Modeling and Control of Hydrosystems

Xavier Litrico • Vincent Fromion

Modeling and Control of Hydrosystems

Springer

Xavier Litrico, PhD
UMR G-EAU, Cemagref
361 rue JF Breton, BP 5095
34196 Montpellier cedex 5
France

Vincent Fromion, PhD
Mathematics, Computer Science
and Genome
INRA, Domaine de Vilvert
78350 Jouy-en-Josas
France

ISBN 978-1-4471-5785-4 ISBN 978-1-84882-624-3 (eBook)
DOI 10.1007/978-1-84882-624-3
Springer Dordrecht Heidelberg London New York

British Library Cataloguing in Publication Data
A catalogue record for this book is available from the British Library

Cover design: eStudioCalamar, Figueres/Berlin

Printed on acid-free paper

Springer is part of Springer Science+Business Media (www.springer.com)

Preface

Hydrosystems are an important element of the hydraulic cycle. In this book we deal with hydrosystems that can be managed, and focus our attention on the management of surface water flows. Other important aspects of the water cycle such as underground water flow are not considered in this book.

With population growth and climatic change impacting water resources, the existing water transport systems need to be modernized, in order to optimize the use of this precious resource.

The methods developed in this book apply to hydrosystems that can be controlled: in irrigation canals, the gates need to be operated in order to deliver water to the user; in navigation waterways, water levels need to be controlled accurately to ensure given water depths along the reaches; in combined sewer systems, overflow may be prevented by opening or closing of gates; hydroelectric power plants in regulated rivers need controllers to maintain the river water levels close to some target value; and regulated rivers used to transport water to users need to have the water release adjusted from the upstream dam to satisfy the water demand. The book mainly focuses on the control of irrigation canals and of regulated rivers, but the techniques presented can easily be adapted to other hydrosystems.

To design automatic controllers for such systems, one needs accurate models describing the open channel dynamics. The Saint-Venant equations are commonly used by hydraulic engineers to describe open channel flow dynamics. These equations are nonlinear partial differential equations, which have a complex behavior in general, can exhibit shocks, and are therefore difficult to study for control design.

In this book we propose to study the linearization of the Saint-Venant equations around realistic steady flow solutions, and to design controllers that stabilize the set of linearizations. This pragmatic approach is closely related to the gain-scheduling approach commonly used in the automatic control community to control real systems such as power plants, airplanes, missiles, etc.

This book uses concepts from automatic control and from hydraulic engineering. We wrote the book in an attempt to bring together both communities. Some developments considered as well-known by one readership may appear rather new to the other one. The book is intended for automatic control researchers or engineers, environmental hydraulics researchers or engineers, but it may also attract the interest of open channels managers and of graduate and undergraduate students in both fields.

The mathematical prerequisites are basic algebra, undergraduate mathematics, and some basic physics. Some sections of the book contain more theoretical material, but they can be skipped by the reader who is more interested by practical applications. We tried to explain the mathematical results by giving physical explanations, including figures and graphs, to illustrate the practical interest of our results.

A set of MATLAB® files are provided (available from www.springer.com/978-1-84882-623-6) so that the interested reader can test the methods developed in the book. These MATLAB® files also use Simulink® and the Control System Toolbox™.

Acknowledgements

This book is the output of several years of fruitful collaboration between the authors, which was supported by various projects. We gratefully acknowledge the financial help of Cemagref and INRA, our home institutions, that launched the AQUAE project (2000–2003), which was the official starting point of our collaboration.

The methods proposed in this book have been validated on an experimental canal located at the University of Évora in Portugal. The collaboration with Professor Manuel Rijo was supported by the French Ministry of Foreign Affairs and GRICES (Gabinete de Relações Internacionais da Ciência e do Ensino Superior), through the French–Portuguese scientific collaboration projects PAI no. 547-B4 in 2000–2003 and PESSOA no. 09685UE in 2005–2007. We thank Professor Rijo for giving us the opportunity to use the canal, and for his hospitality during our stays in Portugal.

The Gignac experimental platform project (2000–2006) financed by Cemagref, Région Languedoc-Roussillon, Agence de l'Eau Rhône-Méditerranée-Corse, Conseil Général de l'Hérault and ASA Gignac enabled us to exchange ideas about canal control during the work on the PhD theses of Iana Guenova and Nadia Bedjaoui.

We also gratefully acknowledge the support of the Région Languedoc-Roussillon, through the project GERHYCO II of the Communauté de Travail des Pyrénées (2002–2004).

Some of the methods included in the book have been used in a short course taught jointly by Pierre-Olivier Malaterre and Xavier Litrico from 2003 to 2007 at the École Nationale Supérieure d'Électronique, Électrotechnique, Informatique, Hydraulique et Télécommunications (ENSEEIHT) in Toulouse, France. This stimulating teaching experience was made possible by Gilles Harran, Alain Sevrain, and Jean-Félix Alquier from ENSEEIHT, who helped us design the practical training experiment, where the students were able to test the controllers on a small-scale canal.

Xavier Litrico would like to thank Cemagref and the University of California at Berkeley for allowing him to spend the academic year 2007–2008 as a Visiting

Scholar in the Department of Civil and Environmental Engineering of UC Berkeley, where most of this book was written. Professor Alex Bayen and his research team were also instrumental in providing a stimulating research atmosphere in Berkeley. The book benefited from numerous discussions with Alex Bayen and his students, including Tarek Rabbani, Saurabh Amin, Qingfang Wu, and Issam Strub.

We are also grateful to our colleagues in France and abroad, whose inputs appear in many parts of this book. The members of the TRANSCAN research team at Cemagref, Jean-Pierre Baume, Gilles Belaud, Pierre-Olivier Malaterre, José Ribot-Bruno, and Pierre-Yves Vion often provided constructive feedback to our new ideas. The research and feedback by our colleagues Jacques Sau, Didier Georges, Georges Bastin, Bert Clemmens, Manuel Rijo, and Erik Weyer had a stimulating impact on our work. The advice given by managers of hydrosystems was also very beneficial. We are grateful to Vincent Kulesza from Gignac Canal, Pierre Rousset, Jean-Luc Deltour, Franck Sanfilippo, and Yann Viala from SCP, and Jean-Luc Trouvat and Patrick Hurand from CACG.

We thank Patrick Saint-Pierre, Gilles Belaud, Jean-Pierre Baume, Jacques Sau, and the anonymous reviewers for their comments on the early drafts of this book.

Special thanks go to Anne Goelzer and Gérard Scorletti for numerous and fruitful discussions on the control aspects of open channel flow.

We thank the students we had the pleasure to interact with, either at Montpellier, Caen, Évora, or at Berkeley: Iana Guenova, Nadia Bedjaoui, Carina Arranja, Miguel Nunes de Almeida, Philippe Pognant-Gros, Marc Dinh, Simon Munier, Florent Di Meglio, Christian Claudel, Dan Work, Ryan Herring, Olli-Pekka Tossavainen, Julie Percelay, Arthur Wiedmer, Jean-Séverin Deckers, and Sébastien Blandin.

Last, but not least, we thank Gabi for her continuous support, and Louise and Octave for their communicative energy.

Montpellier, Xavier Litrico
Jouy-en-Josas, Vincent Fromion

August 2009

Contents

Part III Control of a Multiple-pool Canal

Chapter 1
Introduction

Abstract This chapter gives an overview of the issue of water management. We describe open channel systems, why they are difficult to manage, and why control engineering may provide useful tools to better manage such systems. We also emphasize the difficulties involved in the design of efficient controllers for open channels: they are distributed systems, with large delays between inputs and outputs, few measurements along the system, subject to large and unmeasured perturbations, with a nonlinear behavior. Our approach is based on an engineering background, where pragmatic but efficient solutions can be found to solve practical problems. We limit our study to the linearized Saint-Venant equations, which describe open channel flow around a given equilibrium regime. This is a classical approach in the control engineering community, linked to the gain-scheduling method for designing a controller for a nonlinear system, based on its linearizations.

1.1 Water Management Issues for Hydrosystems

Water is a key factor for life. We need freshwater for domestic, agricultural and industrial purposes. But we also need water to flow in natural streams, to preserve wildlife and biodiversity. Water is a renewable resource, but it is not distributed equally in time or in space.

Since antiquity, human beings have used engineering techniques for water systems: Roman engineers were masters in designing aqueducts and sewer systems. Following these great ancestors, open channels have been built to transport water over long distances, either for irrigation purposes, or for drinking or domestic water supply.

The hydrosystems considered in this book are those where operational water management is necessary and can be improved using control techniques. We briefly describe different types of hydrosystems falling into that category and explain why automatic control may improve their operational management. By operational management, we denote real-time operations on a time scale of minutes to hours, as opposed to tactical management, which works on a time scale of several days or weeks.

1.1.1 Irrigation Canals

It is estimated that irrigation accounts for 80% of the world's total freshwater consumption and that 30% of this water is effectively delivered to the plant. The water efficiency of irrigated systems is therefore very low, and this is partly due to their management [43]. Moreover, the service to users (farmers irrigating their crops) is far from being satisfactory in many cases, and can be greatly improved. These open channel systems therefore have a low efficiency in terms of water resource management and at the same time a low efficiency with respect to the user. This has a great impact on the crop production and on the environment, since diverting water from a natural stream can have large environmental impact.

Irrigated agriculture accounts for 40% of food production. Having to respond to an increased demand linked to population growth, agriculture faces the challenge of increasing crop production while controlling water consumption. Water is also consumed by other users, such as domestic or industrial ones. These factors provide a strong incentive to modernize the management of irrigation canals by using modern control methods.

As an example the Indus Basin Irrigation System, one of the world's largest contiguous irrigation schemes consists of about 60,000 km of open channels irrigating an area of 16 million ha [29]. In West Africa, the Office du Niger in Mali diverts water from the Niger River to irrigate an area of 80,000 ha.

In these traditional systems, water is diverted from main canals to secondary canals, then to distributaries and watercourses. These canals are usually operated manually with a supply-based logic, i.e., the water allocation and distribution is essentially controlled by the supply at the head of the canal. The water efficiency of these systems is usually low due to physical and operational losses. It is estimated that only 25% of the water diverted from the Niger effectively reaches the crop in the Office du Niger [15].

In modern systems, such as the Canal de Provence in Southern France or the Central Arizona Project in the USA, water distribution is supervised with a SCADA (Supervisory Control and Data Acquisition) system, which enables a better monitoring of the system. The systems with the highest efficiency are those with automatically controlled gates, which can react instantaneously when a change is required in the flow at some point of the system.

1.1.2 Navigation Waterways

Open channels have been used as navigation waterways in many countries for a long time. They provide a cheap and environmentally friendly way of transportation. The canals are equipped with gates to control water levels. Water level control is essential in these canals, since a minimum depth is necessary for boats to navigate safely, and the water level should not exceed a maximum value to ensure that boats can pass under the bridges and that the water does not overtop the banks of the canals.

1.1.3 Regulated Rivers

Regulated rivers are used in many countries to sustain summer low flows and supply water for different users (drinking water, industries, irrigation, hydropower, salubrity). In these systems, an upstream dam is used as storage, and the river is used as a channel to convey water to water users. As an example, in southwestern France the Compagnie d'Aménagement des Côteaux de Gascogne (CACG) manages the Neste system, a hydrosystem with 11 watersheds on a total area of 8,000 km^2. The rivers represent a total length of 1,300 km and are regulated using dams with a total volume equal to 50 hm^3. This system is managed using a SCADA system, which centralizes the data collected and also determines the water releases of dams in order to fulfill operational objectives [32].

This use of natural channels prevents important civil engineering work. Moreover, such systems are difficult to manage: there is usually only one control action variable, the discharge released from the dam, and only a few measuring points along the river. The system is subject to large non-measured perturbations (inflows, or withdrawals) and the dynamics of each reach (part of a river between two measuring points) are strongly nonlinear. It is therefore difficult to determine the "optimal" water release of the upstream dam in order to maintain a target discharge downstream. These systems are generally managed manually, but with a low efficiency [33]. The application of automatic control techniques for real-time regulation of these open channel hydraulic systems may lead to large water savings and better service, and may help to maintain a minimum flow for ecological purposes [35].

Other types of regulated rivers are the cascades of hydroelectric powerplants along river reaches [26, 28]. These hydrosystems have been managed for a long time using automatic control techniques in order to optimize production and control water levels [22].

1.1.4 Combined Sewer Systems

Combined sewer systems are present in large cities where sewer pipes have been designed to carry both wastewater and stormwater. During heavy rain that exceeds pipe capacity, the pipes overflow into local water bodies. This is called a combined sewer overflow (CSO), which provides a safety valve that prevents back-ups of untreated wastewater entering homes and businesses, flooding city streets, or bursting underground pipes. An advantage of this system is that, most of the time when rainfall is low to moderate, both the stormwater and wastewater go to the treatment plant. However, the main inconvenience is that for heavy rainfall, wastewater and stormwater are released directly into natural streams without any treatment. In many large cities such as Paris, Barcelona and Montreal, there are underground tanks that can be used to temporarily store water during a heavy rainfall event. There are also sensors distributed on this usually complex network and actuators such as gates. The decision to store water in a given tank can be made based on the knowledge of the dynamics of the system, which are usually described by the Saint-Venant equations of open channel flow.

1.2 Control Engineering for Open Channels

All the previous hydrosystems can be generally described as a network of pools, with each pool representing a portion of the open channel system in between two controlled hydraulic structures (gates or weirs, for example). As a common feature, the dynamic behavior of a pool can be accurately represented by the Saint-Venant equations. This is why this book focuses on the control of systems described by the Saint-Venant equations. We will further restrict our attention to the case of irrigation canals, which allows us to study the main features of a hydrosystem described by the Saint-Venant equations. Some elements about the control of other hydrosystems are provided in Chap. 12 at the end of the book.

In the rest of the book, an irrigation canal is assumed to be described as a series of pools, even if more complex interconnections could easily be handled. We also assume that the control action is the discharge at each cross-structure. Even if in practice canals are controlled using hydraulic structures (gates, weirs), represented by static nonlinear equations, a slave controller on each hydraulic structure can be used to deliver a required discharge (see, e.g., [36] for a comparison of different ways to impose a discharge at a gate). In this context, canal control consists of operational control or regulation of the water levels and flows in the main canal using the available hydraulic structures (gates, moveable weirs, etc.), in order to satisfy real-time requirements, which may vary according to the considered system.

1.2.1 Automatic Control as a Way to Increase Efficiency

Traditionally managed irrigation canals have been shown to have a low efficiency, i.e., the amount of water that reaches the crop represents only a small fraction of the amount of water withdrawn from the environment. On the other hand, modernized canals usually have a better efficiency, and provide a better service to the user. Automatic control therefore appeared as a possible means to improve the water efficiency and at the same time improve the quality of the service (such as timeliness, quantity, flow for a given water delivery). This area has attracted attention from researchers and engineers from the hydraulic community, as attested by the number of research articles published in scientific journals in the 1990s. Historically, apart from a few isolated contributions from the control community, the first large trend of contributions came from the hydraulic community at large, or civil engineers with a background in agronomy or hydraulic design.

In terms of research, the problem is seen differently according to the background of the author. This is clear from the contributions since the 1990s: a hydraulic researcher would tend to favor time domain analysis and to focus on management issues, while an automatic researcher would use a frequency domain approach, or study the stability of the closed-loop system. Research in this area has intensified since the early 1990s, with the ASCE (American Society of Civil Engineers) working groups on canal modeling [4] and canal automation [5]. Results of these working groups were published in the *ASCE Journal of Irrigation and Drainage Engineering*. An interesting result of the group on canal automation is the classification of canal control algorithms, presented in [41]. This paper is based on the PhD thesis [39] and proposes a rigorous classification of canal control methods, giving a clearer view on the state of the art developments at that time.

The working group has also proposed test cases for canal control, which were inspired from real canals [18]. Several control methods have been tested on these test canals (Smith predictor [23], model inversion [37], linear quadratic Gaussian control [40], PI (proportional integral) controllers [17], open-loop control [8]) and were compared using performance indicators [18]. Finally, numerous methods have been tested in simulation on other canals: see, for example, the works presented at the conferences stated in[1, 2, 3].

Even if the definition of benchmarks has offered a good opportunity to compare and rigorously evaluate various control approaches, the authors believe that the automatic control approach is able to provide more in terms of what can be achieved by a control law on a specific canal (e.g., level of performance) and how this goal can be practically reached by a suitable design. In order to achieve such a goal, we have first to discuss the issues associated to the control of irrigation canal pools. Actually, the main control objectives are twofold:

- To ensure a good water management, i.e., minimize the operational water losses
- To provide a good service to the water user, i.e., deliver water in time and in quantity.

It is then interesting to discuss the existing irrigation control in view of these two main objectives.

Most of the existing irrigation canals are operated manually using local upstream control [15]. In this case, the downstream hydraulic structure of a pool is used to control the water level located just upstream. Such a control method imposes many constraints on the user, since water is distributed according to a prespecified schedule (the so-called "water turns"). Also, the discharge is imposed from upstream, and cannot be changed according to the effective demand. This leads to a possible spillage of water.

The main research line to improve this management has been to use automatic control to implement distant downstream control [48]. In that case, the upstream hydraulic structure of a pool is used to control the downstream water level of the pool. The upstream discharge therefore adapts to the water demand and may not use more water than necessary.

It should be noticed that these two main classical control policies for irrigation canals (local upstream control and distant downstream control) lead to completely opposite water management rules. As will be shown in the sequel, the distant downstream control is parsimonious from the water management point of view, but has a low performance with respect to the water user. Indeed, the water released upstream from the pool is directly linked to a downstream water demand, but the time-delay induced by the finite speed of the water transport of the pool necessarily implies that the user will have to wait for some time before his demand is satisfied.

Conversely, the local upstream control policy is water consuming, but leads to a very good performance with respect to the user. In this case, the water demand in a pool is satisfied by reducing the discharge flowing downstream. This implies that there is enough water available and that the user has instantaneous access to the water resource. The local upstream control policy has a high performance from the water user point of view, but is very water consuming since water demands are time-varying, and a large amount of water entering the system is lost at the downstream end without being used.

The main issue at stake for irrigation canal control is therefore a better trade off between water resource management and performance with respect to water users.

This book provides methods to design controllers that implement the desired trade-off between resource management and service to the user.

1.2.2 The Open Channel as a Control Problem

Before explaining the approach developed in this book for the specific canal control problem, we first recall the way generally used by control engineers to handle a control problem. Modern control engineering relies on a computer-based approach. This means that complex control problems can easily be handled if they can be formulated in a way easily handled by computers. In this context, optimization, and more specifically convex optimization is a key ingredient (see, for example [6, 14, 12, 13]). The central role of computational tools and more especially that attached to optimization requires that most of the design requirements be expressed into mathematical criteria involving controlled variables and control action variables. This task is generally difficult, not only because it requires that complex requirements be translated into mathematical criteria, but also because there is a strong interplay between the control objectives and the definition of a simplified model for controller synthesis. Indeed, the controller design methods rely on a model that corresponds to a simplified representation of the considered system, and its consistency with respect to complex requirements can generally only be checked after the synthesis step. Therefore, defining control requirements through mathematical criteria on a particular system is the result of an iterative process where synthesis models and criteria are refined. The main steps of this iterative process are as follows: a synthesis model and a first mathematical formalization of requirements are put together and a controller is then designed. On this basis, the control engineer investigates the features of the obtained controller and generally refines or redefines both the synthesis model and the mathematical criteria until they become compatible. This process is complex and represents the art of control engineers. It can moreover include a refinement of the initial requirements, and even in some cases a strong modification, since some requirements may be infeasible.

As already mentioned, a trade-off has to be obtained between the main control objectives for irrigation control. The associated objectives are quite general and the requirements attached to a specific canal pool have to be defined more specifically. As an example, the control requirements depend on the way water is delivered to the users. If a farmer uses a pump to withdraw water from the canal, the control design requirements will be that the water level should be compatible with the pump functioning. With such a water delivery system, the user can have the desired amount of water whatever the water level in the pool. By contrast, if we assume that the water is obtained through a gravitational delivery offtake, then the control requirement is completely different, since now the delivered discharge directly depends on the water level. In this case, the control has to minimize the water level variations in order that the user have access to his water.

It is obviously unrealistic to consider all the aspects attached to the control of a given canal pool, which is by essence very specific. Therefore, our aim in the sequel is to analyze the main features of the problem that are common to all the canal pool problems and to offer a panel of simple and more advanced controller design methods to handle them.

1.2.3 Control of a Set of Linearizations

From a control point of view, and beyond the control objectives, the main problem considered in this book is to obtain a practical and efficient way to control systems described by the Saint-Venant equations, which are known to provide an accurate mathematical description of open channel flow dynamics. But this highly reliable model consists of a couple of two nonlinear partial differential equations, which is very complex from the control point of view. Very few existing methods can use this model directly for control design purposes (see, e.g., [30, 19, 20, 7]). Moreover, these methods can only handle a small set of design requirements, which are not the most important ones. Therefore, with this model we are not able to define a well-adapted design method, and then another way must be considered. To this end, we use an important property of the system: the hydraulic regimes of a canal change relatively slowly. This explains why most design methods existing in the literature are based on a "rough" description of the system: a linear time invariant linearization around an equilibrium regime [41]. This kind of approach is called the gain-scheduling approach in the automatic control field. It is widespread in many applications, such as the design of controllers for airplanes or nuclear power plants (see [47, 44, 34, 27]).

The main idea of the gain-scheduling approach is to use at different operating points, the linearizations of the nonlinear plant to design a linear time invariant (LTI) controller ensuring locally suitable linear properties assuming that the linearizations are LTI systems. This set of LTI controllers is then interpolated or scheduled as a function of operating points. The main interest of this method is that it allows us to use all the classical and advanced tools developed to control LTI systems. From a theoretical viewpoint, this method also guarantees global stability of the nonlinear system under some technical conditions. For a nonlinear system described by differential equations, if the considered linearizations are those associated to equilibrium regimes, it is possible to show that the exponential stability of its linearizations implies global stability if the variation of the operating points in time remains sufficiently slow (with respect to the dynamics of the system).

This classical justification of the gain-scheduling approach, and the proof of its validity, may certainly be extended to nonlinear partial differential systems such as the Saint-Venant equations. However, this would require advanced mathematical notions, especially since we are working with boundary controlled systems, and is out of the scope of this book. But in most of the cases, it is generally difficult to guarantee that the system with the designed controller is really slowly varying whatever perturbations or inputs are acting on it. Actually, in most applications, the main drawback of the gain-scheduling approach is the lack of theoretical guarantees provided by this approach even concerning the closed-loop stability. In view of such a situation, a posteriori mathematical analysis of the nonlinear closed-loop system properties can be done [27], but generally, a posteriori analysis is often reduced to the validation of the closed-loop through intensive nonlinear simulations. Finally, even if this approach of the nonlinear control problem is heuristic, it can be shown that its goal is fully justified from a control point of view [27].

To summarize, the design approach considered in the sequel is based on the gain-scheduling approach. This leads us to focus on the problem of designing well-suited LTI controllers for a series of linearizations of the system. As it is already known, and will be illustrated in Chap. 11, this approach works surprisingly well in practice, which in turn explains our interest in developing a LTI design approach to control the open channel linearizations that is as sound as possible. It now remains to explain how to design an efficient controller for a set of LTI systems.

1.2.4 From Classical Automatic Control to Robust Control

The main interest of LTI systems that possess a rational transfer function, or a finite state-space representation is related to the existence of an extensive number of design and analysis methods. In this context, two main kinds of approaches exist: the state-space approach and the frequency approach (based on the transfer representation of the linear system). These two approaches are equivalent from a mathematical point of view, since rational transfer function and state-space representations of linear time invariant systems are mathematically equivalent. Nevertheless, from a computer point of view, the state-space approach seems more adequate and forms the basis of most of the available computer-based methods. By contrast, the frequency approach, and more generally the so-called input–output approach, seems more appropriate to handle and frame the control problem and especially adequate for representing uncertainties and robustness problems [50, 45].

From a historical point of view, the frequency approach is associated with the so-called *classical automatic control*, which was developed before World War II through the works of Harold S. Black [9, 10] and Harry Nyquist [42], to cite a few. For more than 50 years now, the classical automatic control approach has constituted a mature method for monovariable control design (see the textbooks of Hendrick W. Bode [11] and I. M. Horowitz [31]). Nevertheless, the application of the classical approach to a specific system often requires a lot of expertise from control engineers ([38]). More importantly, the mathematical nature of problems that are solved by a classical approach remained unknown for a very long period. Indeed, as noted by George Zames in [51], one of the issues of the development of the so-called H_∞ approach was to provide a well-defined optimization problem whose solution would be a LTI controller containing, for example, a classical lead/lag filter effect or which is equivalent to a classical PID (proportional integral derivative). The work of George Zames has also explained the soundness of the frequency approach, especially with respect to the issue of robustness. Indeed, the wide success of classical automatic control is related to its ability to design highly efficient and robust LTI controllers. Zames clearly showed that was not a coincidence. In classical automatic control, the robustness issue is handled by choosing a controller that is able to satisfy the Nyquist criteria for a family of Nyquist plots associated with the set of possible models. This leads to defining the robustness as the ability of the controller to ensure that the neighborhood of the Nyquist plot remains far from the point -1 in the complex plane. Georges Zames then showed in [51] that the use of a weighted H_∞ norm allows formalization of this notion of the Nyquist plot neighborhood as a well-defined optimization problem. He showed that the robustness issue could be handled through the use of a weighted H_∞ norm. In the same way, he also showed that the performance requirement attached to a closed-loop system could be also formulated as a weighted H_∞ norm. To summarize, the valuable contributions of George Zames to control paved the way for an amazing period of automatic control developments where many new results appeared. These results form the basis of *linear robust control*, also called "neo-classic" automatic control.

Today, we have access to a large number of synthesis and analysis tools that allow us, with the help of computers, to develop highly efficient controller design approaches for finite dimension LTI systems. Furthermore, all these developments have been made using deep mathematical treatments, which provide suitable and theoretical answers to some fundamental questions related to the design or the analysis of LTI control systems.

It remains to explain why all these developments and available tools for finite dimensional systems are also useful for controlling the Saint-Venant linearizations, which represent an infinite dimensional system. For this purpose, we use the fact that the Saint-Venant linearizations belong to a specific class of infinite dimensional systems: the Callier–Desoer algebra (investigated in detail in [16, 49, 21]). Surprisingly, most of the results developed for rational transfer functions can be extended to this specific class of LTI systems. The Nyquist type criterion holds true for this class of irrational transfer functions. Let us recall that the Nyquist criterion [42] plays a central role in classical automatic control. Indeed, for systems described by rational transfer functions, it provides a *necessary and sufficient* condition for the stability of the closed-loop stability only based on the properties of the open-loop system, which can be tested graphically. On this basis, all classical tools associated with the frequency approach can be then easily used for this class of irrational transfer functions (more detailed mathematical developments are given in Appendix E). Beyond its role of in classical automatic control, the Nyquist criterion is also the basis of many results attached to linear robust control, such as the developments of the μ/k_m analysis (see [24, 25] and also [45, 46]). This connection between Nyquist and robustness analysis explains why most analysis tools developed in the robust control context are also available for infinite dimensional systems belonging to the Callier–Desoer class. Finally, this class of irrational transfer functions also possesses very interesting properties with respect to the parameterization of all the stabilizing controllers through the use of coprime factors (see [21] and references therein).

Another important issue for designing controllers is related to the possibility to approximate an infinite dimensional system by a finite dimensional one. From a practical point of view, this issue is essential for developing efficient design methods since it allows us to use tools attached to finite dimensional systems. This problem can be considered from two points of view, from a mathematical point of view, or from a more practical one. From a mathematical point of view, the problem of approximation is related to the question of convergence. The main issue is then to know whether the design of a controller based on a series of designs made on a finite dimensional systems of increasing order and which converge to the infinite dimensional system leads to obtaining at the limit the "true" infinite dimensional controller.

The other more practical perspective is related to the fact that the bandwidth of the controller is limited, since the actuator is not able to provide energy in high frequencies. Thus the system and the controller can only be exactly known in a finite range of frequencies, which means that the system can be approximated by a finite order system without any loss of performance.

Following these developments, in this book we present two types of controller design methods:

- First, we follow a classical automatic control approach, by giving general rules for the design of PI controllers. This first step provides simple controllers and enables us to identify the main constraints in the design problem.
- Second, in order to handle more complex situations, we develop a design method based on the H_∞ approach, which can be seen as a natural extension of the classical controller design approach.

1.3 Outline of the Book

The book is divided into three parts:

- **Part I** focuses on open channel flow modeling. After this introductive chapter, we state the Saint-Venant equations and give some of their properties in Chap. 2. The linearized Saint-Venant equations are then studied in the frequency domain in Chap. 3. We analyze the properties of the linearized Saint-Venant transfer matrix, first in the horizontal frictionless case, then in the uniform flow case, and finally in the general nonuniform flow case. Various finite dimensional rational models for control purposes are then presented in Chap. 4. In particular, we use the analysis performed in Chap. 3 to derive rational models based on the poles of the system, and we also study the well-known semi-implicit discretization scheme proposed by Preissmann. Chapter 5 proposes a simplified model that can be used to derive tuning rules for PI controllers of a canal pool.
- **Part II** focuses on the control of a single canal pool. In Chap. 6 we study the behavior of an open channel controlled with hydraulic structures, such as gates, weirs, or hydromechanical automatic gates. Chapter 7 then turns towards the classical feedback control policies of a canal pool: we introduce distant downstream and local upstream control, and propose tuning rules for PI controllers and for H_∞ controllers. Chapter 8 introduces the mixed local upstream/distant downstream control policy, with PI and H_∞ controllers. In Chap. 9 we derive open-loop controllers for a canal pool. This is important when we know in advance the perturbations affecting the channel, such as water withdrawals for irrigation, or discharge perturbations generated at one boundary.
- **Part III** focuses on the control of a multiple-pool canal. Chapter 10 studies decentralized controllers, i.e., controllers which have been designed independently for each pool, and then are put together on the whole canal. We analyze the stability and robustness of the three control schemes (distant downstream, local upstream, and mixed). Chapter 11 presents experimental results obtained on a small-scale canal located in the University of Évora, Portugal. Finally, Chap. 12 shows how the methods presented in the book can be applied to another type of hydrosystem, here a regulated river. Some technical results are given in the Appendix for completeness.

References

[1] (1997) International workshop regulation of irrigation canals: state of the art of research and applications. Ric'97 proceedings, Marrakech-Morocco
[2] (1998) IEEE Conference on Systems, Man and Cybernetics, San Diego, pp. 3850–3920
[3] (1999) European Control Conference, Special Session on Canal Modelling and Control, Karlsruhe
[4] ASCE (1993) Unsteady-flow modeling of irrigation canals. J Irrig Drain Eng 119(4):615–630
[5] ASCE (1998) Special issue on canal automation. J Irrig Drain Eng 124(1)
[6] Başar T, Bernhard P (1991) H_∞-optimal control and related minimax design problems: a dynamical game approach. Birkhauser, Boston MA
[7] Bastin G, Coron JM, d'Andréa Novel B (2009) On Lyapunov stability of linearised Saint-Venant equations for a sloping channel. Networks Heterogeneous Media 4(2):177–187
[8] Bautista E, Clemmens A (1999) Response of ASCE task committee test cases to open-loop control measures. J Irrig Drain Eng 125(4):179–188
[9] Black HS (1934) Stabilized feedback amplifiers. Bell Syst Tech J 13:1–18
[10] Black HS (1977) Inventing the negative feedback amplifier. IEEE Spectrum 14(12):55–60
[11] Bode H (1945) Network analysis and feedback amplifier design. Van Nostrand, New York
[12] Boyd S, Vandenberghe L (1995) Introduction to convex optimization with engineering applications. Lecture Notes, Information Systems Laboratory. Stanford University
[13] Boyd S, Vandenberghe L (2004) Convex optimization. Cambridge University Press
[14] Boyd S, El Ghaoui L, Feron E, Balakrishnan V (1994) Linear matrix inequalities in system and control theory. Studies in Applied Mathematics, SIAM, Philadelphia PA
[15] Burt C, Styles S (1999) Modern water control and management practices in irrigation. Tech. rep., FAO, IPTRID, World Bank
[16] Callier F, Desoer C (1978) An algebra of transfer functions for distributed linear time-invariant systems. IEEE Trans Circuits Syst CAS-25(9):651–662
[17] Clemmens A, Wahlin B (2004) Simple optimal downstream feedback canal controllers: ASCE test case results. J Irrig Drain Eng 130(1):35–46
[18] Clemmens A, Kacerek T, Grawitz B, Schuurmans W (1998) Test cases for canal control algorithms. J Irrig Drain Eng 124(1):23–30
[19] Coron JM, d'Andréa Novel B, Bastin G (2007) A strict lyapunov function for boundary control of hyperbolic systems of conservation laws. IEEE Trans Autom Contr 52(1):2–11
[20] Coron JM, d'Andréa Novel B, Bastin G (2008) Dissipative boundary conditions for one dimensional nonlinear hyperbolic systems. SIAM Journal of Control and Optimization 47:1460–1498
[21] Curtain R, Zwart H (1995) An introduction to infinite dimensional linear systems theory, Text in applied mathematics, vol 21. Springer, Berlin
[22] Dang Van Mien H, Klein F (1992) Robust control for hydraulic power plants. In: Proceedings of the first IEEE conf. on control applications, Dayton, OH
[23] Deltour JL, Sanfilippo F (1998) Introduction of Smith predictor into dynamic regulation. J Irrig Drain Eng 124(4):47–52
[24] Doyle JC (1982) Analysis of feedback systems with structured uncertainties. IEE Proc 129-D(6):242–250
[25] Fan MKH, Tits AL, Doyle JC (1991) Robustness in the presence of mixed parametric uncertainty and unmodeled dynamics. IEEE Trans Autom Contr 36(1):25–38
[26] Foss B, Haug J, Alne J, Aam S (1989) User experience with on-line predictive river flow regulation. IEEE Trans Power Syst 4(3):1089–1094
[27] Fromion V, Scorletti G (2003) A theoretical framework for gain scheduling. Int J Robust Nonlin Contr 13:951–982
[28] Glanzmann G, von Siebenthal M, Geyer T, Papafotiou G, Morari M (2005) Supervisory water level control for cascaded river power plants. In: International conference on hydropower, Stavanger, Norway

[29] Habib Z (2004) Scope of reallocation of river waters for agriculture in the indus basin. PhD thesis, ENGREF, Montpellier

[30] de Halleux J, Prieur C, Coron JM, d'Andréa Novel B, Bastin G (2003) Boundary feedback control in networks of open-channels. Automatica 39:1365–1376

[31] Horowitz I (1963) Synthesis of feedback systems. Academic Press, New York

[32] Hurand P, Kosuth P (1993) Régulations en rivière. La Houille Blanche 2/3:143–149

[33] Kosuth P (1994) Techniques de régulation automatique des systèmes complexes: application aux systèmes hydrauliques à surface libre. PhD thesis, Institut National Polytechnique de Toulouse, (in French)

[34] Leith DJ, Leithead WE (2000) Survey of gain-scheduling analysis and design. International Journal of Control 73:1001–1025

[35] Litrico X (1999) Modélisation, identification et commande robuste de systèmes hydrauliques à surface libre. PhD thesis, ENGREF–Cemagref, (in French)

[36] Litrico X, Malaterre PO, Baume JP, Ribot-Bruno J (2008) Conversion from discharge to gate opening for the control of irrigation canals. J Irrig Drain Eng 134(3):305–314

[37] Liu F, Feyen J, Malaterre PO, Baume JP, Kosuth P (1998) Development and evaluation of canal automation algorithm CLIS. J Irrig Drain Eng 124(1):40–46

[38] Lurie BJ (1986) Feedback maximization. Artech House, Boston, MA

[39] Malaterre PO (1994) Modélisation, analyse et commande optimale LQR d'un canal d'irrigation. PhD thesis, ENGREF–Cemagref

[40] Malaterre PO (1998) Pilote: linear quadratic optimal controller for irrigation canals. J Irrig Drain Eng 124(4):187–194

[41] Malaterre PO, Rogers DC, Schuurmans J (1998) Classification of canal control algorithms. J Irrig Drain Eng 124(1):3–10

[42] Nyquist H (1932) Regeneration theory. Bell Syst Tech J 11:126–147

[43] Plusquellec H, Burt C, Wolter H (1994) Modern water control in irrigation. Tech. Rep. 246, World Bank, Irrigation and Drainage Series

[44] Rugh WJ, Shamma JS (2000) Research on gain scheduling. Automatica 36:1401–1425

[45] Safonov MG (1980) Stability and robustness of multivariable feedback systems. Series in signal processing, optimization, and control; 3, MIT Press, Cambridge

[46] Safonov MG (1982) Stability margins of diagonally perturbed multivariable feedback systems. IEE Proc 129-D:251–256

[47] Shamma JS (1988) Analysis and design of gain scheduled control systems. PhD thesis, Dept. of Mechanical Engineering, MIT, Cambridge, MA

[48] Shand M (1971) Automatic downstream control systems for irrigation canals. PhD thesis, University of California, Berkeley

[49] Vidyasagar M (1992) Nonlinear systems analysis, 2nd edn. Prentice Hall, Englewood Cliffs, NJ

[50] Willems JC (1971) The analysis of feedback systems, Research monographs, vol 62. MIT Press, Cambridge, MA

[51] Zames G (1981) Feedback and optimal sensitivity: Model reference transformations, multiplicative seminorms, and approximate inverses. IEEE Trans Autom Contr 26(2):301–320

Part I
Modeling of Open Channel Flow

Unidirectional open channel flow is classically described by a set of hyperbolic equations expressing conservation of mass and momentum. These equations were first introduced in 1871 by Adhémar Barré de Saint-Venant, and since then they have been classically referred to as the Saint-Venant equations. These equations are introduced in Chap. 2. We then study in Chap. 3 the linearization of these equations around a nonuniform steady flow in the frequency domain, and provide a full characterization of the Saint-Venant transfer matrix in terms of poles and delays. In Chap. 4, we study rational models, and develop a simplified model in Chap. 5, which will be useful for control purposes.

Chapter 2
Modeling of Open Channel Flow

Abstract In this chapter, we present the classical model used to describe open channel hydraulics: the Saint-Venant equations. For completeness, the equations are rigorously derived in Appendix A. We first study some of their mathematical properties, such as the characteristic form. We briefly describe some numerical methods of resolution, and then consider the linearized equations that are valid for small variations around equilibrium regimes. These equations form the basis of all the methods developed in this book.

2.1 Saint-Venant Equations for Open Channel Flow

The Saint-Venant equations were first stated in 1871 in a note to the Comptes-Rendus de l'Académie des Sciences de Paris by Adhémar Barré de Saint-Venant, a French engineer of Ponts et Chaussées [15]. This has remained one of the most popular models among hydraulic engineers to represent the dynamics of open channel flow. It has been applied successfully in many applications, such as river flow forecasting, canal operations, sewer modeling, etc. (see the book by Cunge et al. [6], which is a classical reference for hydraulic engineers).

2.1.1 Saint-Venant Equations

In this section we give the Saint-Venant equations, which express the conservation of mass and momentum for one-dimensional open channel flow. Before stating the equations, we give the hypotheses used in their derivation, and define the notations. The derivation of the equations is detailed in Appendix A for completeness.

2.1.1.1 Hypotheses and Notations

The following assumptions are classically made when deriving the Saint-Venant equations:

1. The flow is one-dimensional, i.e., the velocity is uniform over the cross-section and the water level across the section is horizontal.
2. The streamline curvature is small and vertical accelerations are negligible, hence the pressure is hydrostatic.
3. The effect of boundary friction and turbulence can be accounted for through resistance laws analogous to those used for steady-state flow.
4. The average channel bed slope is small so that the cosine of the angle it makes with the horizontal may be replaced by unity.
5. The variation of channel width along x is small.

We denote by x the longitudinal abscissa and by t the time. We shall use the following notations for the considered variables (see Figs. 2.1 and 2.2): $A(x,t)$ represents the wetted area (m^2), $Q(x,t)$ the discharge (m^3/s) across-section A, $V(x,t)$ the average velocity (m/s) in section A, $Y(x,t)$ the water depth (m), $S_f(x,t)$ the friction slope (m/m), $S_b(x)$ the bed slope (m/m), and g the gravitational acceleration (m/s^2).

We assume that the distributed lateral inflow and outflow are negligible. The methods developed in the book may be readily extended to the case of nonzero lateral inflow or outflow.

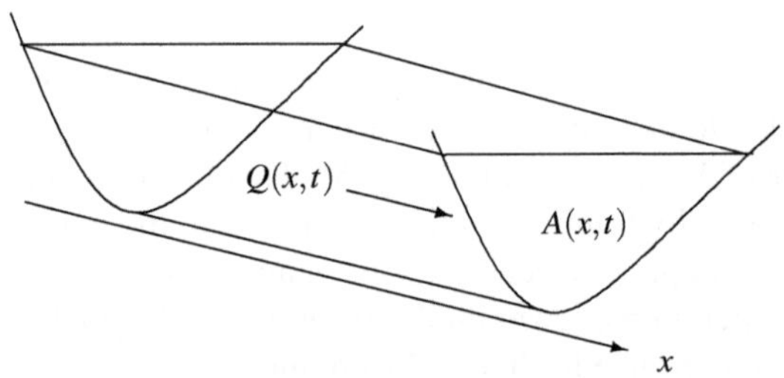

Fig. 2.1 Open channel flow along a longitudinal axis indexed by the abscissa x. The discharge $Q(x,t)$ is defined as $Q(x,t) = A(x,t)V(x,t)$

2.1.1.2 The Equations

The Saint-Venant equations are two coupled partial derivative equations. The first one is the mass conservation equation:

$$\frac{\partial A(x,t)}{\partial t} + \frac{\partial Q(x,t)}{\partial x} = 0, \tag{2.1}$$

and the second one is the momentum conservation equation:

$$\frac{\partial Q(x,t)}{\partial t} + \frac{\partial}{\partial x}\left[\frac{Q^2(x,t)}{A(x,t)}\right] + gA(x,t)\left(\frac{\partial Y(x,t)}{\partial x} + S_f(x,t) - S_b(x)\right) = 0. \quad (2.2)$$

The friction slope S_f is modeled with the classical Manning formula [5]:

$$S_f = \frac{Q^2 n^2}{A^2 R^{4/3}}, \quad (2.3)$$

with n the Manning coefficient ($\mathrm{sm}^{-1/3}$) and R the hydraulic radius (m), defined by $R = A/P$, where P is the wetted perimeter (m) (see Fig. 2.2).

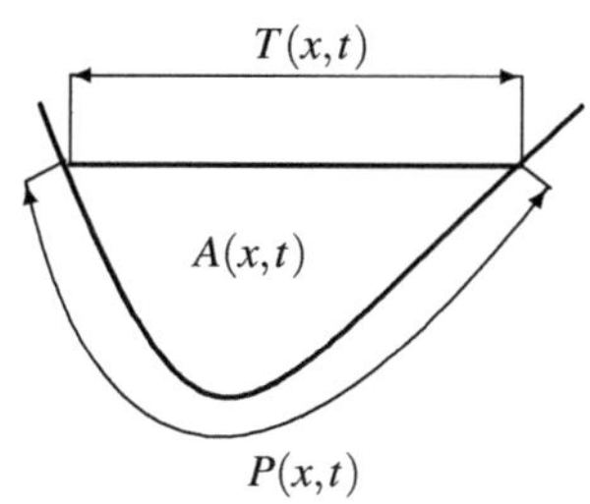

Fig. 2.2 Section of an open channel. Definition of the wetted perimeter P, wetted area A, and top width T

2.1.1.3 Initial and Boundary Conditions

To complete the equations, we need to introduce initial and boundary conditions.

The initial condition is given in terms of $(Q(x,0), Y(x,0))$, for all $x \in [0,L]$, with L the length of the channel.

The choice of the boundary conditions depends on the flow characteristics, and the reverse also holds, since a change in the boundary conditions may change the flow characteristics. This choice will be discussed in Sect. 2.1.4, once the characteristic form of the equations has been introduced.

2.1.1.4 Available Measurements and Controls

Since our objective is to control open channel flow using boundary controls, we need to specify the measurements and controls available. In many practical situations, the available controls are gate openings, and the available measurements are boundary water levels. However, for reasons that will be clarified later, we will consider that we can control the boundary discharges $Q(0,t)$ and $Q(L,t)$, and that we measure the boundary water depths $Y(0,t)$ and $Y(L,t)$. More generally, one may measure water levels at various locations along the canal, but for practical reasons, the boundary levels are easier to measure.

2.1.2 Limitations of Saint-Venant Equations

The Saint-Venant equations are valid as long as the hypotheses stated in Sect. 2.1.1.1 are fulfilled. We mention here some cases where these hypotheses do not hold.

2.1.2.1 Two-dimensional Flow

When the flow is not one-dimensional, which may happen in flood plains or in large rivers, the one-dimensional Saint-Venant equations are inaccurate. In that case, one may consider two-dimensional Saint-Venant equations. We restrict ourselves in this book to flow patterns that can reasonably be modeled with a one-dimensional flow assumption. This is the case of most of open channels that are controlled with hydraulic structures: irrigation or drainage canals, regulated rivers, and sewers.

2.1.2.2 Non-hydrostatic Pressure Distribution

When the pressure distribution is not hydrostatic, the pressure term in the Saint-Venant equations needs to be modified. This is usually linked to hydraulic phenomenon with a small wave length, either due to geometric variations (sharp bend) or hydraulic variations, e.g., undular hydraulic jump, or undular tidal bore.

In this case, a more accurate model is provided by the Boussinesq equations. These equations add a third order derivative term to the momentum conservation of the Saint-Venant equations [21]. Undular bores are positive surges characterized by a train of secondary waves following the surge front. They were studied originally by Favre [8] and are efficiently modeled with the Boussinesq equations [4, 18, 13].

2.1.2.3 Sharp Discontinuities

When the flow encounters sharp discontinuities, such as those provoked by hydraulic structures (weirs or gates), the Saint-Venant equations are no longer applicable. These hydraulic structures are usually treated as internal boundaries modeled with a static algebraic relationship between the flow variables. This static relationship is usually derived from the Bernoulli theorem. In this book, we consider hydraulic structures as passive static controllers acting at the boundaries of the domain. This point will be developed in Chap. 6.

2.1.3 Characteristic Form

The Saint-Venant equations are nonlinear hyperbolic partial differential equations, that can be expressed as a set of four ordinary differential equations in the (x,t) plane using the classical "characteristic form" [1]. These ordinary differential equations are satisfied along the so-called "characteristic curves" in the (x,t) plane. The characteristic form is important because it has long been used as a classical means to solve the equations in simple cases, and because it enables us to understand the physical phenomenon modeled by the hyperbolic equations.

For simplicity, in the following we consider a rectangular prismatic channel, i.e., a channel with a spatially uniform rectangular cross-section.

We first rewrite the equations in terms of the velocity V and the water depth Y. Let T denote the top width (see Fig. 2.2). The area is then equal to $A(x,t) = TY(x,t)$, and the continuity equation then reads:

$$\frac{\partial Y}{\partial t}(x,t) + Y(x,t)\frac{\partial V}{\partial x}(x,t) + V(x,t)\frac{\partial Y}{\partial x}(x,t) = 0. \tag{2.4}$$

For the momentum equation, we first replace $Q(x,t) = TY(x,t)V(x,t)$ in (2.2), apply the product rule of differentiation, and substitute for $\frac{\partial Y}{\partial t}(x,t)$ from the continuity equation (2.4). Finally, dividing by $A(x,t)$ leads to:

$$\frac{\partial V}{\partial t}(x,t) + V(x,t)\frac{\partial V}{\partial x}(x,t) + g\frac{\partial Y}{\partial x}(x,t) = g(S_b - S_f(x,t)). \tag{2.5}$$

Let us introduce the variable $C(x,t) = \sqrt{g\frac{A(x,t)}{T(x,t)}}$ (here $C(x,t) = \sqrt{gY(x,t)}$ due to the rectangular geometry), which is homogeneous to a celerity.

We now express (2.5–2.4) in terms of variables $V(x,t)$ and $C(x,t)$. Using the fact that $\frac{\partial C}{\partial t} = \frac{g}{2C}\frac{\partial Y}{\partial t}$, and dropping the dependence in (x,t) for readability, we get:

$$2\frac{\partial C}{\partial t} + C\frac{\partial V}{\partial x} + 2V\frac{\partial C}{\partial x} = 0, \tag{2.6a}$$

$$\frac{\partial V}{\partial t} + V\frac{\partial V}{\partial x} + 2C\frac{\partial C}{\partial x} = g(S_b - S_f). \tag{2.6b}$$

Now, we observe that adding and subtracting (2.6a) and (2.6b) respectively leads to:

$$\frac{\partial}{\partial t}(V+2C) + (V+C)\frac{\partial}{\partial x}(V+2C) = g(S_b - S_f), \tag{2.7a}$$

$$\frac{\partial}{\partial t}(V-2C) + (V-C)\frac{\partial}{\partial x}(V-2C) = g(S_b - S_f). \tag{2.7b}$$

We have obtained a set of equations with two new variables

$$J_1(x,t) = V(x,t) + 2C(x,t),$$
$$J_2(x,t) = V(x,t) - 2C(x,t).$$

Then, we remark that the left-hand terms of (2.7) involve a derivative along two characteristic curves. A characteristic curve is defined in the (x,t) plane by an ordinary differential equation such as:

$$\frac{dx}{dt} = U(x(t),t). \quad (2.8)$$

This can be depicted as in Fig. 2.3. A characteristic curve described by (2.8) has a local slope equal to $1/U(x_M,t_M)$ at the point of coordinates (x_M,t_M).

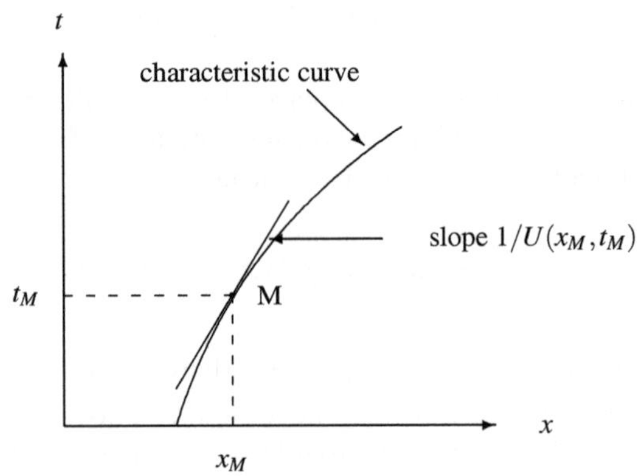

Fig. 2.3 Characteristic line $\frac{dx}{dt} = U(x(t),t)$ in the (x,t) plane

Then, the time derivative of a variable $Z(x(t),t)$ along a characteristic curve defined by (2.8) is given by:

$$\frac{dZ}{dt}(x(t),t) = \frac{\partial Z}{\partial t}(x(t),t) + U(x(t),t)\frac{\partial Z}{\partial x}(x(t),t). \quad (2.9)$$

Applying this property to (2.7), we obtain the following expressions:

$$\frac{\mathrm{d}J_1}{\mathrm{d}t}(x_1,t) = h(x_1, J_1(x_1,t), J_2(x_1,t)), \tag{2.10a}$$

$$\frac{\mathrm{d}x_1}{\mathrm{d}t} = V(x_1,t) + C(x_1,t), \tag{2.10b}$$

$$\frac{\mathrm{d}J_2}{\mathrm{d}t}(x_2,t) = h(x_2, J_1(x_2,t), J_2(x_2,t)), \tag{2.10c}$$

$$\frac{\mathrm{d}x_2}{\mathrm{d}t} = V(x_2,t) - C(x_2,t), \tag{2.10d}$$

with $h(x, J_1(x,t), J_2(x,t)) = g(S_b(x) - S_f(x, J_1(x,t), J_2(x,t)))$, and where the dependence of $x_1(t)$ and $x_2(t)$ on t has been dropped for readability.

The four ordinary differential equations (2.10) are then equivalent to the Saint-Venant equations.

2.1.3.1 Supercritical Versus Subcritical Flow

Such a characteristic form is in fact a very general feature of hyperbolic partial differential equations [7]. It is also useful to understand the fundamental physical behavior of open channel flow: the discharge and water depth (or, equivalently, the velocity and the area) are the result of the interaction of two elementary waves, one traveling downstream, corresponding to the positive characteristic curve C_1, the other one traveling either upstream or downstream, depending on the sign of $V - C$, corresponding to the negative characteristic curve C_2. When $V < C$, the negative characteristic curve C_2 travels upstream, and the flow is called subcritical. When $V > C$, the negative characteristic C_2 curve travels downstream, and the flow is called supercritical (see Fig. 2.4).

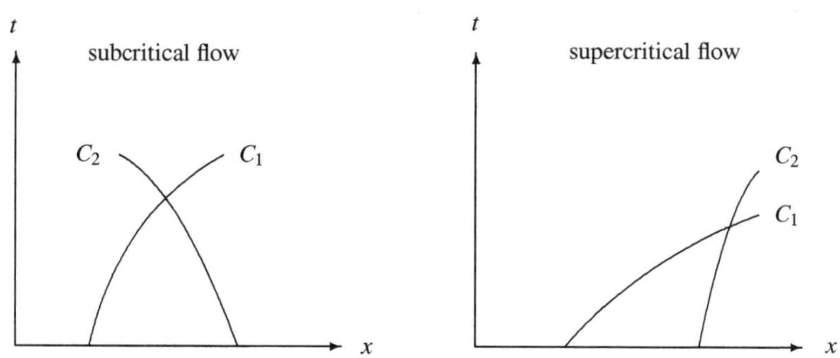

Fig. 2.4 Characteristics in subcritical and supercritical flow

For illustration purposes, we consider below the horizontal frictionless case to give a physical interpretation of these expressions.

2.1.3.2 Horizontal Frictionless Case

In the special case of a horizontal frictionless channel, the slope S_b and the friction slope S_f are identically zero, so that the function $h(x,J_1(x,t),J_2(x,t)) \equiv 0$, which leads to:

$$\frac{\mathrm{d}J_1}{\mathrm{d}t}(x_1,t) = 0 \tag{2.11a}$$

$$\frac{\mathrm{d}x_1}{\mathrm{d}t} = V(x_1,t) + C(x_1,t) \tag{2.11b}$$

$$\frac{\mathrm{d}J_2}{\mathrm{d}t}(x_2,t) = 0 \tag{2.11c}$$

$$\frac{\mathrm{d}x_2}{\mathrm{d}t} = V(x_2,t) - C(x_2,t). \tag{2.11d}$$

In this case, (2.11) can be interpreted as follows:

- The total derivative of the quantity $J_1 = V + 2C$ is zero along the characteristic curve defined by (2.11b). Therefore, this quantity conserves its initial value along the first characteristic curve. This can be displayed in the (x,t) plane, where the solution of (2.11b) is a line denoted C_1, with a positive slope equal to $1/(V+C)$. This corresponds to a wave traveling downstream.
- Similarly, the total derivative of the quantity $J_2 = V - 2C$ is zero along the characteristic curve defined by (2.11d). Therefore, this quantity conserves its initial value along the second characteristic curve. In the (x,t) plane, the solution of (2.11d) is a line denoted C_2, with a slope equal to $1/(V-C)$, which may be negative or positive. When $V > C$, this wave is traveling downstream, otherwise it is traveling upstream.

Therefore, in the horizontal frictionless case, the variables $J_1(x,t)$ and $J_2(x,t)$ are constant along the characteristic curves C_1 and C_2. J_1 and J_2 are called *Riemann invariants* of the hyperbolic system.

In this case, the characteristics method provides a way to derive an analytical solution to the Saint-Venant equations for simple initial conditions. This point will be detailed in Sect. 2.4.2 for the linearized equations, which apply for small variations of depth and discharge around a steady flow.

2.1.3.3 General Case

In the general case, the right-hand term $h(x,J_1(x,t),J_2(x,t))$ is different from zero, and $J_1(x,t)$ and $J_2(x,t)$ are called quasi-invariants, since their values change along the characteristics lines. There is no analytical solution in that case, but the characteristic form can be used to solve the equations numerically. Some classical numerical methods will be presented in Sect. 2.2.

The characteristic form displayed in (2.10) is valid only for rectangular prismatic geometry, but can be generalized to more complex geometry. The expressions of the quasi-invariants are more complex, but they keep the same physical meaning: the flow can be described as the interaction of two waves in the plane (x,t), one that travels at speed $V+C$, the another that travels at speed $V-C$.

2.1.4 Influence of Initial and Boundary Conditions

The characteristics enable us to understand the way an initial condition propagates in the domain. The initial condition will have a range of influence in a triangularly shaped domain in the (x,t) plane (see Fig. 2.5). For the rest of the (x,t) plane, one needs to specify the boundary conditions at $x=0$ and $x=L$.

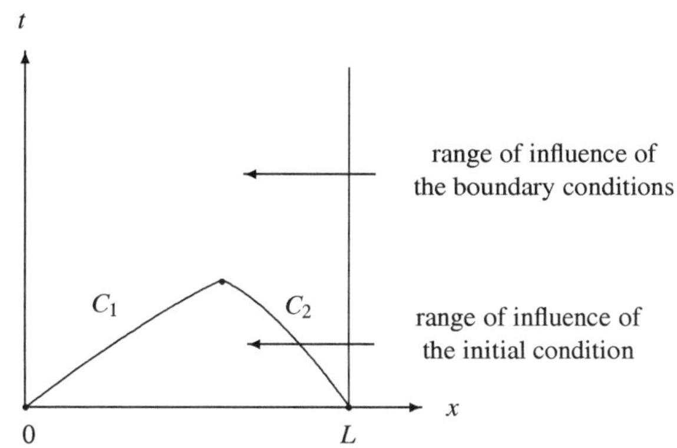

Fig. 2.5 Range of influence of the initial and boundary conditions for subcritical flow

As mentioned above, it is possible to define different boundary conditions according to the flow characteristics. The characteristic form of the equations is useful for understanding the different possible boundary conditions that one may define:

- If the flow is subcritical, two boundary conditions are needed, one upstream and one downstream.
- If the flow is supercritical, two boundary conditions must be defined at the upstream boundary.
- For intermediate situations, i.e., when the flow in the channel is partly subcritical and partly supercritical, one may need to specify one, two, or three boundary

conditions according to the situation. One possibility is then to deal with weak boundary conditions [3, 19]. This complex situation is out of the scope of this book.

We will assume in the rest of the book that the flow is subcritical over the whole channel; therefore we need one condition at each boundary (see Fig. 2.6).

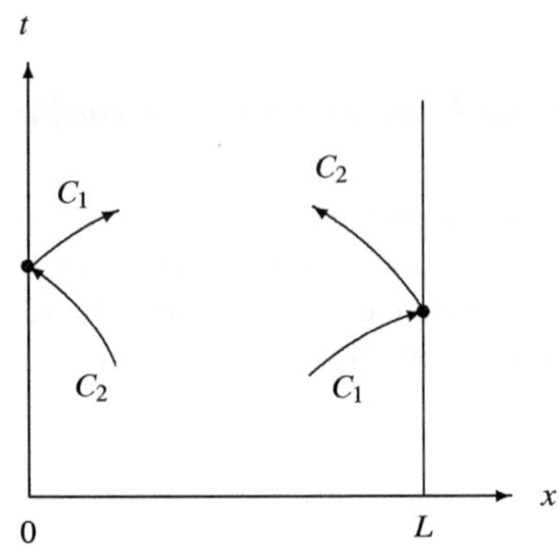

Fig. 2.6 Boundary conditions for subcritical flow

We can either choose the water depth or the discharge as a boundary condition. There are therefore four possibilities:

1. Discharges at both boundaries: $(Q(0,t),Q(L,t))$. This corresponds to the case where two pumps impose the discharges at each boundary, where the canal is closed. This is the case that we will mostly consider in this book, since we will assume that a hydraulic structure such as a gate can deliver a given discharge.
2. Water depths at both boundaries: $(Y(0,t),Y(L,t))$. This corresponds to the case of a body of water connected at each end to very large bodies of water. Such a situation can occur in deltaic zones, where a river can be connected to the sea at one end, and to a lake at the other end.
3. Discharge upstream and water depth downstream: $(Q(0,t),Y(L,t))$. This case corresponds to the case of a river where the upstream discharge is controlled by a dam, and the downstream ends up in a lake.
4. Water upstream and discharge downstream: $(Y(0,t),Q(L,t))$. This unusual case may correspond to a hydroelectric power plant controlling the downstream discharge of a river leaving from a lake.

The developments of this book mainly consider the first case, but can be adapted to the other cases without much difficulty.

2.1.5 *Calibration of the Saint-Venant Model*

When faced with a practical problem of modeling an open channel, a hydraulic engineer usually requires the following data:

- Physical parameters describing the geometry of the open channel: slope, cross-section (shape, dimension, and variation along x).
- Value of the Manning friction coefficient.

Usually, the geometry can be obtained with a measuring campaign, and the Manning coefficient can be either estimated from tables, or calibrated based on steady-state water level and discharge measurements.

2.2 Numerical Methods of Resolution

While this book is not aimed at providing state of the art numerical methods of resolution, for completeness we briefly mention two classical numerical methods of resolution, the method of characteristics and the finite difference Preissmann scheme.

2.2.1 *The Method of Characteristics*

We present two numerical methods based on the characteristics: the original method and the Hartree method.

2.2.1.1 The Original Method

The characteristic form of the Saint-Venant equations that we derived in Sect. 2.1.3 can be used to numerically solve the equations. This method has been detailed extensively by Abbott [1, 2]. In the following we give an illustration of the method in the rectangular case. Let us assume that initial data are known at points A and B (see Fig. 2.7).

The characteristics issuing from these points intersect at the point C. Equations (2.10) can be integrated along the characteristics as follows:

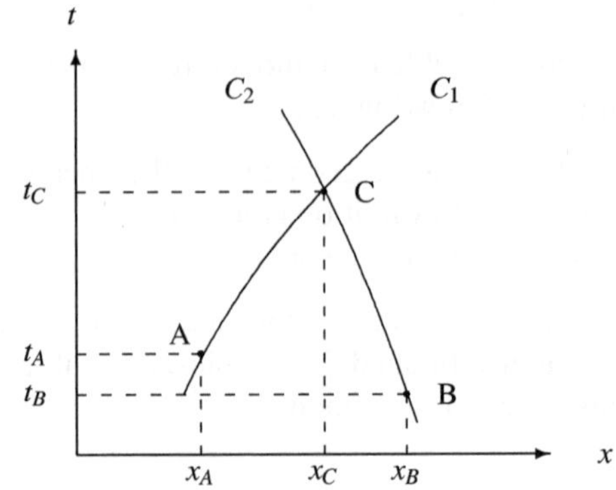

Fig. 2.7 Numerical resolution with the method of characteristics

$$V_C + 2C_C = V_A + 2C_A + \int_{t_A}^{t_C} g(S_b - S_f)\mathrm{d}t \tag{2.12a}$$

$$x_C = x_A + \int_{t_A}^{t_C} (V + C)\mathrm{d}t \tag{2.12b}$$

$$V_C - 2C_C = V_B - 2C_B + \int_{t_B}^{t_C} g(S_b - S_f)\mathrm{d}t \tag{2.12c}$$

$$x_C = x_B + \int_{t_B}^{t_C} (V - C)\mathrm{d}t. \tag{2.12d}$$

These four equations can be solved for the unknowns V_C, C_C, x_C, and t_C. To this end, one needs to evaluate the integral terms. A classical approximation is to use the trapezoidal rule of integration between points A and C, and between points B and C.

The method of characteristics provides an elegant way to turn the two nonlinear partial differential equations (PDE) into a set of four nonlinear ordinary differential equations (ODE), which may be solved numerically. However, the solution is obtained on an irregular grid, which consists of the points of the (x,t) plane located at the intersection of two characteristics: a forward characteristic and a backward characteristic. This requires us to perform an interpolation in order to compute the flow variables at fixed locations.

2.2.1.2 The Hartree Method

The Hartree method enables to use the characteristics method with a fixed grid in the (x,t) plane. Let us assume that the flow properties are known at time $t = (k-1)\Delta t$, and that we want to compute the flow at point $\mathrm{C}(x_C, k\Delta t)$. The idea is to compute the characteristics backwards in time and to interpolate the solution at time t (see Fig. 2.8).

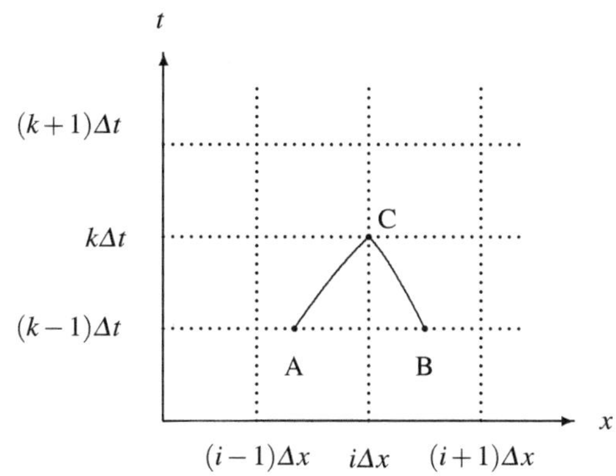

Fig. 2.8 Numerical resolution with the Hartree method

The abscissa of points A and B do not generally coincide with the fixed grid points. It is then necessary to interpolate the velocity and celerity at points A and B between $x = (i-1)\Delta x$ and $x = i\Delta x$, and between $x = i\Delta x$ and $x = (i+1)\Delta x$, respectively. The corresponding discrete characteristic equations are given by:

$$V_C + 2C_C = V_A + 2C_A + g(S_b - S_{fA})\Delta t, \tag{2.13a}$$

$$x_C - x_A = (V_A + C_A)\Delta t, \tag{2.13b}$$

$$V_C - 2C_C = V_B - 2C_B + g(S_b - S_{fB})\Delta t, \tag{2.13c}$$

$$x_C - x_B = (V_B - C_B)\Delta t. \tag{2.13d}$$

The resulting system has four unknowns, V_C, C_C, x_A, and x_B, where the subscripts refer to the points A, B, and C in Fig. 2.8.

The system (2.13) can be solved explicitly using a linear interpolation for the values of V and C at points A and B.

The method of characteristics is an explicit method, which is subject to the Courant–Friedrichs–Levy stability condition:

$$\Delta t < \frac{\Delta x}{|V \pm C|}. \tag{2.14}$$

Therefore, according to the defined space step of the considered problem, the time step size is limited by this condition. This may lead to large computation times.

2.2.2 The Preissmann Implicit Scheme

This implicit scheme was first proposed by Alexandre Preissmann in 1961 when he was hydraulic engineer at SOGREAH, Grenoble, France [14]. In the general Preissmann scheme, a continuous function $f(x,t)$ and its partial derivatives are represented by [6, 22]:

$$f(x,t) = \theta\left[\phi f_{i+1}^{k+1} + (1-\phi) f_i^{k+1}\right] + (1-\theta)\left[\phi f_{i+1}^{k} + (1-\phi) f_i^{k}\right], \tag{2.15a}$$

$$\frac{\partial f}{\partial x}(x,t) = \theta \frac{f_{i+1}^{k+1} - f_i^{k+1}}{\Delta x} + (1-\theta)\frac{f_{i+1}^{k} - f_i^{k}}{\Delta x}, \tag{2.15b}$$

$$\frac{\partial f}{\partial t}(x,t) = \phi \frac{f_{i+1}^{k+1} - f_{i+1}^{k}}{\Delta t} + (1-\phi)\frac{f_{i}^{k+1} - f_i^{k}}{\Delta t}, \tag{2.15c}$$

where i is the space index, k the time index and $\theta \in [0,1]$, $\phi \in [0,1]$ are weighting coefficients. The generalized Preissmann scheme is depicted in Fig. 2.9.

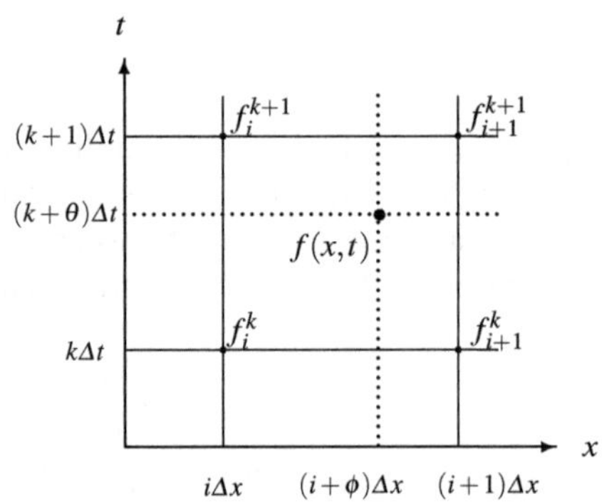

Fig. 2.9 Preissmann implicit discretization scheme

The Preissmann scheme is second-order accurate in both time and space if $\theta = 0.5$ and $\phi = 0.5$, and first-order accurate otherwise. Linear stability analysis shows that the centered scheme ($\phi = 0.5$) is unconditionally stable for $\theta \geq 0.5$. This feature makes it very interesting for practical applications, since contrarily to the case of an explicit scheme, it is not subject to the Courant–Friedrichs–Levy stability condition (2.14) that constrains the time step to small values.

This may explain why the Preissmann scheme has become the standard method for one-dimensional numerical modeling in the field of hydraulic engineering. A stability analysis including convective and friction terms has shown that another condition is necessary for numerical stability in addition to the condition $\theta \geq 0.5$

[16, 22]: the Vedernikov number $\mathscr{V}$ must be smaller than 1, where $\mathscr{V}$ is defined by

$$\mathscr{V} = \frac{a}{b}\frac{A}{R}\frac{\mathrm{d}R}{\mathrm{d}A}F,$$

where a is the exponent on the hydraulic radius and b the exponent on the velocity in the evaluation of the friction slope, A is the cross-sectional area of flow, R the hydraulic radius, and F the Froude number. In the Manning formulation of the friction slope of (2.3), we have $a = \frac{4}{3}$ and $b = 2$.

2.3 Steady Flow Solutions

The steady flow solutions of Saint-Venant equations are obtained by replacing $\frac{\partial}{\partial t}$ by 0 in (2.1) and (2.2). Then, denoting the variables corresponding to the equilibrium regime with a subscript zero ($Q_0(x)$, $Y_0(x)$, etc.), the Saint-Venant equations become:

$$\frac{\mathrm{d}Q_0(x)}{\mathrm{d}x} = 0, \tag{2.16a}$$

$$\frac{\mathrm{d}Y_0(x)}{\mathrm{d}x} = \frac{S_b - S_{f0}(x)}{1 - F_0(x)^2}. \tag{2.16b}$$

F_0 is the Froude number $F_0 = \frac{V_0}{C_0}$ with $C_0 = \sqrt{g\frac{A_0}{T_0}}$, $V_0 = \frac{Q_0}{A_0}$, and T_0 is the top width.

These two equations define an equilibrium regime given by $Q_0(x) = Q_0 = Q_L$ and $Y_0(x)$ solution of the ordinary differential equation (2.16b), for a boundary condition in terms of downstream elevation.

When the right-hand side of (2.16b) is equal to zero, the water depth is constant along the channel. In this case, given $Q_0(x) = Q_0$, the equilibrium solution $Y_0(x) = Y_n$ (also called *normal depth*) can be deduced by solving the following algebraic equation:

$$S_f(Q_0, Y_n) = S_b. \tag{2.17}$$

This specific solution corresponds to the *uniform flow regime*. Equation (2.17) is usually solved numerically with a fixed-point or Newton–Raphson method.

In some cases, the uniform depth can be computed analytically. For large rectangular channels, the hydraulic radius R can be approximated by the water depth Y, and the Manning equation (2.3) reduces to:

$$S_f = \frac{Q^2 n^2}{T^2 Y^{10/3}}. \tag{2.18}$$

Combining (2.17) and (2.18) give the uniform depth Y_n corresponding to a discharge Q_0 in the large rectangular case:

$$Y_n = \left(\frac{Q_0^2 n^2}{T^2 S_b}\right)^{3/10}. \tag{2.19}$$

The uniform flow regime will be specifically developed in the book as an example, since this regime leads to closed-form analytical solutions for the transfer matrix (see Chap. 3). However, we will also show that this specific flow regime is not qualitatively different from realistic nonuniform flow regimes.

Example 2.1. Throughout the book, we will illustrate our results on two trapezoidal prismatic channels, with different characteristics (see Fig. 2.10).

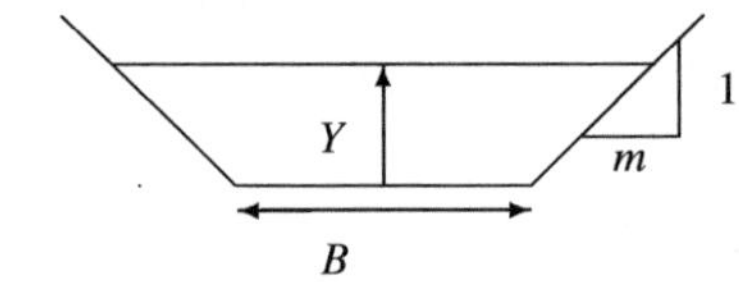

Fig. 2.10 Section of a trapezoidal canal

The channel characteristics are given in Table 2.1, where L is the channel length (m), m the bank slope, B the bed width (m), S_b the bed slope (m/m), n the Manning coefficient ($\mathrm{m}^{-1/3}\mathrm{s}$), and Y_n the normal depth (m) corresponding to the discharge $Q_{\max}$ ($\mathrm{m}^3\mathrm{s}^{-1}$). Canal 1 is a short oscillating canal and canal 2 is a long delayed canal.

Table 2.1 Parameters for the two canals

	L	m	B	S_b	n	Y_n	$Q_{\max}$
Canal 1	3000	1.5	7	0.0001	0.02	2.12	14
Canal 2	6000	1.5	8	0.0008	0.02	2.92	80

Keeping a constant discharge, we compare the backwater curves obtained for different downstream boundary conditions $Y_0(L) = Y_n \times [0.8, 1, 1.2]$ (see Fig. 2.11). We have two types of backwater curves: if $Y_0(L) > Y_n$, then the flow is decelerating along x, this is a so-called "M1" curve (see [20]), and if $Y_0(L) < Y_n$, then the flow is accelerating along x, this is an "M2" curve. The M1 curves are the most widely observed in practice, since they may occur upstream of any kind of obstacle in the flow (hydraulic structure, gate, weir, bridge, etc.). The M2 type curves are observed when there is a sudden drop in the canal or a steep slope downstream.

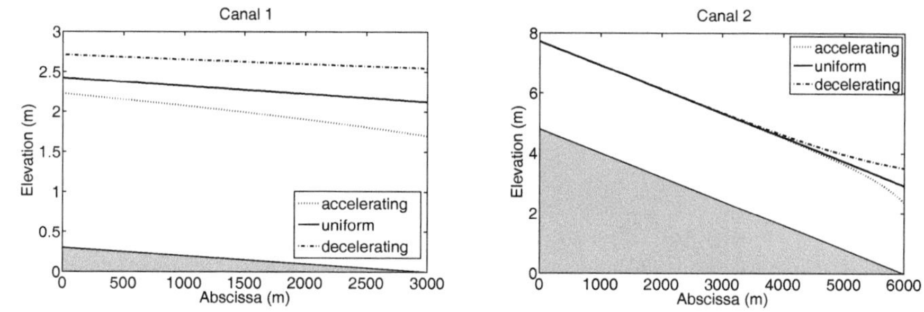

Fig. 2.11 Backwater curves for canals 1 and 2, and various downstream boundary conditions $Y_0(L)$

2.4 Linearized Saint-Venant Equations

We now study the linearized equations around a given steady-state. This is restrictive, since it cannot capture dynamic features such as shocks. Nonetheless, this enables us to use the powerful tools developed for linear systems and, as we will show in Chap. 3, it provides a very detailed analysis of open channel dynamics.

2.4.1 Derivation of the Linearized Equations

The Saint-Venant equations are linearized around an equilibrium steady-state defined by $(Q_0(x), Y_0(x))$. The linearized equations are obtained by putting $Q(x,t) = Q_0(x) + q(x,t)$ and $Y(x,t) = Y_0(x) + y(x,t)$ into (2.1) and (2.2) and expanding in series.

For a given term $f(Q,Y)$ of the Saint-Venant equations, its Taylor expansion can be written as:

$$f(Q,Y) - f(Q_0,Y_0) = f_Q(Q_0,Y_0)q(x,t) + f_Y(Q_0,Y_0)y(x,t) + \text{HOT} \qquad (2.20)$$

where f_Q and f_Y are respectively the partial derivative of f with respect to Q and Y, respectively and HOT stands for higher order terms.

Since we are looking for a *linear* approximation, we keep only the linear terms in y and q in the expansion. All quadratic (and higher order) terms are supposed to be negligible in front of linear terms, even if this is false for large deviations from the equilibrium values. Indeed, the linear approximation is only valid for small deviations from the equilibrium values.

Equation (2.1) then gives:

$$T_0(x)\frac{\partial y}{\partial t} + \frac{\partial q}{\partial x} = 0.$$

After collecting terms in q, y, $\frac{\partial q}{\partial x}$ and $\frac{\partial y}{\partial x}$, (2.2) is linearized as:

$$\frac{\partial q}{\partial t}+2V_0\frac{\partial q}{\partial x}+\delta q+(C_0^2-V_0^2)T_0\frac{\partial y}{\partial x}-\tilde{\gamma}y=0, \tag{2.21}$$

with

$$\tilde{\gamma}=V_0^2\frac{\mathrm{d}T_0}{\mathrm{d}x}+gT_0\Big(\kappa S_{f0}+S_b-(1+2F_0^2)\frac{\mathrm{d}Y_0}{\mathrm{d}x}\Big), \tag{2.22}$$

$$\delta=\frac{2g}{V_0}\Big(S_{f0}-F_0^2\frac{\mathrm{d}Y_0}{\mathrm{d}x}\Big), \tag{2.23}$$

$$\kappa=\frac{7}{3}-\frac{4A_0}{3T_0P_0}\frac{\partial P_0}{\partial Y}, \tag{2.24}$$

and $F_0^2=\frac{V_0^2T_0}{gA_0}$ is the Froude number for the equilibrium regime. Dependence on x is omitted for readability.

The model is obtained by linearization around an equilibrium regime defined by (2.16b). Therefore, substituting the expression of S_{f0} obtained from (2.16b) into (2.22) and (2.23), leads to

$$\tilde{\gamma}=V_0^2\frac{\mathrm{d}T_0}{\mathrm{d}x}+gT_0\Big[(1+\kappa)S_b-(1+\kappa-(\kappa-2)F_0^2)\frac{\mathrm{d}Y_0}{\mathrm{d}x}\Big],$$

$$\delta=\frac{2g}{V_0}\Big(S_b-\frac{\mathrm{d}Y_0}{\mathrm{d}x}\Big).$$

The linearized model of open channel flow is therefore given by the two linear partial differential equations:

$$T_0\frac{\partial y}{\partial t}+\frac{\partial q}{\partial x}=0, \tag{2.25a}$$

$$\frac{\partial q}{\partial t}+2V_0\frac{\partial q}{\partial x}+(C_0^2-V_0^2)T_0\frac{\partial y}{\partial x}+\delta q-\tilde{\gamma}y=0. \tag{2.25b}$$

The initial condition and the boundary conditions are given by:

$$q(x,0)=q_0(x),\quad y(x,0)=y_0(x), \tag{2.26a}$$

$$q(0,t)=u_1(t),\quad q(L,t)=u_2(t). \tag{2.26b}$$

We also assume that the water level deviations are measured at each boundary: $y(0,t)$ and $y(L,t)$.

One may show that with such initial and boundary conditions, the linearized equations are well-posed and satisfy some mathematical properties (see Appendix E for details).

To simplify the notations and to facilitate the forthcoming mathematical analysis, we will use an alternative expression for the linearized Saint-Venant equations, by defining $a(x,t)=T_0(x)y(x,t)$, and expressing the equations as a function of $q(x,t)$

and $a(x,t)$. We also define

$$\alpha(x) = C_0(x) + V_0(x)$$
$$\beta(x) = C_0(x) - V_0(x)$$
$$\gamma(x) = \frac{1}{T_0(x)}\left(\tilde{\gamma}(x) + (C_0^2(x) - V_0^2(x))\frac{\mathrm{d}T_0}{\mathrm{d}x}(x)\right).$$

The linearized equations are then given by:

$$\frac{\partial a}{\partial t} + \frac{\partial q}{\partial x} = 0, \tag{2.27a}$$

$$\frac{\partial q}{\partial t} + (\alpha - \beta)\frac{\partial q}{\partial x} + \alpha\beta\frac{\partial a}{\partial x} + \delta q - \gamma a = 0, \tag{2.27b}$$

with

$$\gamma = \frac{C_0^2}{T_0}\frac{\mathrm{d}T_0}{\mathrm{d}x} + g\left[(1+\kappa)S_b - (1+\kappa-(\kappa-2)F_0^2)\frac{\mathrm{d}Y_0}{\mathrm{d}x}\right].$$

Equations (2.27a–2.27b) are simpler for the mathematical analysis, and are equivalent to (2.25a–2.25b). The solution in terms of variations of water level $y(x,t)$ can readily be deduced by dividing by $T_0(x)$ the solution obtained in terms of variations of wetted area $a(x,t) = T_0(x)y(x,t)$.

Finally, we can rewrite the linearized Saint-Venant equations as follows:

$$\frac{\partial \xi}{\partial t} + \mathbf{A}(x)\frac{\partial \xi}{\partial x} + \mathbf{B}(x)\xi = 0, \tag{2.28}$$

where $\xi(x,t) = (a(x,t), q(x,t))^T : [0,L] \times [0,+\infty) \to \Omega \in \mathbb{R}^2$ is the state of the system and $\mathbf{A}(x) = \begin{pmatrix} 0 & 1 \\ \alpha(x)\beta(x) & \alpha(x)-\beta(x) \end{pmatrix}$, $\mathbf{B}(x) = \begin{pmatrix} 0 & 0 \\ -\gamma(x) & \delta(x) \end{pmatrix}$.

Example 2.2 (Parameters of the linearized Saint-Venant equations for the example canals). Table 2.2 gives the parameters of the example canals at uniform flow.

Table 2.2 Parameters for the two canals at uniform flow

	L	α	β	γ	δ	T_0
Canal 1	3000	4.63	3.33	0.0027	0.003	13.37
Canal 2	6000	6.81	2.39	0.0218	0.0071	16.77

Let us note that the linearized Saint-Venant equations are hyperbolic equations, since the eigenvalues of matrix $\mathbf{A}$, namely $\alpha(x)$ and $-\beta(x)$, are real. In the subcritical flow case, the eigenvalues are of opposite sign.

2.4.2 Characteristic Form

Let us introduce the following change of variable, which corresponds to the characteristic form of the linearized equations:

$$\chi_1(x,t) = q(x,t) + \beta(x)T_0(x)y(x,t), \tag{2.29a}$$
$$\chi_2(x,t) = q(x,t) - \alpha(x)T_0(x)y(x,t). \tag{2.29b}$$

The inverse transform is given by:

$$y(x,t) = \frac{1}{T_0(\alpha+\beta)}[\chi_1(x,t) - \chi_2(x,t)], \tag{2.30a}$$
$$q(x,t) = \frac{1}{\alpha+\beta}[\alpha\chi_1(x,t) + \beta\chi_2(x,t)]. \tag{2.30b}$$

Using a matrix notation and the fact that $\xi = (T_0 y, q)^T$, we have:

$$\chi(x,t) = \mathbf{X}(x)\xi(x,t),$$

with $\mathbf{X}(x) = \begin{pmatrix} \beta(x) & 1 \\ -\alpha(x) & 1 \end{pmatrix}$.

This change of variable enables us to diagonalize matrix $\mathbf{A}(x)$ as follows:

$$\mathbf{A}(x) = \mathbf{X}(x)^{-1}\mathbf{D}(x)\mathbf{X}(x),$$

with

$$\mathbf{D}(x) = \begin{pmatrix} \alpha(x) & 0 \\ 0 & -\beta(x) \end{pmatrix},$$

and

$$\mathbf{X}(x)^{-1} = \frac{1}{\alpha(x)+\beta(x)} \begin{pmatrix} 1 & -1 \\ \alpha(x) & \beta(x) \end{pmatrix}.$$

Equation (2.28) can then be rewritten as:

$$\frac{\partial\chi}{\partial t} + \mathbf{D}(x)\frac{\partial\chi}{\partial x} + \mathbf{E}(x)\chi = 0, \tag{2.31}$$

with $\mathbf{E}(x) = [\mathbf{X}(x)\mathbf{B}(x) - \mathbf{D}(x)\mathbf{X}'(x)]\mathbf{X}(x)^{-1}$, where $f'(x)$ denotes the derivative of f with respect to x.

This change of variable corresponds to the characteristic form of the equations. Indeed, the new variable $\chi(x,t)$ verifies:

$$\frac{d\chi_1}{dt}(x_1,t) = -e_{11}(x_1)\chi_1(x_1,t) - e_{12}(x_1)\chi_2(x_1,t), \tag{2.32a}$$

$$\frac{dx_1}{dt} = \alpha(x_1), \tag{2.32b}$$

$$\frac{d\chi_2}{dt}(x_2,t) = -e_{21}(x_2)\chi_1(x_2,t) - e_{22}(x_2)\chi_2(x_2,t), \tag{2.32c}$$

$$\frac{dx_2}{dt} = -\beta(x_2), \tag{2.32d}$$

with $\mathbf{E}(x) = (e_{ij}(x))$ defined by:

$$e_{11} = \frac{1}{\alpha+\beta}\left[-\gamma+\alpha\delta-\alpha\beta'\right], \tag{2.33a}$$

$$e_{12} = \frac{1}{\alpha+\beta}\left[\gamma+\beta\delta+\alpha\beta'\right], \tag{2.33b}$$

$$e_{21} = \frac{1}{\alpha+\beta}\left[-\gamma+\alpha\delta-\beta\alpha'\right], \tag{2.33c}$$

$$e_{22} = \frac{1}{\alpha+\beta}\left[\gamma+\beta\delta+\beta\alpha'\right], \tag{2.33d}$$

where we have dropped the argument x for readability, and α' and β' denote the derivatives of α and β with respect to x.

This change of variable leads to the characteristic form of the linearized Saint-Venant equations. Indeed, one may show that the variables $\chi_1(x,t)$ and $\chi_2(x,t)$ are closely linked to the characteristic variables $j_1(x,t) = v(x,t) + 2c(x,t)$ and $j_2(x,t) = v(x,t) - 2c(x,t)$, which are the deviations from the characteristic variables J_1 and J_2 around the reference steady flow. To this end, we express $\chi_1(x,t)$ and $\chi_2(x,t)$ in terms of $v(x,t)$ and $c(x,t)$, the variations of velocity and celerity from the steady flow values $V_0(x)$ and $C_0(x)$. Such a link is especially simple in the case of a rectangular channel. Indeed, in that case, we have $v = \frac{q}{T_0 Y_0} - \frac{V_0}{Y_0}y$ and $c = \frac{C_0}{2Y_0}y$, which gives in turn:

$$y = \frac{2Y_0}{C_0}c,$$

$$q = T_0 Y_0\left(v + 2\frac{V_0}{C_0}c\right).$$

Substituting this into the expressions of $\chi(x,t)$ yields:

$$\chi_1 = q + (C_0 - V_0)T_0 y = T_0 Y_0(v+2c),$$

$$\chi_2 = q - (C_0 + V_0)T_0 y = T_0 Y_0(v-2c).$$

Therefore the variables $\chi_1(x,t)$ and $\chi_2(x,t)$ are proportional to the characteristics variables $j_1(x,t)$ and $j_2(x,t)$.

Remark 2.1 (Riemann coordinates in nonuniform flow). From the expression of matrix $\mathbf{E}(x)$, we see that even in the case where the friction and the slope can be neglected ($\delta(x) \approx 0$ and $\gamma(x) \approx 0$), but when the flow is not uniform (e.g., because of a nonuniform geometry), the Riemann coordinates are no longer invariants. This also appears in the general characteristic form, with a supplementary term on the right-hand side due to the nonprismatic geometry.

We now consider simplified models obtained from the Saint-Venant equations.

2.5 Approximate Hydraulic Models for Flow Routing

The Saint-Venant equations require a lot of data (geometry of the channel, longitudinal profile, roughness coefficient) that are not always available in the case of rivers. There are also cases where some terms can be neglected in the equations, leading to simplified models of open channel flow.

2.5.1 *The Diffusive Wave*

Assuming the inertia terms ($\frac{\partial Q}{\partial t} + \frac{\partial Q^2/A}{\partial x}$) are negligible with respect to ($Ag\frac{\partial z}{\partial x}$), the water elevation $z = Y + S_b x$ can be eliminated from the Saint-Venant equations (2.1) and (2.2), leading to the diffusive wave equation [12]:

$$\frac{\partial Q}{\partial t} + C(Q,z,x)\frac{\partial Q}{\partial x} - D(Q,z,x)\frac{\partial^2 Q}{\partial x^2} = 0, \tag{2.34}$$

with $Q(x,t)$ the discharge (m^3/s), $C(Q,z,x)$ the celerity (m/s), and $D(Q,z,x)$ the diffusion (m^2/s).

C and D are given by the following formulas:

$$C(Q,z,x) = \frac{1}{T^2\frac{\partial S_f}{\partial Q}}\left[\frac{\partial T}{\partial x} - \frac{\partial (TS_f)}{\partial z}\right], \tag{2.35}$$

$$D(Q,z,x) = \frac{1}{T\frac{\partial S_f}{\partial Q}}, \tag{2.36}$$

where T stands for the water surface top width.

Supposing uniform geometry and uniform flow regime, one may obtain analytical expressions of C and D in some cases.

With a uniform geometry, (2.35) and (2.36) give $C = 5Q/(3A)$ and $D = Q/(2TS_b)$, where S_b stands for the bed slope of the reach.

In the case of a large rectangular channel, the normal depth is much lower than the surface width $Y \ll T$. The hydraulic radius can be approximated by the normal depth $R \simeq Y$. Manning Strickler equation (2.3) then reads:

$$Q = KA\sqrt{S_b}Y^{2/3}.$$

As $A = TY$, one gets an explicit expression for Y:

$$Y = \left(\frac{Q}{KT\sqrt{S_b}}\right)^{3/5}.$$

C and D can then be expressed as functions of the discharge Q:

$$C = \frac{5K^{3/5}S_b^{3/10}}{3T^{2/5}}Q^{2/5},$$
$$D = \frac{Q}{2TS_b}.$$

Using the variables $\Theta = 1/(3.6C)$ (hydraulic time-lag expressed in h/km) and $Z = 250C/D$ (hydraulic diffusion expressed in km^{-1}) one gets:

$$\Theta = \alpha_\Theta Q^{\beta_\Theta},$$
$$Z = \alpha_Z Q^{\beta_Z},$$

with $\alpha_\Theta = 6T^{-0.4}K^{0.6}S_b^{0.3}$, $\beta_\Theta = -0.4$, $\alpha_Z = 625/3K^{0.6}T^{0.6}S_b^{1.3}$, $\beta_Z = -0.6$.

2.5.2 The Hayami Model

The linear equation obtained with the hypothesis that C and D are constant is called the Hayami equation. This linear hypothesis is valid when the discharge stays within a limited range of variation.

The Hayami equation is a linear partial derivative equation representing the discharge transfer in a river reach around an equilibrium point:

$$\frac{\partial q}{\partial t} + C\frac{\partial q}{\partial x} - D\frac{\partial^2 q}{\partial x^2} = 0, \tag{2.37}$$

where

- q is the discharge (m^3/s,
- C the celerity coefficient (m/s), and
- D the diffusion coefficient (m^2/s).

This parabolic partial differential equation is closely related to the heat equation, which can be controlled using backstepping-type methods [10].

2.6 Summary

In this chapter, we have derived the Saint-Venant equations, which are the classical distributed parameter model for one-dimensional open channel flow. These equations are based on a series of assumptions that proved to be very efficient for describing flow in rivers, canals, and sewers. The equations can be put in the characteristic form, which is useful for understanding the underlying physical phenomenon described by the equations.

We then derived a linearized Saint-Venant model, valid for small variations around an equilibrium flow regime. This model consists of two linear partial differential equations, which will be analyzed in detail and used as a basis for controller design in the following. In fact, it would be desirable to be able to use the nonlinear Saint-Venant equations to design automatic controllers, but this is a very difficult problem, and remains open at the moment. Some preliminary results have been obtained for the horizontal frictionless case using Riemann invariants [9], but much more needs to be done to be able to stabilize the full nonlinear Saint-Venant equations.

The linearized equations are derived around a nonuniform steady flow regime, following the approach developed in [17, 11].

The mathematical analysis of the linearized equations is performed in Appendix E.

References

[1] Abbott M (1966) An introduction to the method of characteristics. Elsevier, New York
[2] Abbott MB (1979) Computational hydraulics. Elements of the theory of free surface flows. Pitman Publishing, London, 324 p
[3] Bardos C, Le Roux A, Nedelec J (1979) First order quasilinear equations with boundary conditions. Commun in Part Diff Eqs 4(9):1017–1034
[4] Boussinesq VJ (1877) Essai sur la théorie des eaux courantes, vol 23. Mémoires présentés par divers savants à l'Académie des Sciences, Paris
[5] Chow V (1988) Open-channel hydraulics. McGraw-Hill, New York, 680 p
[6] Cunge J, Holly F, Verwey A (1980) Practical aspects of computational river hydraulics. Pitman Advanced Publishing Program
[7] Evans L (1998) Partial differential equations, Graduate studies in mathematics, vol 19. American Mathematical Society, Providence, RI
[8] Favre H (1935) Etude théorique et expérimentale des ondes de translation dans les canaux découverts. Dunod, Paris
[9] de Halleux J, Prieur C, Coron JM, d'Andréa Novel B, Bastin G (2003) Boundary feedback control in networks of open-channels. Automatica 39:1365–1376
[10] Krstic M, Smyshlyaev A (2008) Boundary control of PDEs: A course on backstepping designs. SIAM
[11] Litrico X, Fromion V (2004) Frequency modeling of open channel flow. J Hydraul Eng 130(8):806–815
[12] Miller W, Cunge J (1975) Simplified equations of unsteady flow, in Mahmood K, Yevjevich V (eds.), *Unsteady flow in open channels*. Water Resources Publications, Fort Collins, CO, pp 183–257

[13] Mohapatra PK, Chaudhry MH (2004) Numerical solution of Boussinesq equations to simulate dam-break flows. J Hydraul Eng 130(2):156–159
[14] Preissmann A (1961) Propagation des intumescences dans les canaux et rivières. In: 1er Congrès de l'Association Française de Calcul, Grenoble, France, pp 433–442
[15] Barré de Saint-Venant A (1871) Théorie du mouvement non-permanent des eaux avec application aux crues des rivières et à l'introduction des marées dans leur lit. Comptes rendus Acad Sci Paris 73:148–154, 237–240
[16] Samuels P, Skeels C (1990) Stability limits for Preissmann's scheme. J Hydraul Eng 116(8):997–1012
[17] Schuurmans J (1997) Control of water levels in open-channels. PhD thesis, ISBN 90-9010995-1, Delft University of Technology
[18] Soares-Frazao S, Zech Y (2002) Undular bores and secondary waves - experiments and hybrid finite-volume modelling. J Hydr Res 40(1):33–43
[19] Strub I, Bayen A (2006) Weak formulation of the boundary condition for scalar conservation laws: an application to highway traffic modelling. Int J Robust Nonlin Contr 16(16):733–748
[20] Sturm T (2001) Open-channel hydraulics. McGraw-Hill, New York
[21] Treske A (1994) Undular bores (Favre waves) in open channels – experimental studies. J Hydr Res 32(3):355–370
[22] Venutelli M (2002) Stability and accuracy of weighted four-point implicit finite difference schemes for open channel flow. J Hydraul Eng 128(3):281–288

Chapter 3
Frequency Domain Analysis of Open Channel Flow

Abstract In this chapter, we analyze the linearized Saint-Venant equations. First the horizontal frictionless case, then the uniform flow case, and finally the nonuniform flow case are treated in depth, with a complete characterization of the Saint-Venant transfer matrix in terms of poles, delay, and series expansion. For the two former cases, the analysis of the system properties is possible due to the existence of closed-form expressions for the transfer functions. By contrast, for the latter case, in order to bypass the absence of closed-form expressions, we provide a complete and new approach based on the use of efficient and convenient numerical schemes.

3.1 Introduction

Our objective in this chapter is to fully characterize the dynamic behavior of the Saint-Venant equations linearized around any steady flow regime. We first study the specific cases of horizontal frictionless channel and uniform flow conditions, where analytical expressions can be obtained. These results enable us to obtain a very detailed description of the open channel flow dynamics. However, most realistic flow patterns are not uniform, which is why it is very important to be able to extend the analysis to nonuniform flow cases.

To this end, we develop an efficient numerical method to compute the frequency response of an open channel around nonuniform flow. This method also enables us to extend to the nonuniform case most of the results obtained for the uniform case, in terms of characterization of the transfer matrix.

These results provide an essential tool to control the open channel flow. Having the complete frequency response for any flow regime enables us to use Bode plots, Nyquist plots, and Nichols charts to design and analyze controllers.

We will propose methods to derive rational and simplified models for control purposes in Chaps. 4 and 5.

We begin with the horizontal frictionless case, which enables us to present in a simple way the key elements of the analysis, and shows the connection between the transfer function and the characteristic.

3.2 Horizontal Frictionless Channel Case

Let us recall here the linearized Saint-Venant equations (2.28) in the horizontal frictionless case:

$$\frac{\partial \xi}{\partial t} + \mathbf{A}\frac{\partial \xi}{\partial x} = 0, \tag{3.1}$$

where $\xi(x,t) = (T_0 y(x,t), q(x,t))^T$ is the state of the system, and $\mathbf{A} = \begin{pmatrix} 0 & 1 \\ \alpha\beta & \alpha - \beta \end{pmatrix}$, $\alpha = C_0 + V_0$, and $\beta = C_0 - V_0$ are constant.

We now use the Laplace transform to derive the transfer matrix. We denote in the following $\mathscr{L}(f) = \hat{f}(s)$ the Laplace transform of a function $f(t)$. The Laplace transform is defined by:

$$\mathscr{L}(f) = \hat{f}(s) = \int_0^\infty e^{-st} f(t)\mathrm{d}t, \tag{3.2}$$

and has the following useful property:

$$\mathscr{L}(f'(t)) = s\hat{f}(s) - f(0), \tag{3.3}$$

where $f'(t)$ denotes the time derivative of $f(t)$, and $f(0)$ is the value of f at time $t = 0$.

3.2.1 Derivation of the Transfer Matrix

We first derive the state-transition matrix, then use the boundary conditions to compute the poles of the system.

3.2.1.1 State-transition Matrix

Following the results of Appendix E, (3.1) is well-defined, and thus Laplace transformable[1]. Applying the Laplace transform on (3.1), and using property (3.3) leads to:

$$\frac{\partial \hat{\xi}(x,s)}{\partial x} = -s\mathbf{A}^{-1}\hat{\xi}(x,s) + \mathbf{A}^{-1}\xi_0(x), \tag{3.4}$$

[1] A more detailed justification can be found in [9], p. 507.

where $\xi_0(x) = \xi(x,0)$ is the initial condition at $t = 0$.

The general solution of this ordinary differential equation in x is then given by:

$$\hat{\xi}(x,s) = \Phi(x,s)\hat{\xi}(0,s) + \Phi(x,s)\int_0^x \Phi(v,s)^{-1}\mathbf{A}^{-1}\xi_0(v)\mathrm{d}v, \tag{3.5}$$

with $\Phi(x,s)$ the state-transition matrix, given by $\Phi(x,s) = e^{-s\mathbf{A}^{-1}x}$.

The exponential of a matrix is not easy to compute in general, but when the matrix is diagonal, then its exponential is easily obtained as follows:

$$\exp(\mathrm{diag}(\lambda_i)) = \mathrm{diag}(\exp(\lambda_i)).$$

Let us therefore diagonalize matrix $\mathbf{A}$. We have

$$\mathbf{A} = \mathbf{X}^{-1}\mathbf{D}\mathbf{X},$$

with $\mathbf{D} = \begin{pmatrix} \alpha & 0 \\ 0 & -\beta \end{pmatrix}$, $\mathbf{X} = \begin{pmatrix} \beta & 1 \\ -\alpha & 1 \end{pmatrix}$, and $\mathbf{X}^{-1} = \frac{1}{\alpha+\beta}\begin{pmatrix} 1 & -1 \\ \alpha & \beta \end{pmatrix}$.

Then, the state-transition matrix $\Phi(x,s)$ can be computed as follows:

$$\Phi(x,s) = e^{-sx\mathbf{A}^{-1}} = \mathbf{X}^{-1}e^{-sx\mathbf{D}^{-1}}\mathbf{X}. \tag{3.6}$$

The elements of matrix $\Phi(x,s)$ are then given by:

$$\phi_{11}(x,s) = \frac{\beta e^{-\frac{sx}{\alpha}} + \alpha e^{\frac{sx}{\beta}}}{\alpha+\beta}, \tag{3.7a}$$

$$\phi_{12}(x,s) = \frac{e^{-\frac{sx}{\alpha}} - e^{\frac{sx}{\beta}}}{\alpha+\beta}, \tag{3.7b}$$

$$\phi_{21}(x,s) = \frac{\alpha\beta}{\alpha+\beta}\left(e^{-\frac{sx}{\alpha}} - e^{\frac{sx}{\beta}}\right), \tag{3.7c}$$

$$\phi_{22}(x,s) = \frac{\alpha e^{-\frac{sx}{\alpha}} + \beta e^{\frac{sx}{\beta}}}{\alpha+\beta}. \tag{3.7d}$$

Let us note that this diagonalization corresponds to the change of variable used in Sect. 2.4.2 to obtain the characteristics variables.

3.2.1.2 Boundary Conditions

For simplicity, we assume zero initial conditions in the following. The boundary conditions have a great impact on the dynamics of the distributed system. In our case, the boundary conditions are the deviations in upstream and downstream discharge.

Let us apply (3.5) at $x = 0$ to express $\hat{y}(0,s)$ as a function of the boundary controls $\hat{q}(0,s)$ and $\hat{q}(L,s)$:

$$T_0\hat{y}(0,s) = -\frac{\phi_{22}(L,s)}{\phi_{21}(L,s)}\hat{q}(0,s) + \frac{1}{\phi_{21}(L,s)}\hat{q}(L,s). \tag{3.8}$$

This algebraic manipulation is valid as long as $\phi_{21}(L,s) \neq 0$. In fact, the values of s such that $\phi_{21}(L,s) = 0$ are the poles of the system.

3.2.1.3 Poles

The poles of a transfer function $P(s)$ are the values of s such that $P(s)$ is singular. The poles characterize the dynamic behavior of a linear system. Moreover, as we will show hereafter, there is a strong link between the time and spatial behavior of the open channel. The poles will be useful to illustrate this coupling between time and space.

In our case, the poles of the Saint-Venant transfer matrix in the horizontal frictionless case are obtained as solutions of $\phi_{21}(L,s) = 0$, or:

$$e^{-(\tau_1+\tau_2)s} = 1,$$

with $\tau_1 = \frac{L}{\alpha}$ and $\tau_2 = \frac{L}{\beta}$, which leads to:

$$(\tau_1 + \tau_2)s = \pm 2jk\pi,$$

with $k \in \mathbb{N}$.

Therefore the poles are given by:

$$p_{\pm k} = \pm\frac{2jk\pi}{\tau_1 + \tau_2}. \tag{3.9}$$

There is an infinite number of poles, which is linked to the fact that the system is described by a partial differential equation, therefore it is an infinite dimensional system. All the poles are located on the imaginary axis. There is a pole at zero (the open channel acts as an integrator for low frequencies) and the other poles are complex conjugate, with a zero real part. The poles are evenly distributed along the imaginary axis, separated by $\frac{2\pi}{\tau_1+\tau_2}$.

Example 3.1 (Poles of a horizontal frictionless channel). As an example, we consider an open channel with the following parameters: $\alpha = 4.63$ m/s, $\beta = 3.33$ m/s, and $L = 3000$ m. These parameters correspond to the linearized Saint-Venant equations representing a horizontal frictionless channel with trapezoidal geometry (bed width of 7 m, side slope of 1.5 m/m), around a stationary regime corresponding to a discharge $Q_0 = 14$ m^3/s and a water depth $Y_0 = 2.12$ m.

In that case, the delays are given by $\tau_1 = 648$ s and $\tau_2 = 900$ s. The poles of the canal are therefore given by $p_{\pm k} = \pm\frac{jk\pi}{774} = \pm jk4 \times 10^{-4}$ rad/s. The first 21 poles are depicted in Fig. 3.1. The first pole corresponds to the resonant frequency of the first oscillating mode in Figs. 3.2 and 3.3.

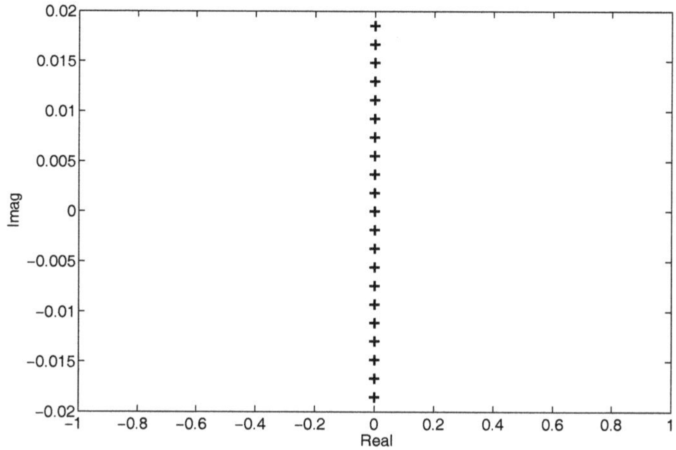

Fig. 3.1 Poles of a horizontal frictionless channel in the complex plane

In fact, one may derive an equation for $q(x,t)$ by eliminating the water depth $y(x,t)$ in (2.25b) using the mass conservation equation (2.25a). This yields the following wave equation:

$$\left(\frac{\partial}{\partial x}+\frac{1}{\alpha}\frac{\partial}{\partial t}\right)\left(\frac{\partial}{\partial x}-\frac{1}{\beta}\frac{\partial}{\partial t}\right)q=0,$$

with boundary conditions $q(0,t)$ and $q(L,t)$.

This is another way of showing that the oscillating modes correspond to the interaction of two gravity waves, one traveling downstream at speed $\alpha = V_0 + C_0$ and one traveling upstream at speed $\beta = C_0 - V_0$.

3.2.1.4 Input–Output Transfer Matrix

Let $P(s)$ denote the Saint-Venant transfer matrix relating the inputs $\hat{q}(0,s)$ and $\hat{q}(L,s)$ to the outputs $\hat{y}(0,s)$ and $\hat{y}(L,s)$.

Using (3.8) and the state-transition matrix (3.7), we obtain the input–output transfer matrix $P(s)$:

$$\begin{pmatrix}\hat{y}(0,s)\\ \hat{y}(L,s)\end{pmatrix} = P(s)\begin{pmatrix}q(0,s)\\ q(L,s)\end{pmatrix}, \tag{3.10}$$

with $P(s) = \Big(p_{ij}(s)\Big)$, and

$$p_{11}(s) = \frac{\beta + \alpha e^{-(\tau_1+\tau_2)s}}{\alpha\beta T_0\left(1 - e^{-(\tau_1+\tau_2)s}\right)}, \quad (3.11a)$$

$$p_{12}(s) = -\frac{(\alpha+\beta)e^{-\tau_2 s}}{\alpha\beta T_0\left(1 - e^{-(\tau_1+\tau_2)s}\right)}, \quad (3.11b)$$

$$p_{21}(s) = \frac{(\alpha+\beta)e^{-\tau_1 s}}{\alpha\beta T_0\left(1 - e^{-(\tau_1+\tau_2)s}\right)}, \quad (3.11c)$$

$$p_{22}(s) = -\frac{\alpha + \beta e^{-(\tau_1+\tau_2)s}}{\alpha\beta T_0\left(1 - e^{-(\tau_1+\tau_2)s}\right)}. \quad (3.11d)$$

Example 3.2 (Bode plots of a horizontal frictionless open channel). We consider the canal in Example 3.1. Figures 3.2 and 3.3 depict the Bode plots of transfer functions $p_{ij}(s)$ for this horizontal frictionless channel. The transfers $p_{21}(s)$ and $p_{12}(s)$ include a time-delay, as can be seen from the phase, which decreases towards $-\infty$. The other transfers seem to be minimum phase (this will be shown in the following). The poles are clearly visible in the frequency responses, with peaks located at multiples of a resonance frequency equal to about 4×10^{-3} rad/s. This frequency corresponds to the total travel time of a perturbation to reach its origin after reflection on both boundaries.

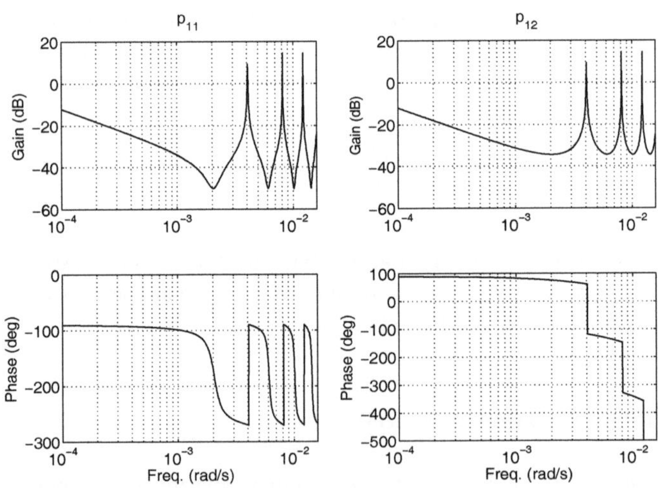

Fig. 3.2 Bode plot of the transfer functions $p_{11}(s)$ and $p_{12}(s)$ of a horizontal frictionless channel

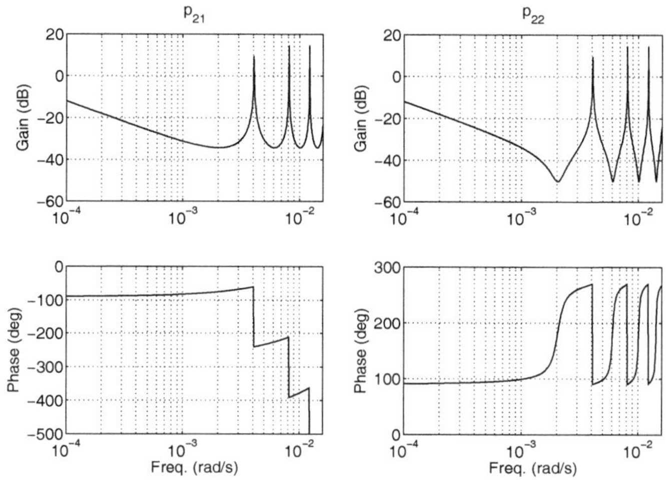

Fig. 3.3 Bode plot of the transfer functions $p_{21}(s)$ and $p_{22}(s)$ of a horizontal frictionless channel

3.2.1.5 Distributed Transfer Matrix

From the state-transition matrix $\Phi(x,s)$, one may also derive the distributed transfer matrix relating the water depth $\hat{y}(x,s)$ and the discharge $\hat{q}(x,s)$ at any point x in the canal pool to the upstream and downstream discharges. This transfer matrix, denoted by $G(x,s)$ is given by:

$$\begin{pmatrix} \hat{y}(x,s) \\ \hat{q}(x,s) \end{pmatrix} = G(x,s) \begin{pmatrix} \hat{q}(0,s) \\ \hat{q}(L,s) \end{pmatrix}, \tag{3.12}$$

with $g_{11}(x,s) = \frac{1}{T_0}\left[\phi_{12}(x,s) - \phi_{11}(x,s)\frac{\phi_{22}(L,s)}{\phi_{21}(L,s)}\right]$, $g_{12}(x,s) = \frac{\phi_{11}(x,s)}{T_0\phi_{21}(L,s)}$, $g_{21}(x,s) = \phi_{22}(x,s) - \phi_{21}(x,s)\frac{\phi_{22}(L,s)}{\phi_{21}(L,s)}$, and $g_{22}(x,s) = \frac{\phi_{21}(x,s)}{\phi_{21}(L,s)}$.

This leads to the following explicit expressions:

$$g_{11}(x,s) = \frac{\beta + \alpha e^{-(1-\frac{x}{L})(\tau_1+\tau_2)s}}{\alpha\beta T_0(1 - e^{-(\tau_1+\tau_2)s})} e^{-\frac{x}{\alpha}s}, \tag{3.13a}$$

$$g_{12}(x,s) = -\frac{\alpha + \beta e^{-\frac{x}{L}(\tau_1+\tau_2)s}}{\alpha\beta T_0(1 - e^{-(\tau_1+\tau_2)s})} e^{-\frac{L-x}{\beta}s}, \tag{3.13b}$$

$$g_{21}(x,s) = \frac{1 - e^{-(1-\frac{x}{L})(\tau_1+\tau_2)s}}{1 - e^{-(\tau_1+\tau_2)s}} e^{-\frac{x}{\alpha}s}, \tag{3.13c}$$

$$g_{22}(x,s) = \frac{1 - e^{-\frac{x}{L}(\tau_1+\tau_2)s}}{1 - e^{-(\tau_1+\tau_2)s}} e^{-\frac{(L-x)}{\beta}s}. \tag{3.13d}$$

Example 3.3 (Spatial Bode plot of a horizontal frictionless channel). Figure 3.4 depicts the Bode plots of the distributed transfer functions $g_{ij}(s)$ for the horizontal frictionless channel of Example 3.1. We observe that there is an integrator in low frequencies, whose gain is constant along x, and we clearly see nodes and anti-nodes of oscillation. These oscillations are linked to the poles of the transfer matrix, which come in complex conjugate pairs. The first pair of complex poles (corresponding to $k=1$) leads to a node of oscillation close to the middle of the pool, the second pair of poles ($k=2$) leads to two nodes of oscillation, the third pair of poles to three nodes, etc. We will give a more detailed explanation of this phenomenon when studying the modal decomposition of the distributed transfer matrix in Sect. 3.2.2.2.

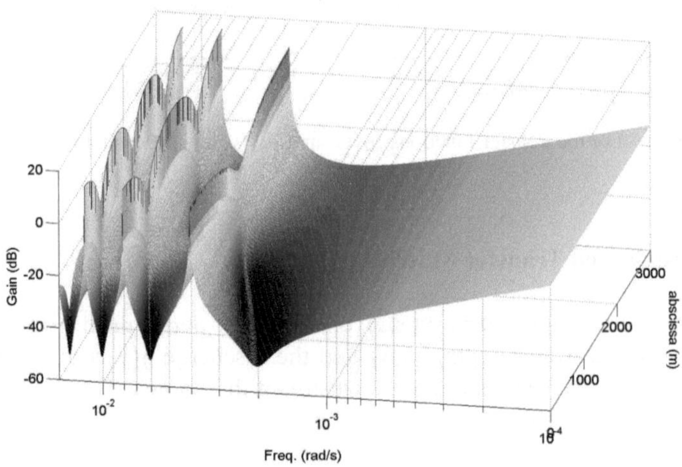

Fig. 3.4 Spatial Bode plot of the transfer function $g_{11}(x,s)$ of a horizontal frictionless channel

3.2.1.6 Link with the Characteristics

We show here that the transfer matrix associated with the characteristics contains only time-delays.

Expressing (3.5) in terms of the characteristic variable $\chi(x,t) = \mathbf{X}\xi(x,t)$ introduced in (2.29), we get:

$$\hat{\chi}(x,s) = e^{-s\mathbf{D}^{-1}x}\hat{\chi}(0,s) + \bar{\chi}_0(x,s), \tag{3.14}$$

with $e^{-s\mathbf{D}^{-1}x} = \begin{pmatrix} e^{-\frac{sx}{\alpha}} & 0 \\ 0 & e^{\frac{sx}{\beta}} \end{pmatrix}$ and $\bar{\chi}_0(x,s) = \int_0^x e^{-s\mathbf{D}^{-1}(x-v)}\mathbf{D}^{-1}\chi_0(v)\mathrm{d}v$.

The components of the solution can then be expressed as:

$$\hat{\chi}_1(x,s) = e^{-s\frac{x}{\alpha}}\hat{\chi}_1(0,s) + \bar{\chi}_{10}(x,s), \tag{3.15a}$$

$$\hat{\chi}_2(x,s) = e^{-s\frac{L-x}{\beta}}\hat{\chi}_2(L,s) + \bar{\chi}_{20}(x,s), \tag{3.15b}$$

with

$$\bar{\chi}_{10}(x,s) = \frac{1}{\alpha}\int_0^x e^{-s\frac{x-v}{\alpha}}\chi_{10}(v)\mathrm{d}v, \tag{3.16a}$$

$$\bar{\chi}_{20}(x,s) = \frac{1}{\beta}\int_x^L e^{s\frac{x-v}{\beta}}\chi_{20}(v)\mathrm{d}v. \tag{3.16b}$$

Let us now define τ_1 and τ_2, the delays for downstream and upstream propagation, respectively:

$$\tau_1 = \frac{L}{\alpha} = \frac{L}{C_0+V_0}, \tag{3.17a}$$

$$\tau_2 = \frac{L}{\beta} = \frac{L}{C_0-V_0}. \tag{3.17b}$$

Using (3.15), we can now express the input–output transfer matrix in terms of the characteristics variables at the boundaries:

$$\begin{pmatrix}\hat{\chi}_1(L,s)\\ \hat{\chi}_2(0,s)\end{pmatrix} = \begin{pmatrix} e^{-\tau_1 s} & 0 \\ 0 & e^{-\tau_2 s}\end{pmatrix}\begin{pmatrix}\hat{\chi}_1(0,s)\\ \hat{\chi}_2(L,s)\end{pmatrix} + \begin{pmatrix}\bar{\chi}_{10}(L,s)\\ \bar{\chi}_{20}(0,s)\end{pmatrix}.$$

We see that in the horizontal frictionless case, the transfer matrix is diagonal in the characteristics variables and only contains delays. This is consistent with the physical interpretation of the characteristics: the variable $\chi_1(x,t)$ can be expressed as a delayed function of its value at the upstream boundary $x=0$, and the variable $\chi_2(x,t)$ can be expressed as a delayed function of its value at the downstream boundary $x=L$. The flow results from the superposition of these two traveling waves.

In order to recover the transfer matrix $P(s)$, we need to introduce the physical boundary conditions using (2.29) as follows:

$$\hat{\chi}_1(0,s) = \hat{q}(x,s) + \beta T_0\hat{y}(x,s), \tag{3.18a}$$

$$\hat{\chi}_2(x,s) = \hat{q}(x,s) - \alpha T_0\hat{y}(x,s), \tag{3.18b}$$

and express the outputs $\hat{y}(0,s)$ and $\hat{y}(L,s)$ as functions of the inputs $\hat{q}(0,s)$ and $\hat{q}(L,s)$. This yields the input–output transfer matrix $P(s)$ in (3.11).

3.2.2 Properties of the Transfer Matrix

3.2.2.1 Delays

It is well-known that open channel flow is a delayed process. We show in the following that the transfer functions of linearized Saint-Venant equations can be factorized as the product of a pure delay and a delay-free part.

Input–Output Transfer Matrix

For the input–output transfer matrix, this leads to the following factorization for $p_{12}(s)$ and $p_{21}(s)$:

$$\begin{aligned} p_{11}(s) &= \tilde{p}_{11}(s), \\ p_{12}(s) &= \tilde{p}_{12}(s)e^{-\tau_2 s}, \\ p_{21}(s) &= \tilde{p}_{21}(s)e^{-\tau_1 s}, \\ p_{22}(s) &= \tilde{p}_{22}(s), \end{aligned}$$

with $\tau_1 = \frac{L}{C_0+V_0}$, $\tau_2 = \frac{L}{C_0-V_0}$, and $\tilde{p}_{21}(s) = -\tilde{p}_{12}(s) = \frac{(\alpha+\beta)}{\alpha\beta T_0(1-e^{-(\tau_1+\tau_2)s})}$.

Distributed Transfer Matrix

For the distributed transfer matrix, the delays are obtained directly as $e^{-\frac{x}{\alpha}s}$ for $g_{11}(x,s)$ and $g_{21}(x,s)$ and $e^{-\frac{L-x}{\beta}s}$ for $g_{12}(x,s)$ and $g_{22}(x,s)$. The transfer functions can therefore be factorized as follows:

$$\begin{aligned} g_{11}(x,s) &= \tilde{g}_{11}(x,s)e^{-\frac{x}{\alpha}s}, \\ g_{12}(x,s) &= \tilde{g}_{12}(x,s)e^{-\frac{L-x}{\beta}s}, \\ g_{21}(x,s) &= \tilde{g}_{21}(x,s)e^{-\frac{x}{\alpha}s}, \\ g_{22}(x,s) &= \tilde{g}_{22}(x,s)e^{-\frac{L-x}{\beta}s}, \end{aligned}$$

where the delay-free parts $\tilde{g}_{ij}(x,s)$ are defined by

$$\tilde{g}_{11}(x,s) = \frac{\beta + \alpha e^{-(1-\frac{x}{L})(\tau_1+\tau_2)s}}{\alpha\beta T_0(1-e^{-(\tau_1+\tau_2)s})},$$

$$\tilde{g}_{12}(x,s) = -\frac{\alpha + \beta e^{-\frac{x}{L}(\tau_1+\tau_2)s}}{\alpha\beta T_0(1-e^{-(\tau_1+\tau_2)s})},$$

$$\tilde{g}_{21}(x,s) = \frac{1-e^{-(1-\frac{x}{L})(\tau_1+\tau_2)s}}{1-e^{-(\tau_1+\tau_2)s}},$$

$$\tilde{g}_{22}(x,s) = \frac{1-e^{-\frac{x}{L}(\tau_1+\tau_2)s}}{1-e^{-(\tau_1+\tau_2)s}}.$$

This factorization extends that of the input–output transfer matrix.

We can prove in fact that the transfer functions $\tilde{g}_{ij}(x,s)$ are minimum phase, i.e., that they do not contain any other noninvertible element. This point is important for control purposes, and will be studied in Chap. 9. In fact, the factorization corresponds to the so-called inner-outer factorization [11]. The delay-free part corresponds to the outer part of the system, and the inner part corresponds to the delay. The outer part has a stable and causal inverse, which allows us to claim that this part is then free of delay, since a delay does not possess a stable and causal inverse.

3.2.2.2 Modal Decompositions

Rational Modal Decomposition

It is also possible to derive a modal decomposition for the distributed transfer functions (3.13), based on the explicit expression of the poles.

Indeed, the Cauchy residues theorem implies that each transfer function $g_{ij}(x,s)$ of (3.13) can be decomposed as an infinite sum:

$$g_{ij}(x,s) = b_{ij}(x) + \frac{a_{ij}^{(0)}(x)}{s} + \sum_{k=-\infty,k\neq 0}^{\infty} \frac{a_{ij}^{(k)}(x)s}{p_k(s-p_k)}, \tag{3.19}$$

with

$$a_{ij}^{(k)}(x) = \lim_{s\to p_k}(s-p_k)g_{ij}(x,s), \tag{3.20}$$

and

$$b_{ij}(x) = \frac{\partial}{\partial s}[sg_{ij}(x,s)]|_{s=0}. \tag{3.21}$$

The coefficient $a_{ij}^{(k)}(x)$ is the residue of transfer function $g_{ij}(x,s)$ at the pole p_k.

Proof. The proof can be easily adapted from that in Appendix B.1. □

This decomposition shows a specific feature of the dynamics of linearized open channel flow: the time and space components can be partially decoupled, as depicted in Fig. 3.5. The response to an input can be computed first by a series of time domain filters (corresponding to the poles p_k), and then this response is modulated by the coefficients $a_{ij}^{(k)}(x)$ to compute the output at the abscissa x.

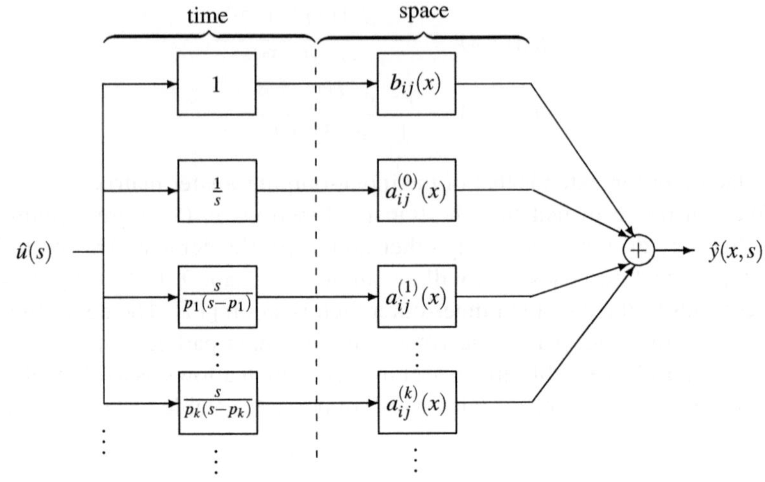

Fig. 3.5 Schematic representation of the modal decomposition of transfer function $g_{ij}(x,s)$

The residues $a_{ij}^{(k)}(x)$ can be computed analytically with (3.20), leading to

$$a_{11}^{(k)}(x) = \frac{e^{\frac{\alpha-\beta}{\alpha+\beta}\frac{jk\pi x}{L}}}{T_0 L}\left[\cos\left(\frac{k\pi x}{L}\right) + \frac{\alpha-\beta}{\alpha+\beta} j \sin\left(\frac{k\pi x}{L}\right)\right], \tag{3.22a}$$

$$a_{12}^{(k)}(x) = (-1)^{k+1}\frac{e^{\frac{\beta-\alpha}{\alpha+\beta}\frac{jk\pi(L-x)}{L}}}{T_0 L}\left[\cos\left(\frac{k\pi x}{L}\right) + \frac{\alpha-\beta}{\alpha+\beta} j \sin\left(\frac{k\pi x}{L}\right)\right], \tag{3.22b}$$

$$a_{21}^{(k)}(x) = \frac{e^{\frac{\alpha-\beta}{\alpha+\beta}\frac{jk\pi x}{L}}}{\tau_1+\tau_2}\sin\left(\frac{k\pi x}{L}\right), \tag{3.22c}$$

$$a_{22}^{(k)}(x) = (-1)^{k+1}\frac{e^{\frac{\beta-\alpha}{\alpha+\beta}\frac{jk\pi(L-x)}{L}}}{\tau_1+\tau_2}\sin\left(\frac{k\pi x}{L}\right), \tag{3.22d}$$

and $b_{ij}(x)$ are given by:

$$b_{11}(x) = \frac{(\beta-\alpha)}{2\alpha\beta T_0}\left(1-2\frac{x}{L}\right), \tag{3.23a}$$

$$b_{12}(x) = \frac{(\alpha-\beta)}{2\alpha\beta T_0}\left(1-2\frac{x}{L}\right), \tag{3.23b}$$

$$b_{21}(x) = 1-\frac{x}{L}, \tag{3.23c}$$

$$b_{22}(x) = \frac{x}{L}. \tag{3.23d}$$

The modal factors $a_{ij}^{(k)}(x)$ appear to be a combination of sine and cosine functions, modified by an exponential term. This fundamental aspect of the transfer function decomposition will be recovered in the uniform flow case, where slope and friction are nonzero.

Example 3.4 (Modal factors of a horizontal frictionless channel). The function $a_{11}^{(k)}(x)$ of the distributed modal decomposition of $g_{11}(x,s)$ of the horizontal frictionless channel is depicted in Fig. 3.6.

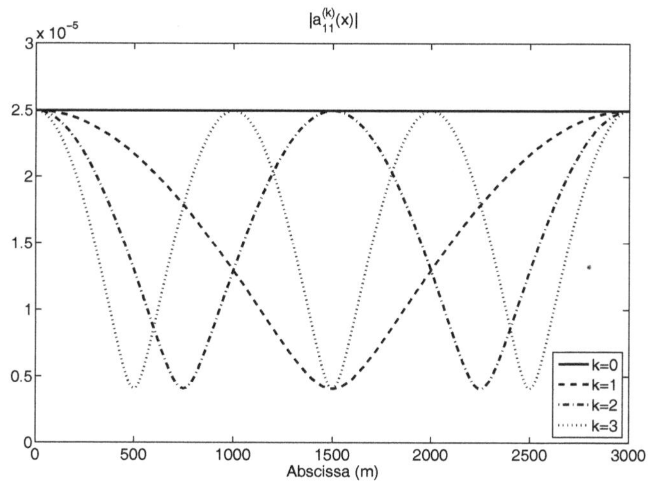

Fig. 3.6 Function $a_{11}^{(k)}(x)$ of the distributed modal decomposition for a horizontal frictionless channel

We see that the integrator gain is constant; it does not depend on x. This is due to the fact that the canal is horizontal, with zero friction. The first mode, corresponding to $k=1$ is clearly visible, with a node of oscillation in the middle of the pool. The second mode corresponding to $k=2$ leads to two nodes of oscillation, located at $L/4$ and $3L/4$. We recover here the feature already noticed in the spatial Bode plot of the distributed transfer function $g_{11}(x,s)$ in Fig. 3.4.

Modal Decomposition of the Delay-free Part

Using the same idea, one may also compute the modal decomposition on the delay-free part of the transfer functions. The delay-free part $\tilde{g}_{ij}(x,s)$ of (3.13) can be decomposed as an infinite sum:

$$\tilde{g}_{ij}(x,s) = \tilde{b}_{ij}(x) + \frac{\tilde{a}_{ij}^{(0)}}{s} + \sum_{k=-\infty,k\neq 0}^{\infty} \frac{s\tilde{a}_{ij}^{(k)}(x)}{p_k(s-p_k)}, \tag{3.24}$$

with

$$\tilde{a}_{ij}^{(k)}(x) = \lim_{s\to p_k} (s-p_k)\tilde{g}_{ij}(x,s), \tag{3.25}$$

and

$$\tilde{b}_{ij}(x) = \frac{\partial}{\partial s}[s\tilde{g}_{ij}(x,s)]|_{s=0}. \tag{3.26}$$

The residues can be computed analytically with (3.25), leading to

$$\tilde{a}_{11}^{(k)}(x) = e^{\frac{\beta}{\alpha+\beta}\frac{jk\pi x}{L}} a_{11}^{(k)}(x), \tag{3.27a}$$

$$\tilde{a}_{12}^{(k)}(x) = e^{\frac{\alpha}{\alpha+\beta}\frac{jk\pi(L-x)}{L}} a_{12}^{(k)}(x), \tag{3.27b}$$

$$\tilde{a}_{21}^{(k)}(x) = e^{\frac{\beta}{\alpha+\beta}\frac{jk\pi x}{L}} a_{21}^{(k)}(x), \tag{3.27c}$$

$$\tilde{a}_{22}^{(k)}(x) = e^{\frac{\alpha}{\alpha+\beta}\frac{jk\pi(L-x)}{L}} a_{22}^{(k)}(x). \tag{3.27d}$$

$\tilde{b}_{ij}(x)$ are given by:

$$\tilde{b}_{11}(x) = \frac{\beta-\alpha}{2\alpha\beta T_0} + \frac{x}{\beta T_0 L}, \tag{3.28a}$$

$$\tilde{b}_{12}(x) = \frac{\beta-\alpha}{2\alpha\beta T_0} - \frac{(L-x)}{\alpha T_0 L}, \tag{3.28b}$$

$$\tilde{b}_{21}(x) = 1 - \frac{x}{L}, \tag{3.28c}$$

$$\tilde{b}_{22}(x) = \frac{x}{L}. \tag{3.28d}$$

These expressions directly stem from the definitions of $\tilde{b}_{ij}(x)$ and $\tilde{a}_{ij}^{(k)}(x)$. Indeed, for $g_{11}(x,s)$, we have:

$$g_{11}(x,s) = \tilde{g}_{11}(x,s)e^{-s\frac{x}{\alpha}},$$

therefore $\tilde{b}_{11}(x)$ is obtained by differentiating $s\tilde{g}_{11}(x,s)$:

$$\begin{aligned}\frac{\mathrm{d}}{\mathrm{d}s}[s\tilde{g}_{11}(x,s)] &= \frac{\mathrm{d}}{\mathrm{d}s}[se^{s\frac{x}{\alpha}}g_{11}(x,s)] \\ &= b_{11}(x) + \frac{x}{\alpha}a_{11}^{(0)}(x).\end{aligned}$$

The other terms are obtained in a similar way.

3.2.3 Time Domain Expressions

We now study the system in the time domain and provide explicit expressions for the linearized Saint-Venant equations using 1) a method based on the characteristics, which can be used to solve the equations step by step, since one needs to specify the value of the boundary conditions, 2) a method based on the input–output transfer matrix, using an infinite series expansion, and 3) a method based on the modal factorization.

3.2.3.1 Explicit Expressions for the Characteristics

The solution (3.14) can be expressed in the time domain using the changes of variables $t_1 = \frac{x-v}{\alpha}$ for the first integral of (3.16) and $t_2 = \frac{v-x}{\beta}$ for the second one. This yields:

$$\bar{\chi}_{10}(x,s) = \int_0^{\frac{x}{\alpha}} e^{-st_1}\chi_{10}(x-\alpha t_1)\mathrm{d}t_1, \tag{3.29a}$$

$$\bar{\chi}_{20}(x,s) = \int_0^{\frac{L-x}{\beta}} e^{-st_2}\chi_{20}(x+\beta t_2)\mathrm{d}t_2. \tag{3.29b}$$

We now observe that $\bar{\chi}_{10}(x,s)$ and $\bar{\chi}_{20}(x,s)$ are the Laplace transforms of $\chi_{10}(x)$ and $\chi_{20}(x)$ on a truncated time interval, $[0, \frac{x}{\alpha}]$ for $\chi_{10}(x)$ and $[0, \frac{L-x}{\beta}]$ for $\chi_{20}(x)$. This is related to the fact that the initial condition influences a given sector of the (x,t) plane. Since we also know that the inverse Laplace transform of $e^{-\tau s}\hat{u}(s)$ is the delayed function $u(t-\tau)$, we obtain the following expressions in the time domain:

$$\chi_1(x,t) = \begin{cases}\chi_{10}(x-\alpha t) & \text{if } x \geq \alpha t,\\ \chi_1\left(0, t-\frac{x}{\alpha}\right) & \text{if } x < \alpha t,\end{cases} \tag{3.30a}$$

$$\chi_2(x,t) = \begin{cases}\chi_{20}(x+\beta t) & \text{if } L-x \geq \beta t,\\ \chi_2\left(L, t-\frac{L-x}{\beta}\right) & \text{if } L-x < \beta t.\end{cases} \tag{3.30b}$$

Let us note that these expressions can be obtained directly using the characteristic form (see (E.13) in Appendix E.2.2.2).

In (3.30), we have assumed that the solution remains inside the domain $[0,L]$, which is valid as long as $t < \tau_1$ for the first characteristics $\chi_1(x,t)$, and as long as $t < \tau_2$ for $\chi_2(x,t)$. It remains to investigate the effect of the boundary conditions. In that case, we need to express the $\chi_1(0,t)$ as a function of $\chi_2(0,t)$ and $\chi_2(L,t)$ as a function of $\chi_1(L,t)$. Then, the solution is once again provided by the propagation given by (3.30), but shifted in time.

Doing this computation in an iterative manner, we end up with an infinite series of terms linked to the reflections on the boundaries. This result can be obtained directly by using an infinite series expansion of the transfer matrix.

3.2.3.2 Infinite Series Expansion

The transfer functions $p_{ij}(s)$ also have an explicit time domain expression as an infinite series expansion. To this end, we use the classical expansion

$$\frac{1}{1-z} = \sum_{k=0}^{\infty} z^k, \tag{3.31}$$

with $z = e^{-(\tau_1+\tau_2)s}$, which yields the following series expansions for $p_{ij}(s)$:

$$p_{11}(s) = \frac{1}{\alpha\beta T_0}\sum_{k=0}^{\infty}\left[\beta + \alpha e^{-(\tau_1+\tau_2)s}\right]e^{-k(\tau_1+\tau_2)s}, \tag{3.32a}$$

$$p_{12}(s) = -\frac{(\alpha+\beta)}{\alpha\beta T_0}\sum_{k=0}^{\infty} e^{-(k\tau_1+(k+1)\tau_2)s}, \tag{3.32b}$$

$$p_{21}(s) = \frac{(\alpha+\beta)}{\alpha\beta T_0}\sum_{k=0}^{\infty} e^{-((k+1)\tau_1+k\tau_2)s}, \tag{3.32c}$$

$$p_{22}(s) = -\frac{1}{\alpha\beta T_0}\sum_{k=0}^{\infty}\left[\alpha + \beta e^{-(\tau_1+\tau_2)s}\right]e^{-k(\tau_1+\tau_2)s}. \tag{3.32d}$$

Going back to the time domain, we get:

$$\begin{aligned} y(0,t) = \frac{1}{\alpha\beta T_0}\sum_{k=0}^{\infty}\Big[&\beta q(0,t-k(\tau_1+\tau_2)) + \alpha q(0,t-(k+1)(\tau_1+\tau_2)) \\ &-(\alpha+\beta)q(L,t-\tau_2-k(\tau_1+\tau_2))\Big], \end{aligned} \tag{3.33a}$$

$$\begin{aligned} y(L,t) = \frac{1}{\alpha\beta T_0}\sum_{k=0}^{\infty}\Big[&(\alpha+\beta)q(0,t-\tau_1-k(\tau_1+\tau_2)) \\ &-\alpha q(L,t-k(\tau_1+\tau_2)) - \beta q(L,t-(k+1)(\tau_1+\tau_2))\Big]. \end{aligned} \tag{3.33b}$$

The water levels at the boundaries can therefore be expressed as a series of the delayed values of the discharges at each boundary. This expresses the way the characteristics reflect on the boundaries. These expressions also provide an explicit solution to the linearized Saint-Venant equations in that case.

As an example, we provide here the variations of water levels corresponding to unit step inputs of the boundary discharges. This can be computed using (3.33). The step responses of each transfer function $p_{ij}(s)$ are given by:

$$y_{11}(t) = \frac{1}{\alpha\beta T_0}\sum_{k=0}^{\infty}[\beta\mathbf{1}(t-k(\tau_1+\tau_2))+\alpha\mathbf{1}(t-(k+1)(\tau_1+\tau_2))], \tag{3.34a}$$

$$y_{12}(t) = -\frac{(\alpha+\beta)}{\alpha\beta T_0}\sum_{k=0}^{\infty}\mathbf{1}(t-\tau_2-k(\tau_1+\tau_2)), \tag{3.34b}$$

$$y_{21}(t) = \frac{(\alpha+\beta)}{\alpha\beta T_0}\sum_{k=0}^{\infty}\mathbf{1}(t-\tau_1-k(\tau_1+\tau_2)), \tag{3.34c}$$

$$y_{22}(t) = -\frac{1}{\alpha\beta T_0}\sum_{k=0}^{\infty}[\alpha\mathbf{1}(t-k(\tau_1+\tau_2))+\beta\mathbf{1}(t-(k+1)(\tau_1+\tau_2))], \tag{3.34d}$$

where $\mathbf{1}(t) = \begin{cases} 0 & \text{if } t<0 \\ 1 & \text{if } t\geq 0 \end{cases}$ is the unit step function.

The water level deviations are then directly obtained as a combination of delayed functions of the boundary discharges.

Example 3.5 (Step responses of a horizontal frictionless channel). The step responses computed using (3.34) are depicted in Fig. 3.7 for the example horizontal frictionless channel. In this figure we can see that the boundary water levels are influenced by the boundary discharges and by the reflections of the discharge variations on the boundaries.

For the step response of transfer function $p_{11}(s)$, we see that the effect of the discharge $q(0,t)$ on the water level $y(0,t)$ is instantaneous, and then, after $\tau_1+\tau_2 =$ 1548 s, we see the effect of the reflected step on the downstream boundary condition. For the step response of $p_{12}(s)$, the effect of the discharge $q(L,t)$ on the water level $y(0,t)$ is delayed by $\tau_2 = 900$ s, and then, at time $\tau_1+2\tau_2 = 2448$ s, we see the effect of the reflected step on the upstream boundary condition. For the step response of $p_{21}(s)$, the effect of the discharge $q(0,t)$ on the water level $y(L,t)$ is delayed by $\tau_1 = 648$ s, and then, at time $t = 2\tau_1+\tau_2 = 2196$ s, we see the effect of the step that has reflected on both the downstream and the upstream boundary conditions. For the step response of $p_{22}(s)$, the effect of the discharge $q(L,t)$ on the water level $y(L,t)$ is instantaneous, and after $\tau_1+\tau_2 = 1548$ s, we see the effect of the step that has reflected on the upstream boundary condition.

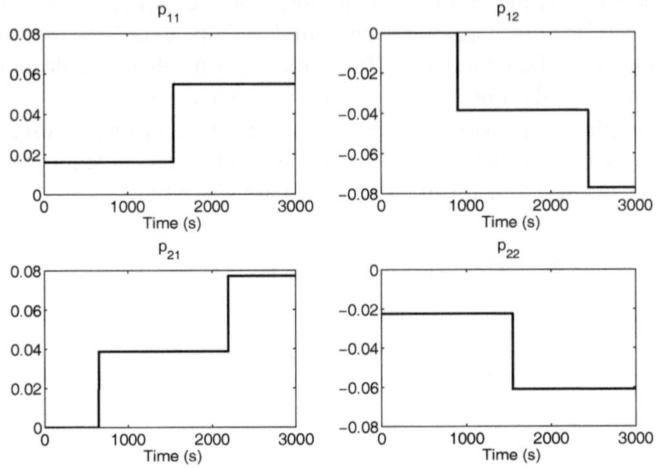

Fig. 3.7 Step responses of Saint-Venant transfer functions $p_{11}(s)$, $p_{12}(s)$, $p_{21}(s)$, and $p_{22}(s)$ for a horizontal frictionless channel

3.2.3.3 State Space Expressions

Expression Based on Rational Modal Decomposition

The expressions with infinite series (3.19) and (3.24) give solutions where time and space are decoupled. Indeed, the solution of the linearized Saint-Venant equations can be obtained as the linear combination of linear time invariant systems with space dependent parameters for the observation equation.

Let us first consider the case of the integrator with two inputs, the upstream discharge $q(0,t)$ and the downstream discharge $q(L,t)$. The system's transfer function is then given by:

$$\hat{y}(x,s) = \frac{1}{s}[a_{11}^{(0)}(x)\hat{q}(0,s) + a_{12}^{(0)}(x)\hat{q}(L,s)].$$

The integrator coefficients have opposite signs (see (3.22) for $k=0$) and therefore the system can be cast in the following state-space representation:

$$\begin{aligned}\frac{\mathrm{d}z(t)}{\mathrm{d}t} &= (1 \;\; -1)u(t),\\ y(x,t) &= a_{11}^{(0)}(x)z(t),\end{aligned}$$

with $u(t) = \begin{pmatrix} q(0,t) \\ q(L,t) \end{pmatrix}$. In this representation, the state is first updated, and the output $y(x,t)$ is then proportional to the state, with a coefficient that depends on x. This enables us to decouple the time and the spatial behavior of the system.

The same can be done for each pair of complex poles p_k and p_{-k}, when $k \neq 0$. In this case, p_k and p_{-k} are either complex conjugate, or both negative real.

Let us first notice that the modal factors verify:

$$
\begin{aligned}
a_{12}^{(k)}(x) &= \mu_k a_{11}^{(k)}(x),\\
a_{22}^{(k)}(x) &= \mu_k a_{21}^{(k)}(x),
\end{aligned}
$$

with $\mu_k = (-1)^{k+1} e^{jk\frac{\beta-\alpha}{\alpha+\beta}}$.

Then, the second-order system is given by:

$$
\begin{aligned}
\hat{y}(x,s) &= \frac{a_{11}^{(k)}(x)s}{p_k(s-p_k)}(\hat{q}(0,s)+\mu_k\hat{q}(L,s)) + \frac{a_{11}^{(-k)}(x)s}{p_{-k}(s-p_{-k})}(\hat{q}(0,s)+\mu_k\hat{q}(L,s))\\
&= \left(\frac{a_{11}^{(k)}(x)}{s-p_k} + \frac{a_{11}^{(k)}(x)}{p_k}\right)(\hat{q}(0,s)+\mu_k\hat{q}(L,s))\\
&\quad + \left(\frac{a_{11}^{(-k)}(x)}{s-p_{-k}} + \frac{a_{11}^{(-k)}(x)}{p_{-k}}\right)(\hat{q}(0,s)+\mu_k\hat{q}(L,s)).
\end{aligned}
$$

This second-order system can be represented by the following state-space system:

$$
\begin{aligned}
\frac{\mathrm{d}z(t)}{\mathrm{d}t} &= \begin{pmatrix} 0 & 1\\ a_1 & a_2 \end{pmatrix} z(t) + \begin{pmatrix} 0 & b_1\\ 1 & b_2 \end{pmatrix} u(t),\\
y(x,t) &= \begin{pmatrix} c_1(x) & c_2(x) \end{pmatrix} z(t) + \begin{pmatrix} d_1(x) & d_2(x) \end{pmatrix} u(t),
\end{aligned}
\tag{3.35}
$$

where $u(t) = [q(0,t), q(L,t)]^T$ is the input, $z(t)$ the state, $y(x,t)$ the measured output of the system, and the parameters are given by:

$$
\begin{aligned}
a_1 &= -p_k p_{-k},\\
a_2 &= p_k + p_{-k},\\
b_1 &= \frac{\mu_k - \mu_{-k}}{p_k - p_{-k}},\\
b_2 &= \frac{\mu_k p_k - \mu_{-k} p_{-k}}{p_k - p_{-k}},\\
c_1(x) &= -(a_{11}^{(-k)}(x)p_k + a_{11}^{(k)}(x)p_{-k}),\\
c_2(x) &= a_{11}^{(k)}(x) + a_{11}^{(-k)}(x),\\
d_1(x) &= \frac{a_{11}^{(k)}(x)}{p_k} + \frac{a_{11}^{(-k)}(x)}{p_{-k}},\\
d_2(x) &= \frac{a_{12}^{(k)}(x)}{p_k} + \frac{a_{12}^{(-k)}(x)}{p_{-k}},
\end{aligned}
$$

where the parameters $a_{ij}^{(k)}(x)$ are given by (3.22).

Let us note that when p_k and p_{-k} are complex conjugate, so are $a_{ij}^{(k)}$ and $a_{ij}^{(-k)}$. Therefore the parameters $a_1, a_2, b_1, b_2, c_1(x), c_2(x), d_1(x)$, and $d_2(x)$ are real.

Similarly, the discharge $q(x,t)$ can be expressed as an output of the dynamical system (3.35), replacing the coefficients by $a_{21}^{(\pm k)}$ and $a_{22}^{(\pm k)}$.

Finally, gathering all the contributions of the poles, the linearized Saint-Venant equations can be represented by the following infinite dimensional linear system:

$$\begin{aligned} \frac{\mathrm{d}z(t)}{\mathrm{d}t} &= Az(t) + Bu(t), \\ \xi(x,t) &= C(x)z(t) + D(x)u(t), \end{aligned} \tag{3.36}$$

with the input $u(t) = \begin{pmatrix} q(0,t) \\ q(L,t) \end{pmatrix}$, the state $z(t)$, and the output $\xi(x,t) = \begin{pmatrix} y(x,t) \\ q(x,t) \end{pmatrix}$.

The matrix A is given by

$$A = \begin{pmatrix} 0 & 0 & 0 & \cdots & \cdots \\ 0 & A^{(1)} & 0 & \cdots & \cdots \\ 0 & 0 & A^{(2)} & 0 & \cdots \\ \vdots & \vdots & 0 & \ddots & \ddots \\ 0 & \vdots & \vdots & \ddots & \ddots \end{pmatrix},$$

where $A^{(k)} = \begin{pmatrix} 0 & 1 \\ -p_k p_{-k} & p_k + p_{-k} \end{pmatrix}$ and B is given by:

$$B = \begin{pmatrix} B^{(0)} \\ B^{(1)} \\ B^{(2)} \\ \vdots \end{pmatrix},$$

where $B^{(0)} = \begin{pmatrix} 1 & -1 \end{pmatrix}$ and $B^{(k)} = \begin{pmatrix} 0 & \frac{\mu_k - \mu_{-k}}{p_k - p_{-k}} \\ 1 & \frac{\mu_k p_k - \mu_{-k} p_{-k}}{p_k - p_{-k}} \end{pmatrix}$.

The observation equation parameters depend on x. The term $C(x)$ is given by:

$$C(x) = \begin{pmatrix} C_1(x) \\ C_2(x) \end{pmatrix},$$

with

$$C_i(x) = \begin{pmatrix} a_{i1}^{(0)}(x) & C_i^{(1)}(x) & C_i^{(2)}(x) & \cdots & C_i^{(k)}(x) & \cdots \end{pmatrix}$$

and where

$$C_i^{(k)}(x) = \begin{pmatrix} -(a_{i1}^{(-k)}(x) p_k + a_{i1}^{(k)}(x) p_{-k}) & a_{i1}^{(k)}(x) + a_{i1}^{(-k)}(x) \end{pmatrix}.$$

The direct term $D(x)$ is given by

$$D(x) = \begin{pmatrix} D_{11}(x) & D_{12}(x) \\ D_{21}(x) & D_{22}(x) \end{pmatrix},$$

with

$$D_{ij}(x) = b_{ij}(x) + \sum_{k=1}^{\infty} \left(\frac{a_{ij}^{(k)}(x)}{p_k} + \frac{a_{ij}^{(-k)}(x)}{p_{-k}} \right),$$

where $b_{ij}(x)$ is given by (3.23).

Therefore the time domain response of the linearized Saint-Venant equations can be obtained as the output of a linear system, which state depends only on time, not on x. We will see that this remarkable feature is conserved for the uniform flow and nonuniform flow cases.

Expressions Based on the Delay-free Modal Decomposition

Using the delay-free modal decomposition given in (3.24) and following the same line as above, one can derive an input delayed system equivalent to the linearized Saint-Venant equations:

$$\begin{aligned} \frac{d\tilde{z}(t)}{dt} &= \tilde{A}\tilde{z}(t) + \tilde{B}(x) \begin{pmatrix} u_1(t - \frac{x}{\alpha}) \\ u_2(t - \frac{L-x}{\beta}) \end{pmatrix}, \\ \xi(x,t) &= \tilde{C}(x)z(t) + \tilde{D}(x) \begin{pmatrix} u_1(t - \frac{x}{\alpha}) \\ u_2(t - \frac{L-x}{\beta}) \end{pmatrix}, \end{aligned} \tag{3.37}$$

where the matrices $\tilde{A}$, $\tilde{B}(x)$, $\tilde{C}(x)$, and $\tilde{D}(x)$ are obtained in the same way as matrices A, B, $C(x)$, and $D(x)$, except that the parameters $a_{ij}^{(k)}(x)$ and $b_{ij}(x)$ are replaced by $\tilde{a}_{ij}^{(k)}(x)$ and $\tilde{b}_{ij}(x)$ given by (3.25) and (3.26), respectively, and the term μ_k is replaced by:

$$\tilde{\mu}_k(x) = \mu_k e^{\left(\frac{L-x}{\beta} - \frac{x}{\alpha}\right)p_k}.$$

In this case, the input matrix $\tilde{B}(x)$ depends explicitly on x. Therefore, when we consider the delay-free system, it is no longer possible to express it as a series of a linear time invariant system, where the space variable x only appears in the output equation.

The horizontal frictionless case is rather unrealistic, but enabled us to illustrate the physical behavior captured by the Saint-Venant equations. We now consider a more realistic case, where the canal has nonzero slope and friction.

3.3 Uniform Flow Case

In the uniform flow case, the nonzero slope and friction modify the characteristics: they induce a coupling that appears as a source term in the characteristics equations. But, as we will show below, it is still possible to eliminate this coupling by using a *dynamic* change of variable, which can easily be obtained in the Laplace domain.

We recall the linearized Saint-Venant equations given by (2.28), written here for the uniform flow case:

$$\frac{\partial \xi}{\partial t} + \mathbf{A}\frac{\partial \xi}{\partial x} + \mathbf{B}\xi = 0, \tag{3.38}$$

where $\xi(x,t) = (a(x,t), q(x,t))^T : [0,L] \times [0,+\infty) \to \Omega \in \mathbb{R}^2$ is the state of the system and $\mathbf{A} = \begin{pmatrix} 0 & 1 \\ \alpha\beta & \alpha-\beta \end{pmatrix}$, $\mathbf{B} = \begin{pmatrix} 0 & 0 \\ -\gamma & \delta \end{pmatrix}$.

In the uniform flow case, the parameters α, β, γ and δ are constant and given by:

$$\begin{aligned} \alpha &= C_0 + V_0, \\ \beta &= C_0 - V_0, \\ \gamma &= g(1+\kappa)S_b, \\ \delta &= \frac{2gS_b}{V_0}, \end{aligned}$$

with $\kappa = \frac{7}{3} - \frac{4A_0}{3T_0P_0}\frac{\partial P_0}{\partial Y}$.

3.3.1 Derivation of the Transfer Matrix

We apply the Laplace transform to the linear partial differential equations (3.38), using the classical relation $\hat{\frac{df}{dt}} = s\hat{f}(s) - f(0)$, which yields:

$$\frac{\partial \hat{\xi}(x,s)}{\partial x} = \mathscr{A}(s)\hat{\xi}(x,s) + \mathscr{B}\xi(x,0), \tag{3.39}$$

with $\mathscr{A}(s) = -\mathbf{A}^{-1}(sI + \mathbf{B})$ and $\mathscr{B} = \mathbf{A}^{-1}$, i.e.:

$$\begin{aligned} \mathscr{A}(s) &= \frac{1}{\alpha\beta}\begin{pmatrix} (\alpha-\beta)s+\gamma & -s-\delta \\ -\alpha\beta s & 0 \end{pmatrix}, \\ \mathscr{B} &= \frac{1}{\alpha\beta}\begin{pmatrix} (\beta-\alpha) & 1 \\ \alpha\beta & 0 \end{pmatrix}. \end{aligned}$$

Let us diagonalize matrix $\mathscr{A}(s)$:

$$\mathscr{A}(s) = \mathscr{X}(s)^{-1}\mathscr{D}(s)\mathscr{X}(s),$$

with

$$\mathscr{X}(s) = \begin{pmatrix} \frac{s}{\lambda_2(s)} & 1 \\ \frac{s}{\lambda_1(s)} & 1 \end{pmatrix}, \tag{3.40a}$$

$$\mathscr{D}(s) = \begin{pmatrix} \lambda_1(s) & 0 \\ 0 & \lambda_2(s) \end{pmatrix}, \tag{3.40b}$$

$$\mathscr{X}(s)^{-1} = \frac{\lambda_1(s)\lambda_2(s)}{s(\lambda_1(s)-\lambda_2(s))} \begin{pmatrix} 1 & -1 \\ -\frac{s}{\lambda_1(s)} & \frac{s}{\lambda_2(s)} \end{pmatrix}, \tag{3.40c}$$

and where $\lambda_1(s)$ and $\lambda_2(s)$ are the eigenvalues of $\mathscr{A}(s)$, solutions of the equation:

$$\det(\lambda I - \mathscr{A}(s)) = 0.$$

This equation is a second-order polynomial equation in λ:

$$\alpha\beta\lambda^2 - [(\alpha-\beta)s+\gamma]\lambda - s(s+\delta) = 0, \tag{3.41}$$

which has, in general, two solutions:

$$\lambda_1(s) = \frac{(\alpha-\beta)s+\gamma-\sqrt{d(s)}}{2\alpha\beta}, \tag{3.42a}$$

$$\lambda_2(s) = \frac{(\alpha-\beta)s+\gamma+\sqrt{d(s)}}{2\alpha\beta}, \tag{3.42b}$$

with $d(s) = (\alpha+\beta)^2 s^2 + 2[(\alpha-\beta)\gamma + 2\alpha\beta\delta]s + \gamma^2$.

In the following, we will factorize $d(s)$ as follows:

$$d(s) = (\alpha+\beta)^2[(s+b)^2 - a^2],$$

with

$$a^2 = \frac{4\alpha\beta(\alpha\delta-\gamma)(\gamma+\beta\delta)}{(\alpha+\beta)^4},$$

$$b = \frac{(\alpha-\beta)\gamma+2\alpha\beta\delta}{(\alpha+\beta)^2}.$$

Let us show that $\gamma < \alpha\delta$, which ensures that parameter a is always real since $a^2 > 0$. Indeed, $\gamma < \alpha\delta$ is equivalent to:

$$\frac{1+\kappa}{2}V_0 < V_0 + C_0 \Leftrightarrow F_0 < \frac{2}{\kappa-1}.$$

Now, we show that $\frac{2}{\kappa-1}$ is always larger than 1, which means that the above inequality is always verified for subcritical flow where $F_0 < 1$.

Parameter κ is defined by (2.24), therefore, since $\frac{\partial P_0}{\partial Y} > 0$, we have

$$\kappa - 1 = \frac{4}{3}\left(1 - \frac{A_0}{T_0 P_0}\frac{\partial P_0}{\partial Y}\right) < 2.$$

Finally, we have shown that parameter a is real and strictly positive. It is also easy to prove that for subcritical flow, we have $a < b$.

3.3.1.1 State-transition Matrix

Once matrix $\mathscr{A}(s)$ is diagonalized, the differential equation (3.39) can be solved analytically:

$$\hat{\xi}(x,s) = \Phi(x,s)\hat{\xi}(0,s) + \bar{\xi}_0(x,s), \tag{3.43}$$

with

$$\Phi(x,s) = \mathscr{X}(s)^{-1}e^{\mathscr{D}(s)x}\mathscr{X}(s), \tag{3.44}$$

and $\bar{\xi}_0(x,s) = \Phi(x,s)\int_0^x \Phi(v,s)^{-1}\mathscr{B}\xi(v,0)dv$.

The elements of matrix $\Phi(x,s)$ are given by:

$$\phi_{11}(x,s) = \frac{\lambda_1(s)e^{\lambda_1(s)x} - \lambda_2(s)e^{\lambda_2(s)x}}{\lambda_1(s) - \lambda_2(s)}, \tag{3.45a}$$

$$\phi_{12}(x,s) = \frac{\lambda_1(s)\lambda_2(s)(e^{\lambda_1(s)x} - e^{\lambda_2(s)x})}{s(\lambda_1(s) - \lambda_2(s))}, \tag{3.45b}$$

$$\phi_{21}(x,s) = \frac{s(e^{\lambda_2(s)x} - e^{\lambda_1(s)x})}{\lambda_1(s) - \lambda_2(s)}, \tag{3.45c}$$

$$\phi_{22}(x,s) = \frac{\lambda_1(s)e^{\lambda_2(s)x} - \lambda_2(s)e^{\lambda_1(s)x}}{\lambda_1(s) - \lambda_2(s)}. \tag{3.45d}$$

$\Phi(x,s)$ is the state-transition matrix for the differential equation (3.39). For simplicity, we assume zero initial conditions for the following developments, therefore $\xi_0(x,s) = 0$.

3.3.1.2 Boundary Conditions

Specifying the boundary conditions leads us to express the state as a function of the boundary discharges. To this end, we use (3.43) for $x = 0$, and get the equality:

$$T_0\hat{y}(0,s) = -\frac{\phi_{22}(L,s)}{\phi_{21}(L,s)}\hat{q}(0,s) + \frac{1}{\phi_{21}(L,s)}\hat{q}(L,s). \tag{3.46}$$

Once again, this is valid provided that $\phi_{21}(L,s)$ is not equal to zero. The values of s such that $\phi_{21}(L,s) = 0$ correspond to the poles of the Saint-Venant transfer matrix.

3.3.1.3 Poles

The poles are obtained as the solutions of:

$$\phi_{21}(L,s) = 0, \tag{3.47}$$

where $\phi_{21}(x,s)$ is given by (3.45c).

If s_0 is such that $\lambda_1(s_0) = \lambda_2(s_0)$, we have:

$$\phi_{21}(L,s_0) = -s_0 L e^{\lambda_1(s_0)L},$$

therefore s_0 is not a pole.

When $\lambda_1(s) \neq \lambda_2(s)$, (3.47) is equivalent to

$$s(e^{\lambda_2(s)L} - e^{\lambda_1(s)L}) = 0.$$

There is a pole in zero $p_0 = 0$, which means that the canal pool acts as an integrator and the other poles verify the equation:

$$\lambda_1(s) - \lambda_2(s) = \frac{2jk\pi}{L},$$

with $k \in \mathbb{N}^*$. The case $k = 0$ leads to $\lambda_1(s) = \lambda_2(s)$, therefore $s = s_0$, which is not a pole. Finally, the poles are solutions of the second-order equation:

$$d(s) = -\frac{4\alpha^2\beta^2k^2\pi^2}{L^2},$$

with $k \in \mathbb{N}^*$.

The poles $(p_{\pm k})_{k\in\mathbb{N}^*}$ are then given by:

$$p_{\pm k} = -\frac{(\alpha-\beta)\gamma + 2\alpha\beta\delta}{(\alpha+\beta)^2} \pm \frac{2\alpha^2\beta^2}{(\alpha+\beta)^2}\sqrt{\Delta(k)}, \tag{3.48}$$

with $\Delta(k) = \frac{(\alpha+\beta)^2}{\alpha^2\beta^2}\left[\frac{(\alpha\delta-\gamma)(\beta\delta+\gamma)}{\alpha\beta(\alpha+\beta)^2} - \frac{k^2\pi^2}{L^2}\right]$.

Let $k_m \in \mathbb{N}^*$ be the greatest integer such that $\Delta(k_m) \geq 0$. Then the poles obtained for $0 < k \leq k_m$ are negative real, and those obtained for $k > k_m$ are complex conjugate, with a constant real part equal to $-\frac{(\alpha-\beta)\gamma+2\alpha\beta\delta}{(\alpha+\beta)^2}$, which means that they are located on a vertical line in the left half plane. Canal pools with a dominant oscillating behavior correspond to $\Delta(1) < 0$. Let us note that (3.48) recovers the case of zero slope and frictionless canal, where the poles are located on the imaginary axis.

Example 3.6 (Poles of the example canals). The poles of the example canals 1 and 2 are depicted in Fig. 3.8. Canal 1 has an oscillating behavior, all its poles being complex conjugate. Canal 2 has two negative real poles for $k_m = 1$, and the other ones are complex conjugate. The complex poles of canal 2 have a larger negative real part than those of canal 1. Canal 2 should therefore exhibit a damped behavior compared to canal 1.

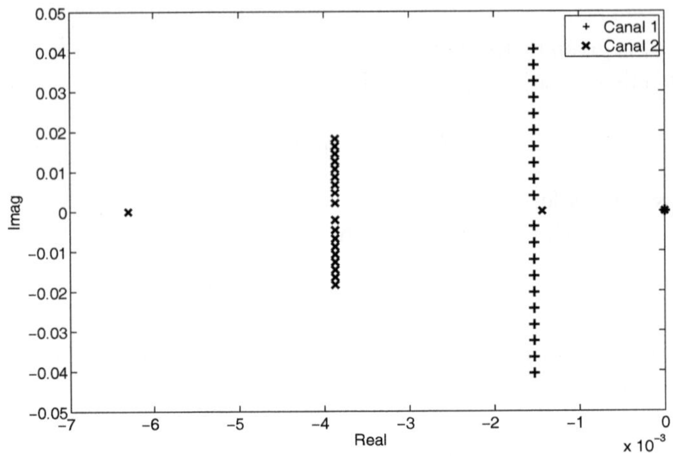

Fig. 3.8 Poles of canals 1 and 2 in the complex plane

3.3.1.4 Input–Output Transfer Matrix

In the Laplace domain, $P(s) = (p_{ij}(s))$ denotes the 2×2 transfer matrix relating the inputs to the outputs. Using the algebraic relation (3.46) and the state-transition matrix (3.45), we obtain the input–output transfer matrix:

$$\begin{pmatrix} \hat{y}(0,s) \\ \hat{y}(L,s) \end{pmatrix} = \begin{pmatrix} p_{11}(s) & p_{12}(s) \\ p_{21}(s) & p_{22}(s) \end{pmatrix} \begin{pmatrix} \hat{q}(0,s) \\ \hat{q}(L,s) \end{pmatrix}, \tag{3.49}$$

with $p_{11}(s) = -\frac{\phi_{22}(L,s)}{T_0\phi_{21}(L,s)}$, $p_{12}(s) = \frac{1}{T_0\phi_{21}(L,s)}$, $p_{21}(s) = \frac{\phi_{12}(L,s)}{T_0} - \frac{\phi_{11}(L,s)\phi_{22}(L,s)}{T_0\phi_{21}(L,s)}$ and $p_{22}(s) = \frac{\phi_{11}(L,s)}{T_0\phi_{21}(L,s)}$.

This leads to the final expressions:

$$p_{11}(s) = \frac{\lambda_2(s)e^{\lambda_1(s)L} - \lambda_1(s)e^{\lambda_2(s)L}}{T_0 s\left(e^{\lambda_2(s)L} - e^{\lambda_1(s)L}\right)}, \tag{3.50a}$$

$$p_{12}(s) = \frac{\lambda_1(s) - \lambda_2(s)}{T_0 s\left(e^{\lambda_2(s)L} - e^{\lambda_1(s)L}\right)}, \tag{3.50b}$$

$$p_{21}(s) = \frac{(\lambda_2(s) - \lambda_1(s))e^{(\lambda_1(s)+\lambda_2(s))L}}{T_0 s\left(e^{\lambda_2(s)L} - e^{\lambda_1(s)L}\right)}, \tag{3.50c}$$

$$p_{22}(s) = \frac{\lambda_1(s)e^{\lambda_1(s)L} - \lambda_2(s)e^{\lambda_2(s)L}}{T_0 s\left(e^{\lambda_2(s)L} - e^{\lambda_1(s)L}\right)}. \tag{3.50d}$$

This is similar to the expressions obtained by several other authors [6, 7, 21, 3].

Example 3.7 (Bode plots of the input output transfer matrix). Bode plots of transfer functions $p_{21}(s)$ and $p_{22}(s)$ for canals 1 and 2 are depicted in Figs. 3.9 and 3.10. We observe that the low frequency behavior of the system is dominated by the integrator; that there is a delay in transfer function $p_{21}(s)$. In high frequencies, the oscillating modes of canal 1 are clearly visible, while canal 2 is damped and tends towards a constant gain.

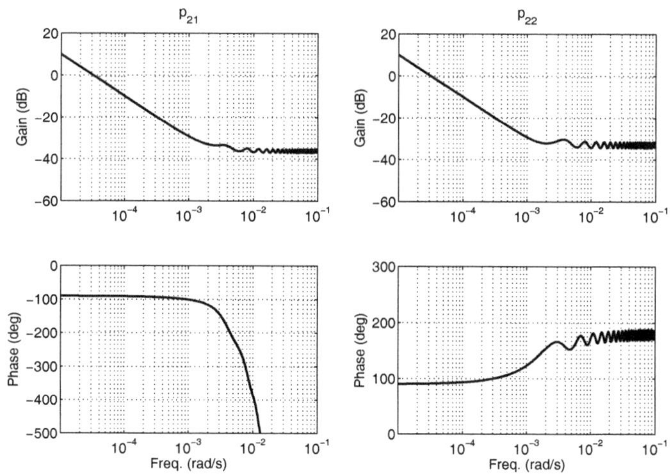

Fig. 3.9 Bode plots of transfer functions $p_{21}(s)$ and $p_{22}(s)$ for canal 1

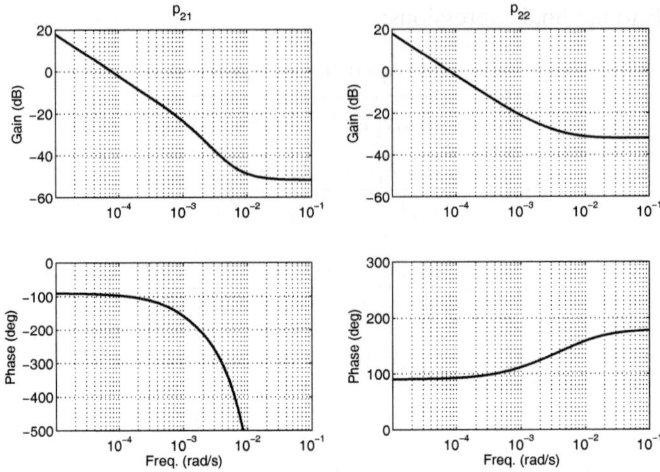

Fig. 3.10 Bode plots of transfer functions $p_{21}(s)$ and $p_{22}(s)$ for canal 2

3.3.1.5 Distributed Transfer Matrix

After elementary manipulations, we obtain the distributed transfer matrix, relating the water depth $\hat{y}(x,s)$ and the discharge $\hat{q}(x,s)$ at any point x in the canal pool to the upstream and downstream discharges:

$$\begin{pmatrix} \hat{y}(x,s) \\ \hat{q}(x,s) \end{pmatrix} = G(x,s) \begin{pmatrix} \hat{q}(0,s) \\ \hat{q}(L,s) \end{pmatrix}, \tag{3.51}$$

with

$$g_{11}(x,s) = \frac{\lambda_2(s)e^{\lambda_2(s)x+\lambda_1(s)L} - \lambda_1(s)e^{\lambda_1(s)x+\lambda_2(s)L}}{T_0 s(e^{\lambda_2(s)L} - e^{\lambda_1(s)L})}, \tag{3.52a}$$

$$g_{12}(x,s) = \frac{\lambda_1(s)e^{\lambda_1(s)x} - \lambda_2(s)e^{\lambda_2(s)x}}{T_0 s(e^{\lambda_2(s)L} - e^{\lambda_1(s)L})}, \tag{3.52b}$$

$$g_{21}(x,s) = \frac{e^{\lambda_1(s)x+\lambda_2(s)L} - e^{\lambda_2(s)x+\lambda_1(s)L}}{e^{\lambda_2(s)L} - e^{\lambda_1(s)L}}, \tag{3.52c}$$

$$g_{22}(x,s) = \frac{e^{\lambda_2(s)x} - e^{\lambda_1(s)x}}{e^{\lambda_2(s)L} - e^{\lambda_1(s)L}}. \tag{3.52d}$$

These expressions generalize those obtained in the horizontal frictionless case. The input–output transfer matrix $P(s)$ can be obtained from the distributed transfer matrix $G(x,s)$ as follows: $p_{11}(s) = g_{11}(0,s)$, $p_{12}(s) = g_{11}(L,s)$, $p_{21}(s) = g_{12}(0,s)$, and $p_{22}(s) = g_{12}(L,s)$.

Example 3.8 (Spatial Bode plot of the distributed transfer matrix). The spatial Bode plots of the distributed transfer function $g_{11}(x,s)$ are depicted in Figs. 3.11 and 3.12 for example canals 1 and 2. The amplitude of the transfer functions is depicted in dB versus a log scale for the frequency.

The spatial Bode plot of canal 1 is similar to that of the horizontal frictionless channel, even if the modes are much more damped in this case. One can see the integrator in low frequencies. For higher frequencies, the oscillating modes can be seen with the corresponding nodes and anti-nodes of oscillations along the channel.

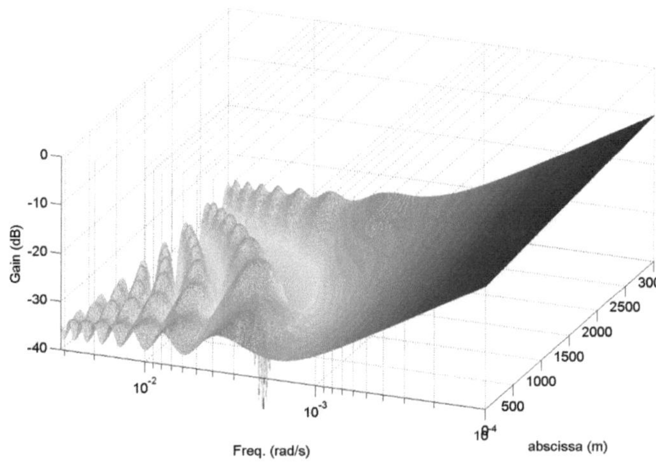

Fig. 3.11 Spatial Bode plots of transfer function $g_{11}(x,s)$ for canal 1

Canal 2 is much more damped than canal 1, but there are oscillating modes at the downstream end of the canal. This means that the water level response to the upstream discharge may exhibit local oscillations close to the downstream end of the canal. The surprising thing here is that these oscillations do not appear in the input–output transfer functions (see Fig. 3.10).

3.3.1.6 Link with the Characteristics Form

Let us denote by ζ the variable defined by:

$$\hat{\zeta}(x,s) = \mathscr{X}(s)\hat{\xi}(x,s), \tag{3.53}$$

with $\mathscr{X}(s)$ the matrix given by (3.40a).

We observe that in the case where $\gamma = \delta = 0$, which corresponds to a frictionless horizontal channel, we have $\frac{s}{\lambda_2(s)} = \beta$ and $\frac{s}{\lambda_1(s)} = -\alpha$. Therefore, we have in

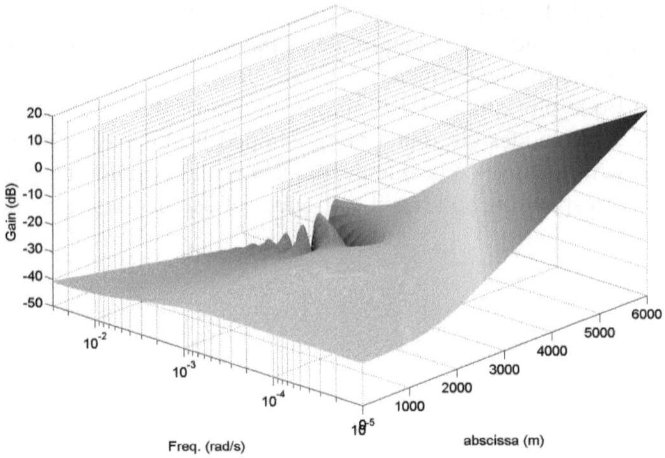

Fig. 3.12 Spatial Bode plots of transfer function $g_{11}(x,s)$ for canal 2

this case $\mathscr{X}(s) = \mathbf{X}$, where $\mathbf{X}$ is the matrix used to convert the equations into the characteristic form.

The change of variable of (3.53) can therefore be considered as an extension of the change of variable $\chi(x,t) = \mathbf{X}\xi(x,t)$ leading to the characteristic form. Indeed, both changes of variable enable us to diagonalize the system of equations, one in the horizontal frictionless case, the other one in the uniform flow case. However, the change of variable $\hat{\zeta}(x,s) = \mathscr{X}(s)\hat{\xi}(x,s)$ is a *dynamic* one, since $\mathscr{X}(s)$ depends on the Laplace variable s.

Using (3.43), we are able to express the solutions in terms of the generalized characteristic variables as follows:

$$\hat{\zeta}_1(x,s) = e^{\lambda_1(s)x}\hat{\zeta}_1(0,s) \tag{3.54a}$$

$$\hat{\zeta}_2(x,s) = e^{-\lambda_2(s)(L-x)}\hat{\zeta}_2(L,s). \tag{3.54b}$$

This diagonal form can also be interpreted as traveling waves: one traveling downstream at "speed" $-\frac{s}{\lambda_1(s)}$, the other one traveling upstream at "speed" $\frac{s}{\lambda_2(s)}$.

We now analyze in detail the Saint-Venant transfer matrix obtained above. We first characterize the delays of the transfer matrix, which are important for control purposes, as a delay limits the available performance of the closed-loop system [2, 14]. We then study the poles of the model obtained and derive various series decompositions for the transfer matrix.

3.3.2 Properties of the Transfer Matrix

3.3.2.1 Delays

Input–Output Transfer Matrix

One may show that the input–output Saint-Venant transfer matrix includes pure time-delays. Transfer function $p_{21}(s)$ includes a pure delay $e^{-\tau_1 s}$, and $p_{12}(s)$ includes a pure delay $e^{-\tau_2 s}$. The transfer functions $p_{11}(s)$ and $p_{22}(s)$ have no delay.

We finally get the following factorization of all elements of the transfer matrix:

$$\begin{aligned} p_{11}(s) &= \tilde{p}_{11}(s), \\ p_{12}(s) &= \tilde{p}_{12}(s)e^{-\tau_2 s}, \\ p_{21}(s) &= \tilde{p}_{21}(s)e^{-\tau_1 s}, \\ p_{22}(s) &= \tilde{p}_{22}(s), \end{aligned}$$

with

$$\begin{aligned} \tilde{p}_{12}(s) &= \frac{(\lambda_1(s)-\lambda_2(s))}{T_0 s\left(1-e^{(\lambda_1(s)-\lambda_2(s))L}\right)} e^{\tau_2 s-\lambda_2(s)L}, \\ \tilde{p}_{21}(s) &= \frac{(\lambda_2(s)-\lambda_1(s))}{T_0 s\left(1-e^{(\lambda_1(s)-\lambda_2(s))L}\right)} e^{\tau_1 s+\lambda_1(s)L}. \end{aligned}$$

Let us note that when $|s| \to \infty$, we have

$$\begin{aligned} \lambda_1(s)L &= -\tau_1 s + O(1), \\ \lambda_2(s)L &= \tau_2 s + O(1). \end{aligned}$$

This explains why the terms $e^{\tau_1 s+\lambda_1(s)L}$ and $e^{\tau_2 s-\lambda_2(s)L}$ are delay-free.

This result shows that the Saint-Venant transfer matrix includes delays only linked to the wave propagations, which correspond to the delays obtained by the characteristics.

Distributed Transfer Matrix

As in the horizontal frictionless case, the delays are obtained directly as $e^{-\frac{x}{\alpha}s}$ for the distributed transfer functions $g_{11}(x,s)$ and $g_{21}(x,s)$, and $e^{-\frac{L-x}{\beta}s}$ for the distributed transfer functions $g_{12}(x,s)$ and $g_{22}(x,s)$.

The transfer functions can therefore be factorized as

$$\begin{aligned}
g_{11}(x,s) &= \tilde{g}_{11}(x,s)e^{-\frac{x}{\alpha}s},\\
g_{12}(x,s) &= \tilde{g}_{12}(x,s)e^{-\frac{L-x}{\beta}s},\\
g_{21}(x,s) &= \tilde{g}_{21}(x,s)e^{-\frac{x}{\alpha}s},\\
g_{22}(x,s) &= \tilde{g}_{22}(x,s)e^{-\frac{L-x}{\beta}s},
\end{aligned}$$

where the delay-free parts $\tilde{g}_{ij}(x,s)$ are given by:

$$\begin{aligned}
\tilde{g}_{11}(x,s) &= \frac{\lambda_2(s)e^{(\lambda_1(s)-\lambda_2(s))(L-x)} - \lambda_1(s)}{T_0 s(1-e^{(\lambda_1(s)-\lambda_2(s))L})}e^{\left(\frac{s}{\alpha}+\lambda_1(s)\right)x},\\
\tilde{g}_{12}(x,s) &= \frac{\lambda_1(s)e^{(\lambda_1(s)-\lambda_2(s))x} - \lambda_2(s)}{T_0 s(1-e^{(\lambda_1(s)-\lambda_2(s))L})}e^{\left(\frac{s}{\beta}-\lambda_2(s)\right)(L-x)},\\
\tilde{g}_{21}(x,s) &= \frac{1-e^{(\lambda_1(s)-\lambda_2(s))(L-x)}}{1-e^{(\lambda_1(s)-\lambda_2(s))L}}e^{\left(\frac{s}{\alpha}+\lambda_1(s)\right)x},\\
\tilde{g}_{22}(x,s) &= \frac{1-e^{(\lambda_1(s)-\lambda_2(s))x}}{1-e^{(\lambda_1(s)-\lambda_2(s))L}}e^{\left(\frac{s}{\beta}-\lambda_2(s)\right)(L-x)}.
\end{aligned}$$

3.3.2.2 Asymptotic Estimate of the Poles

The following proposition provides an asymptotic estimate of the poles for high frequencies.

Proposition 3.1. *For $k \in \mathbb{N}$, $k \gg k_h = \frac{L}{\pi(\alpha+\beta)}\sqrt{\frac{(\alpha\delta-\gamma)(\beta\delta+\gamma)}{\alpha\beta}}$ the poles of the Saint-Venant transfer matrix (3.49) can be approximated by:*

$$p_{\pm k} \approx -\frac{(r_1+r_2)L}{\tau_1+\tau_2} \pm \frac{2jk\pi}{\tau_1+\tau_2}, \tag{3.55}$$

with $r_1 = \frac{\alpha\delta-\gamma}{\alpha(\alpha+\beta)}$ and $r_2 = \frac{\beta\delta+\gamma}{\beta(\alpha+\beta)}$, and $\tau_1 = \frac{L}{\alpha}$ the delay for downstream propagation, and $\tau_2 = \frac{L}{\beta}$ the delay for upstream propagation.

Proof. When $k \gg k_h$, the eigenvalues $\lambda_1(s)$ and $\lambda_2(s)$ can be approximated by:

$$\begin{aligned}
\lambda_1(s) &\approx -r_1 - \frac{s}{\alpha},\\
\lambda_2(s) &\approx r_2 + \frac{s}{\beta}.
\end{aligned}$$

Then, (3.55) is obtained from straightforward manipulations of (3.48) for $k \gg k_h$.

□

One can show that this high frequency approximation leads to a relative error smaller than 10% when $k > 5.7 \times k_h$ and smaller than 5% when $k > 10.7 \times k_h$.

This proposition shows that for high frequencies the poles of the Saint-Venant transfer matrix are close to those of the following damped wave equation:

$$\left(\frac{\partial}{\partial x}+r_1+\frac{1}{\alpha}\frac{\partial}{\partial t}\right)\left(\frac{\partial}{\partial x}-r_2-\frac{1}{\beta}\frac{\partial}{\partial t}\right)q=0,$$

with boundary conditions $q(0,t)$ and $q(L,t)$. Using the Laplace transform, this equation reduces to an ODE in x, with eigenvalues equal to $-r_1-\frac{s}{\alpha}$ and $r_2+\frac{s}{\beta}$. The obtained transfer function has a denominator $1-e^{-(r_1+r_2)L-(\tau_1+\tau_2)s}$, whose roots coincide with the poles' approximation (3.55). This shows that the oscillating modes correspond to the interaction of two gravity waves, one traveling downstream at speed $\alpha=V_0+C_0$ with attenuation factor r_1, and one traveling upstream at speed $\beta=C_0-V_0$ with attenuation factor r_2.

Example 3.9 (Asymptotic estimates for the example canals). Table 3.1 compares the exact value of the poles obtained from (3.48) with their high frequency approximation obtained from (3.55) for the two example canals at uniform flow.

The high frequency approximation is rather good for canal 1, even for small values of k. Indeed, for this canal, $k_h=0.36$, which means that the approximation error is smaller than 10% for $k\geq 5.7\times 0.36=2.07$. For canal 2, one sees that the accuracy of the poles approximation increases as index k increases. For this canal, $k_h=1.65$, which means that the approximation error is smaller than 10% for $k\geq 5.7\times 1.65=9.5$.

Table 3.1 Exact value and high frequency approximation for the poles of the two canals

k	Canal 1		Canal 2	
	Exact	Approx.	Exact	Approx.
± 1	$-0.0015\pm 0.0038j$	$-0.0015\pm 0.0041j$	$-0.0014,\ -0.0063$	$-0.0039\pm 0.0019j$
± 2	$-0.0015\pm 0.0080j$	$-0.0015\pm 0.0081j$	$-0.0039\pm 0.0021j$	$-0.0039\pm 0.0037j$
± 3	$-0.0015\pm 0.0121j$	$-0.0015\pm 0.0122j$	$-0.0039\pm 0.0046j$	$-0.0039\pm 0.0056j$
± 4	$-0.0015\pm 0.0162j$	$-0.0015\pm 0.0162j$	$-0.0039\pm 0.0068j$	$-0.0039\pm 0.0074j$
± 5	$-0.0015\pm 0.0202j$	$-0.0015\pm 0.0203j$	$-0.0039\pm 0.0088j$	$-0.0039\pm 0.0093j$
± 10	$-0.0015\pm 0.0406j$	$-0.0015\pm 0.0406j$	$-0.0039\pm 0.0183j$	$-0.0039\pm 0.0185j$

3.3.2.3 Modal Decompositions

Rational Series Expansion

For simplicity, we assume in the following that the poles have single multiplicity, i.e., that $\Delta(k)\neq 0$, but the solution can easily be extended to the case where $\Delta(k)=0$.

Then, the Cauchy residues theorem implies that each transfer function $g_{ij}(x,s)$ can be decomposed as an infinite sum:

$$g_{ij}(x,s) = b_{ij}(x) + \frac{a_{ij}^{(0)}(x)}{s} + \sum_{k=-\infty,k\neq 0}^{\infty} \frac{a_{ij}^{(k)}(x)s}{p_k(s-p_k)}, \tag{3.56}$$

with

$$a_{ij}^{(k)}(x) = \lim_{s\to p_k}(s-p_k)g_{ij}(x,s), \tag{3.57}$$

and

$$b_{ij}(x) = \frac{\partial}{\partial s}[sg_{ij}(x,s)]|_{s=0}. \tag{3.58}$$

The coefficient $a_{ij}^{(k)}(x)$ is the residue of transfer function $g_{ij}(x,s)$ at the pole p_k.

Proof. The proof can be easily adapted from that in Appendix B.1. □

The residues can be computed analytically with (3.57), leading to, for $k=0$:

$$a_{11}^{(0)}(x) = \frac{\gamma}{\alpha\beta T_0}\frac{e^{\frac{\gamma x}{\alpha\beta}}}{e^{\frac{\gamma L}{\alpha\beta}}-1}, \tag{3.59a}$$

$$a_{12}^{(0)}(x) = -a_{11}^{(0)}(x), \tag{3.59b}$$

$$a_{21}^{(0)}(x) = 0, \tag{3.59c}$$

$$a_{22}^{(0)}(x) = 0, \tag{3.59d}$$

and, for $k \neq 0$:

$$a_{11}^{(k)}(x) = -sgn(k)\frac{2k^2\pi^2 e^{\frac{((\alpha-\beta)p_k+\gamma)}{2\alpha\beta}x}}{T_0L^3p_k\sqrt{\Delta(k)}}\times \left[\cos\left(\frac{k\pi x}{L}\right)+\frac{((\alpha-\beta)p_k+\gamma)L}{2\alpha\beta k\pi}\sin\left(\frac{k\pi x}{L}\right)\right], \tag{3.60a}$$

$$a_{12}^{(k)}(x) = sgn(k)(-1)^k\frac{2k^2\pi^2 e^{\frac{((\alpha-\beta)p_k+\gamma)}{2\alpha\beta}(x-L)}}{T_0L^3p_k\sqrt{\Delta(k)}}\times \left[\cos\left(\frac{k\pi x}{L}\right)+\frac{((\alpha-\beta)p_k+\gamma)L}{2\alpha\beta k\pi}\sin\left(\frac{k\pi x}{L}\right)\right], \tag{3.60b}$$

$$a_{21}^{(k)}(x) = -sgn(k)\frac{2k\pi e^{\frac{((\alpha-\beta)p_k+\gamma)}{2\alpha\beta}x}}{L^2p_k\sqrt{\Delta(k)}}\sin\left(\frac{k\pi x}{L}\right), \tag{3.60c}$$

$$a_{22}^{(k)}(x) = sgn(k)(-1)^k\frac{2k\pi e^{\frac{((\alpha-\beta)p_k+\gamma)}{2\alpha\beta}(x-L)}}{L^2p_k\sqrt{\Delta(k)}}\sin\left(\frac{k\pi x}{L}\right), \tag{3.60d}$$

and $b_{ij}(x)$ are given by:

$$b_{11}(x) = \frac{e^{\psi\frac{x}{L}}}{T_0(e^{\psi}-1)^2}\Bigg[\frac{\alpha-\beta}{\alpha\beta}\Big(e^{\psi}(1+\psi(\frac{x}{L}-1))-1-\psi\frac{x}{L}\Big)$$
$$+\frac{\delta}{\gamma}\Big(e^{\psi(1-\frac{x}{L})}(e^{\psi}-1)+e^{\psi}(1+\psi(\frac{x}{L}-2))-1-\psi\frac{x}{L}\Big)\Bigg], \tag{3.61a}$$

$$b_{12}(x) = \frac{1}{T_0(e^{\psi}-1)^2}\Bigg[\frac{\alpha-\beta}{\alpha\beta}e^{\psi\frac{x}{L}}\Big(\psi e^{\psi}+(1-e^{\psi})(1+\psi\frac{x}{L})\Big)$$
$$+\frac{\delta}{\gamma}\Big((1+e^{\psi})\psi e^{\psi\frac{x}{L}}+(1-e^{\psi})(1+e^{\psi\frac{x}{L}}(1+\psi\frac{x}{L}))\Big)\Bigg], \tag{3.61b}$$

$$b_{21}(x) = \frac{e^{\psi}-e^{\psi(\frac{x}{L}-1)}}{e^{\psi}-1}, \tag{3.61c}$$

$$b_{22}(x) = \frac{e^{\psi\frac{x}{L}}-1}{e^{\psi}-1}, \tag{3.61d}$$

with $\psi = \frac{\gamma L}{\alpha\beta}$.

The modal factors $a_{ij}^{(k)}(x)$ are a combination of sine and cosine functions of $\frac{k\pi x}{L}$, modified by a multiplicative exponential term that depends on the difference $\alpha-\beta$ and the coefficient γ. This feature generalizes the result already obtained for the horizontal frictionless case. Coefficients γ and δ are directly linked to the slope of the open channel. The larger the slope, the larger their influence on the modal factors. Let us note that we recover the horizontal frictionless case by letting γ and δ go to zero, and δ/γ be finite.

Rational Decomposition of the Delay-free Part

We have shown that the transfer functions $g_{ij}(x,s)$ can be factorized as a delay times a delay-free part. Therefore, one can find an approximation of $g_{ij}(x,s)$ by applying the Cauchy residues theorem on transfer functions $\tilde{g}_{ij}(x,s)$, and then adding the delay. Finally, each transfer function $\tilde{g}_{ij}(x,s)$ can be decomposed as:

$$\tilde{g}_{ij}(x,s) = \tilde{b}_{ij}(x) + \frac{\tilde{a}_{ij}^{(0)}(x)}{s} + \sum_{k=-\infty,k\neq 0}^{\infty}\frac{\tilde{a}_{ij}^{(k)}(x)s}{p_k(s-p_k)}, \tag{3.62}$$

with

$$\tilde{a}_{ij}^{(k)}(x) = \lim_{s\to p_k}(s-p_k)\tilde{g}_{ij}(x,s), \tag{3.63}$$

and

$$\tilde{b}_{ij}(x) = \frac{\partial}{\partial s}[s\tilde{g}_{ij}(x,s)]|_{s=0}. \tag{3.64}$$

The coefficient $\tilde{a}_{ij}^{(k)}(x)$ is the residue of transfer function $\tilde{g}_{ij}(x,s)$ at the pole p_k.

Proof. The proof can be easily adapted from that in Appendix B.1. □

The residues can be computed analytically with (3.63), leading to, for $k = 0$:

$$\tilde{a}_{ij}^{(0)}(x) = a_{ij}^{(0)}(x), \tag{3.65}$$

and, for $k \neq 0$:

$$\tilde{a}_{11}^{(k)}(x) = e^{\frac{x}{\alpha}p_k} a_{11}^{(k)}(x), \tag{3.66a}$$
$$\tilde{a}_{12}^{(k)}(x) = e^{\frac{L-x}{\beta}p_k} a_{12}^{(k)}(x), \tag{3.66b}$$
$$\tilde{a}_{21}^{(k)}(x) = e^{\frac{x}{\alpha}p_k} a_{21}^{(k)}(x), \tag{3.66c}$$
$$\tilde{a}_{22}^{(k)}(x) = e^{\frac{L-x}{\beta}p_k} a_{22}^{(k)}(x), \tag{3.66d}$$

and $\tilde{b}_{ij}(x)$ are given by:

$$\tilde{b}_{11}(x) = b_{11}(x) + \frac{x}{\alpha} a_{11}^{(0)}(x), \tag{3.67a}$$
$$\tilde{b}_{12}(x) = b_{12}(x) + \frac{(L-x)}{\beta} a_{12}^{(0)}(x), \tag{3.67b}$$
$$\tilde{b}_{21}(x) = b_{21}(x), \tag{3.67c}$$
$$\tilde{b}_{22}(x) = b_{22}(x). \tag{3.67d}$$

Example 3.10 (Modal factors of the example canals). The functions $a_{11}^{(k)}(x)$ of the distributed modal decomposition of $g_{11}(x,s)$ of the example canals 1 and 2 are depicted in Figs. 3.13 and 3.14.

For canal 1, which is flat and has oscillating modes, we recover the behavior already observed for the horizontal frictionless case: the modes of oscillation are clearly visible. The integrator in this case is slightly modified by the slope and the friction, whereas it was constant in the horizontal frictionless case. We clearly see the oscillating modes and the associated nodes and anti-nodes of oscillation. A node corresponds to a point where the traveling waves interfere negatively with each other. An anti-node corresponds to a positive interference: the perturbations are in phase and act additively.

For canal 2, which is steep, we see that the dominant term is the integrator, which is greatly modified by the slope. The poles $p_{\pm 1}$ are negative real, therefore not oscillating, and those obtained for $|k| > 1$ are oscillating. The corresponding modal factors are strongly modified by the slope, compared to those of canal 1.

Figure 3.15 depicts the value of coefficients $b_{11}(x)$ and $b_{12}(x)$ of the distributed modal decomposition of the transfer functions $g_{11}(x,s)$ and $g_{12}(x,s)$ for canals 1 and 2.

The term $b_{ij}(x)$ corresponds to a direct term in transfer function $g_{ij}(x,s)$. We see that this term varies a lot with x for both canals. It even changes sign, which

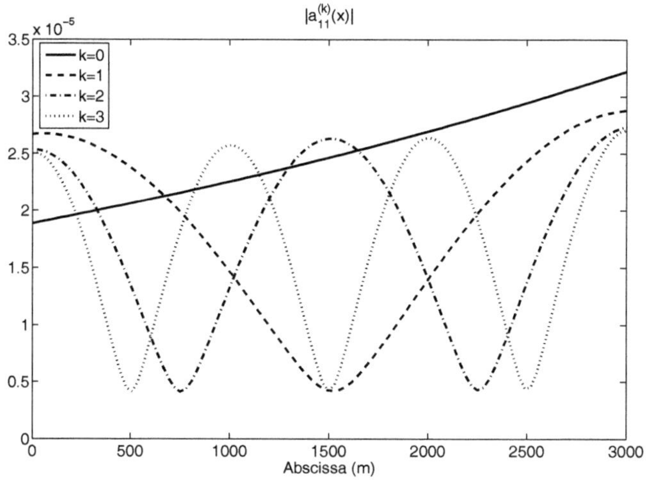

Fig. 3.13 Function $|a_{11}^{(k)}(x)|$ of the distributed modal decomposition for canal 1

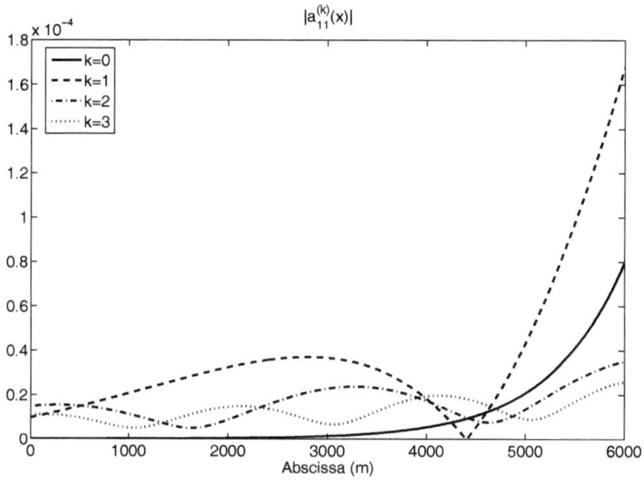

Fig. 3.14 Function $|a_{11}^{(k)}(x)|$ of the distributed modal decomposition for canal 2

is a surprising result: this means that the water level change resulting from a step discharge input contains a direct negative term when x is larger than a certain value.

The decomposition (3.56) will be used in Chap. 4 to obtain rational approximations of open channel flow.

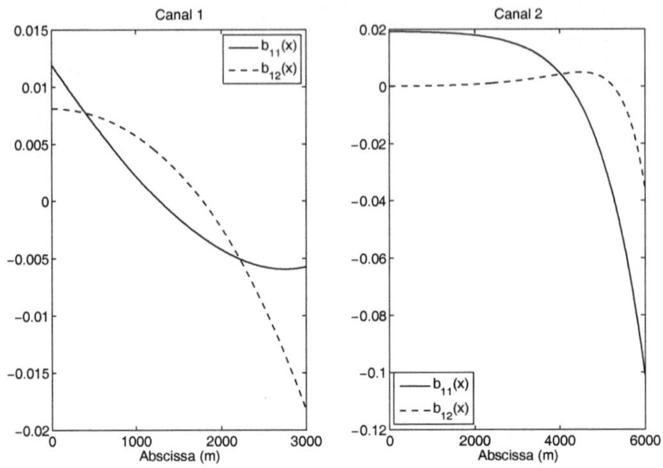

Fig. 3.15 Functions $b_{11}(x)$ and $b_{12}(x)$ of the distributed modal decomposition for canals 1 and 2

3.3.3 Time Domain Expressions

We now study the system in the time domain and provide explicit expressions for the linearized Saint-Venant equations using 1) a method based on the characteristics, which generalizes that already obtained for the horizontal frictionless channel case, 2) a method based on the input–output transfer matrix, using an infinite series expansion, and 3) a method based on the modal factorization.

3.3.3.1 Explicit Expressions for the Characteristics

The change of variable $\hat{\zeta}(x,s) = \mathscr{X}(s)\hat{\xi}(x,s)$ enabled us to diagonalize the system, and the solution in terms of the new characteristics variables has been obtained directly in the frequency domain using (3.54). These expressions can also be translated into the time domain. Let us introduce the irrational transfer functions $\hat{F}(s)$ and $\hat{H}(l,s)$ defined by:

$$\hat{F}(s) = s + b - \sqrt{(s+b)^2 - a^2}, \tag{3.68}$$

$$\hat{H}(l,s) = e^{l\hat{F}(s)} - 1. \tag{3.69}$$

Using inverse Laplace transform tables [1], pp. 1025–1027, one can check that:

$$F(t) = \mathscr{L}^{-1}\left\{\hat{F}(s)\right\} = \frac{ae^{-bt}}{t} I_1(at), \tag{3.70}$$

$$H(l,t) = \mathscr{L}^{-1}\left\{\hat{H}(l,s)\right\} = \frac{ale^{-bt}}{\sqrt{t^2-l^2}} I_1\left(a\sqrt{t^2-l^2}\right), \tag{3.71}$$

with $I_1(t)$ the modified Bessel function[2] of order 1.

Using (3.42) and (3.68), the eigenvalues $\lambda_1(s)$ and $\lambda_2(s)$ can be expressed as:

$$\lambda_1(s) = -r_1 - \frac{s}{\alpha} + \theta\hat{F}(s), \tag{3.72a}$$

$$\lambda_2(s) = r_2 + \frac{s}{\beta} - \theta\hat{F}(s), \tag{3.72b}$$

with $r_1 = \frac{\alpha\delta-\gamma}{\alpha(\alpha+\beta)}$, $r_2 = \frac{\beta\delta+\gamma}{\beta(\alpha+\beta)}$ and $\theta = \frac{(\alpha+\beta)}{2\alpha\beta}$.

Therefore, we have:

$$e^{\lambda_1(s)x} = e^{-r_1x}e^{-\frac{x}{\alpha}s}\left[1+\hat{H}(\theta x,s)\right],$$

$$e^{-\lambda_2(s)(L-x)} = e^{-r_2(L-x)}e^{-\frac{L-x}{\beta}s}\left[1+\hat{H}(\theta(L-x),s)\right].$$

Applying the inverse Laplace transform to (3.54) enables us to express the generalized characteristics as a function of their values at the boundary:

$$\zeta_1(x,t) = e^{-r_1x}\left[\zeta_1\left(0,t-\frac{x}{\alpha}\right) + \int_{\frac{x}{\alpha}}^{t} H\left(\theta x, t-\tau\right)\zeta_1\left(0,\tau-\frac{x}{\alpha}\right)\mathrm{d}\tau\right], \tag{3.73a}$$

$$\begin{aligned}\zeta_2(x,t) = e^{-r_2(L-x)}\Bigg[&\zeta_2\left(L,t-\frac{(L-x)}{\beta}\right)\\ &+\int_{\frac{L-x}{\beta}}^{t} H\left(\theta(L-x),t-\tau\right)\zeta_2\left(L,\tau-\frac{(L-x)}{\beta}\right)\mathrm{d}\tau\Bigg].\end{aligned} \tag{3.73b}$$

The expressions (3.73) generalize those already obtained in the horizontal frictionless case. The variable $\zeta_1(x,t)$ is obtained as the sum of its delayed value at $x=0$ plus its delayed convolution with an infinite dimensional kernel $H(l,t)$, multiplied by a damping coefficient e^{-r_1x}. The same behavior occurs for $\zeta_2(x,t)$.

These expressions can be used to derive time domain expressions for the physical variables q and y. Details can be found in Appendix C.

---

[2] For integer values of n, the modified Bessel functions can be computed using the following series expansion:

$$I_n(t) = \sum_{k=0}^{\infty} \frac{1}{k!(n+k)!}\left(\frac{t}{2}\right)^{2k+n}.$$

3.3.3.2 State-space Expressions

The state-space expressions already obtained in the horizontal frictionless case are also valid in the uniform case. For the rational state-space expressions, the expressions for matrices $A, B, C(x)$, and $D(x)$ remain the same, with the parameters $a_{ij}^{(k)}(x)$ and $b_{ij}(x)$ given by (3.57) and (3.58), respectively. The parameter μ_k is in this case given by:

$$\mu_k = (-1)^{k+1} e^{\left(\frac{(\alpha-\beta)p_k+\gamma}{2\alpha\beta}\right)L}.$$

For the delayed state-space expressions, the expressions for $\tilde{C}(x)$ and $\tilde{D}(x)$ remain the same, with the parameters $\tilde{a}_{ij}^{(k)}(x)$ and $\tilde{b}_{ij}(x)$ given by (3.63) and (3.64), respectively. The parameter $\tilde{\mu}_k$ is in this case given by:

$$\tilde{\mu}_k(x) = \mu_k e^{\left(\frac{L-x}{\beta} - \frac{x}{\alpha}\right)p_k}.$$

3.4 Nonuniform Flow

We now consider the general case of nonuniform flow, where analytical results are no longer possible. We first show that the computation of the transfer matrix is equivalent to solving a specific ordinary differential equation parameterized by the complex Laplace variable s. The application of classical numerical integration methods, such as Runge–Kutta, faces a main difficulty. Due to structural properties of the considered differential equation, the classical compromise between precision of the approximate solution and integration step size necessarily leads to a large computational time for each value of s. The frequency domain of interest goes from the low frequency behavior of the canal (typically $\omega_r/100$ with ω_r the resonant frequency $\omega_r = 2\pi/(\tau_1 + \tau_2)$ with $\tau_1 = L/(V_0 + C_0)$ and $\tau_2 = L/(C_0 - V_0)$, respectively, the downstream and upstream propagation time-delays) to the high frequency behavior (typically 10 times the resonant frequency). Since it is important to have a good representation in between these extreme frequencies (and especially to correctly reproduce the resonant modes), it is necessary to compute a large number of frequency points (typically 500 points) [15]. To this end, it is therefore essential to have an efficient numerical method to compute the frequency response.

3.4.1 Computation of the Transfer Matrix

In the general nonuniform flow case, the linearized Saint-Venant equations are given by (2.28):

$$\frac{\partial \xi}{\partial t} + \mathbf{A}(x)\frac{\partial \xi}{\partial x} + \mathbf{B}(x)\xi = 0, \tag{3.74}$$

where $\xi(x,t) = (a(x,t), q(x,t))^T : [0,L] \times [0,+\infty) \rightarrow \Omega \in \mathbb{R}^2$ is the state of the system and $\mathbf{A}(x) = \begin{pmatrix} 0 & 1 \\ \alpha(x)\beta(x) & \alpha(x)-\beta(x) \end{pmatrix}$, $\mathbf{B}(x) = \begin{pmatrix} 0 & 0 \\ -\gamma(x) & \delta(x) \end{pmatrix}$.

In the nonuniform flow case, it is no longer possible to diagonalize the system. Indeed, applying the Laplace transform to (3.74) and after elementary manipulations, we obtain the following differential equation, parameterized by the complex Laplace variable s:

$$\frac{\partial \hat{\xi}(x,s)}{\partial x} = \mathscr{A}(x,s)\hat{\xi}(x,s) + \mathscr{B}(x)\xi(x,0), \tag{3.75}$$

with $\xi(x,t) = (T_0(x)y(x,t), q(x,t))^T$ and where $\mathscr{A}(x,s)$ and $\mathscr{B}(x)$ depend on x:

$$\mathscr{A}(x,s) = \frac{1}{\alpha(x)\beta(x)} \begin{pmatrix} (\alpha(x)-\beta(x))s+\gamma(x) & -s-\delta(x) \\ -\alpha(x)\beta(x)s & 0 \end{pmatrix},$$

$$\mathscr{B}(x) = \frac{1}{\alpha(x)\beta(x)} \begin{pmatrix} (\beta(x)-\alpha(x)) & 1 \\ \alpha(x)\beta(x) & 0 \end{pmatrix}.$$

To simplify the exposition, we assume in the following that $\xi(x,0) = 0$.

One may show that the change of variable that diagonalizes $\mathscr{A}(x,s)$ introduces a supplementary term due to the derivative of matrix $\mathscr{X}(x,s)$ with respect to x. Indeed, let us introduce $\hat{\zeta}(x,s) = \mathscr{X}(x,s)\hat{\xi}(x,s)$, where $\mathscr{X}(x,s)$ is such that $\mathscr{A}(x,s) = \mathscr{X}(x,s)^{-1}\mathscr{D}(x,s)\mathscr{X}(x,s)$. Plugging this variable into (3.75) leads to:

$$\frac{\partial \hat{\zeta}(x,s)}{\partial x} = \left(\mathscr{D}(x,s) + \frac{\partial \mathscr{X}(x,s)}{\partial x}\mathscr{X}(x,s)^{-1}\right)\hat{\zeta}(x,s). \tag{3.76}$$

The matrix $\mathscr{D}(x,s) + \frac{\partial \mathscr{X}(x,s)}{\partial x}\mathscr{X}(x,s)^{-1}$ is not diagonal because of the term $\frac{\partial \mathscr{X}(x,s)}{\partial x}\mathscr{X}(x,s)^{-1}$. Therefore we can no longer provide an analytical solution to this equation.

However, since the differential equation (3.75) is linear, we know that its general solution always exists, is unique and is given by [13], p. 598:

$$\hat{\xi}(x,s) = \Phi(x,s)\hat{\xi}(0,s) = \begin{pmatrix} \phi_{11}(x,s) & \phi_{12}(x,s) \\ \phi_{21}(x,s) & \phi_{22}(x,s) \end{pmatrix} \hat{\xi}(0,s), \tag{3.77}$$

where $\Phi(x,s)$ is the state-transition matrix associated to the differential equation (3.75).

As a matter of fact, one may think that, as in the scalar case, the state-transition matrix could be written as $\Phi(x,s) = e^{\int_0^x \mathscr{A}(v,s)\mathrm{d}v}$. However, this is not true in general. Indeed, using the definition of $e^A \triangleq \sum_{k=0}^{\infty} \frac{A^k}{k!}$, we have:

$$e^{\int_0^x \mathscr{A}(v,s)\mathrm{d}v} = I + \int_0^x \mathscr{A}(v,s)\mathrm{d}v + \frac{1}{2}\int_0^x \mathscr{A}(v,s)\mathrm{d}v \int_0^x \mathscr{A}(v,s)\mathrm{d}v + \dots$$

and therefore:

$$\begin{aligned}\frac{\partial}{\partial x}\left(e^{\int_0^x \mathscr{A}(v,s)\mathrm{d}v}\right) &= \mathscr{A}(x,s) + \frac{\mathscr{A}(x,s)}{2}\int_0^x \mathscr{A}(v,s)\mathrm{d}v \\ &\quad + \int_0^x \mathscr{A}(v,s)\mathrm{d}v\frac{\mathscr{A}(x,s)}{2} + \ldots \\ &\neq \mathscr{A}(x,s)e^{\int_0^x \mathscr{A}(v,s)\mathrm{d}v}\,!\end{aligned}$$

The only case where this holds is when $\mathscr{A}(x,s)$ and $\int_0^x \mathscr{A}(v,s)\mathrm{d}v$ commute. This is true, e.g., when $\mathscr{A}(x,s)$ is constant with respect to x, i.e., in the uniform flow case. In the nonuniform flow case, this is generally not true, this is why we need to find a numerical way to compute the state-transition matrix.

Once the state-transition matrix is obtained, the transfer matrix corresponding to the original differential equation (3.75) is given by:

$$\begin{pmatrix}\hat{y}(0,s)\\ \hat{y}(L,s)\end{pmatrix} = \begin{pmatrix}p_{11}(s) & p_{12}(s)\\ p_{21}(s) & p_{22}(s)\end{pmatrix}\begin{pmatrix}\hat{q}(0,s)\\ \hat{q}(L,s)\end{pmatrix}, \tag{3.78}$$

with $p_{11}(s) = -\frac{\phi_{22}(L,s)}{T_0(0)\phi_{21}(L,s)}$, $p_{12}(s) = \frac{1}{T_0(0)\phi_{21}(L,s)}$, $p_{21}(s) = \frac{\phi_{12}(L,s)}{T_0(L)} - \frac{\phi_{11}(L,s)\phi_{22}(L,s)}{T_0(L)\phi_{21}(L,s)}$, and $p_{22}(s) = \frac{\phi_{11}(L,s)}{T_0(L)\phi_{21}(L,s)}$, provided that $\phi_{21}(L,s)$ is not equal to zero. As in the uniform flow case, the values of s such that $\phi_{21}(L,s) = 0$ correspond to the poles of the Saint-Venant transfer matrix.

The problem is therefore to solve the differential equation (3.75) and compute the state-transition matrix $\Phi(x,s)$. This problem is not easy to solve, due to its oscillatory and unstable nature.

3.4.1.1 Difficulty of Numerical Integration

Let us analyze the numerical problem, with a more technical approach. In the uniform case, the integration of the ODE (3.76) can be restated as the integration of two complex ODEs given by

$$\frac{\mathrm{d}\zeta_i}{\mathrm{d}x} = \lambda_i(s)\zeta_i, \quad i = 1,2, \tag{3.79}$$

with $\zeta(x) := \mathscr{X}(s)\xi(x) = (\zeta_1(x), \zeta_2(x))^T$.

The general solution of the complex first-order ODE (3.79) is given by

$$\zeta_i(x) = e^{\lambda_i(s)x}\zeta_i(0). \tag{3.80}$$

In this case, for high frequencies $s = j\omega$, the real part $\Re(\lambda_i(j\omega))$ remains small, while the imaginary part $\Im(\lambda_i(j\omega))$ is proportional to ω. Then the qualitative behavior of the solution (3.80) is close to a sinusoidal response of frequency $\omega_i = |\lambda_i(j\omega)|$ (i.e., $v_i(x) \approx e^{\Re(\lambda_i(j\omega))x}\cos(\omega_i x)$) (see Fig. 3.16). This is the reason why classical numerical methods necessitate a very small step size to solve ODE (3.79) with good precision.

As an example, let us consider the case of the Euler numerical method to integrate (3.79). In this case, this first-order method leads to approximate $e^{\lambda_i(s)x}$ by its Taylor expansion of order 1:

$$e^{\lambda_i(s)x} \approx 1 + \lambda_i(s)x.$$

The Euler method leads us to approximate a sinusoidal response at a frequency proportional to ω by a linear approximation. In order to have good precision a very small step size is needed (typically less than $1/\omega$).

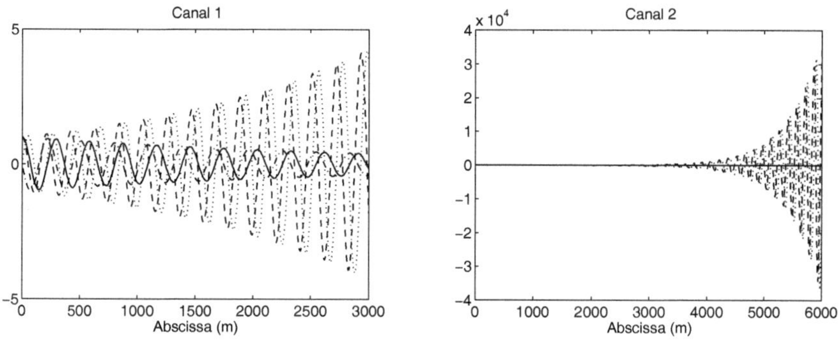

Fig. 3.16 Real and imaginary parts of $e^{\lambda_1(j\omega)x}$ and $e^{\lambda_2(j\omega)x}$ for $\omega = 0.1$ rad/s for canals 1 and 2, uniform flow case

More generally, it is well-known that when systems have highly oscillatory solutions, classical numerical integration schemes (e.g., Runge–Kutta) lead to inefficient algorithms with respect to the computation time. As pointed out by [12], this is due to the fact that the numerical error is explicitly linked to the value of the derivatives of the solution $\dot{\zeta}$, $\ddot{\zeta}$, ..., which are very large in cases of poorly damped systems.

A way to avoid this difficulty is to use an exponential-type method (see [4]). Rather surprisingly, in the case of constant matrix $\mathscr{A}(s)$this solution exactly corresponds to the above computation associated to the uniform case.

3.4.1.2 Numerical Solution for the State-transition Matrix

The proposed numerical scheme is based on a the following idea: since the solution can be obtained analytically for a pool with constant water depth, a pool under backwater flow conditions can be separated into small parts where the water depth is almost constant.

Let x_k be a space discretization of interval $[0,L]$ into n subintervals:

$$0 = x_0 < x_1 < \cdots < x_k < \cdots < x_n = L, \quad x_{k+1} = x_k + h_k. \tag{3.81}$$

The link between $\hat{\xi}(x_k,s) = (T_0(x_k)\hat{y}(x_k,s), \hat{q}(x_k,s))^T$ at the downstream end of a part and $\hat{\xi}(x_{k-1},s)$ at the upstream end can be obtained using (3.43). This process can be propagated downstream, leading to a multiplication of the state-transition matrices.

The numerical scheme and its convergence properties are detailed in Appendix B.3. This numerical scheme is at least of order 1 when the derivative of $\mathscr{A}(x,s)$ with respect to x is bounded.

The proposed numerical scheme leads to the following approximation of the state-transition matrix of (3.75):

$$\Phi(x_n,s) \approx \prod_{k=n-1}^{0} e^{\mathscr{A}(x_k,s)h_k}, \tag{3.82}$$

which can be interpreted as an approximation of nonuniform regime by a series of constant water depth regimes associated with the value of $\mathscr{A}(x,s)$ at x_k.

This is represented in Fig. 3.17 with the real water profile (—) and the corresponding approximation of water depth (– –) used to compute the solution. This method can be interpreted as the equivalent of the well-known rectangle method to compute an integral.

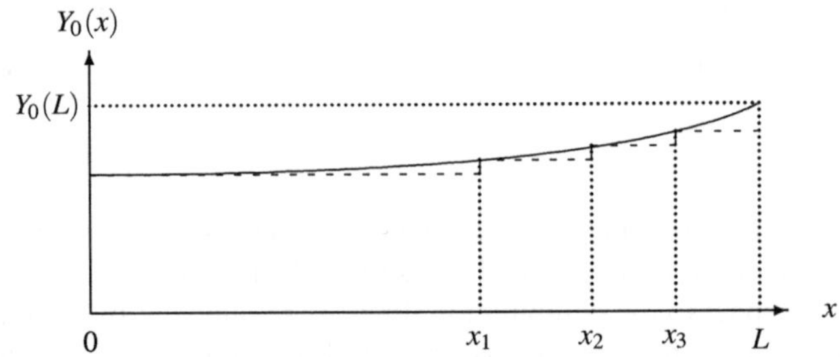

Fig. 3.17 Schematic representation of the integration method approximation of the water depth; – is the real water profile and - - is the corresponding approximation

Example 3.11 (Parameters of the linearized Saint-Venant equations for the example canals). We focus on the two example canals, with the backwater profiles already depicted in Fig. 2.11. The discharge is kept constant, and we investigate the effect of accelerating or decelerating flow by changing the downstream boundary condition Y_L from 0.8 to 1.2 times the normal depth. Tables 3.2 gives the variation of the parameters of the example canals for nonuniform flow.

The variation of parameters $\alpha(x)$ and $\beta(x)$ is depicted in Fig. 3.18 for canals 1 and 2. We see that parameters for canal 1 are much more sensitive to nonuniform flow conditions than for canal 2, where for most of the channel length, the parameters value are close to the uniform flow case.

Table 3.2 Parameters for the two canals in nonuniform flow

		$Y_0(L)$	$T_0(L)$	$\alpha(L)$	$\beta(L)$	$\gamma(L)$	$\delta(L)$
Accelerating	Canal 1	1.70	12.09	4.49	2.76	0.0028	0.0023
	Canal 2	2.34	15.02	7.16	1.22	0.0220	0.0053
Decelerating	Canal 1	2.55	14.64	4.80	3.79	0.0027	0.0039
	Canal 2	3.51	18.52	6.68	3.24	0.0216	0.0091

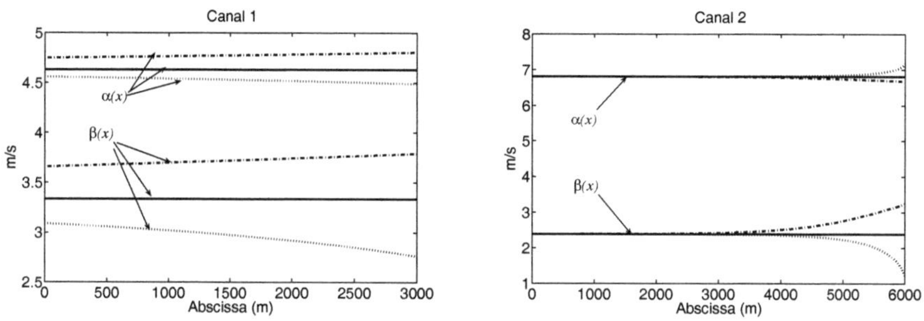

Fig. 3.18 Variation of parameters $\alpha(x)$ and $\beta(x)$ for canals 1 and 2 in uniform (—), accelerating (···), and decelerating (– · –) flows

3.4.1.3 Pole Computation

As we have seen, the poles correspond to the values of s such that $\phi_{21}(L,s) = 0$. In the nonuniform case, the poles are obtained numerically, by finding the zeros of (3.47). Since we can compute the transition matrix numerically, we know how to compute $\phi_{21}(L,s)$ for any $s \in \mathbb{C}$ and we can look for values of s such that $\phi_{21}(L,s) = 0$.

Practically, we seek the value of s that minimizes the modulus of $\phi_{21}(L,s)$. This search is done starting from the values obtained analytically for the uniform regime.

We used this numerical method to find the values of s that minimize the modulus of $\phi_{21}(L,s)$. This has been done for the first seven pairs of poles for the canals in Example 2.1. The starting guess was provided by the high frequency poles approximation (3.91).

The results are depicted in Figs. 3.19 and 3.20 for canals 1 and 2, respectively. In both figures, we see that an accelerating flow tends to dampen the oscillating poles, while a decelerating flow tends to reduce their damping. However, the first negative real pole of canal 2 reacts differently: its damping increases for decelerating flow.

This confirms our results from the asymptotic study in high frequencies, but it does not necessarily apply for low frequencies.

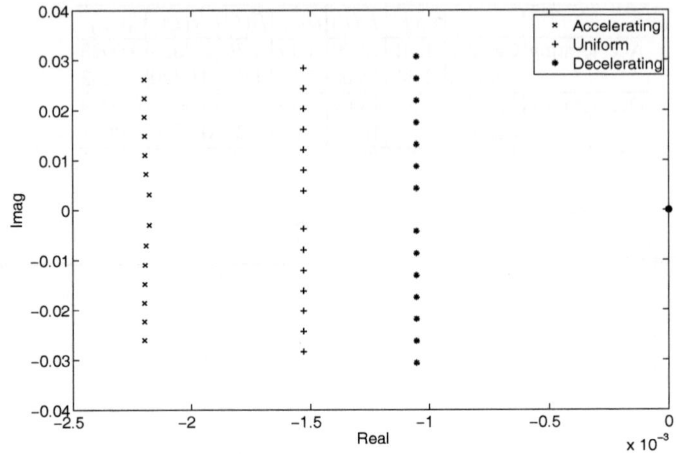

Fig. 3.19 Location of the poles of Saint-Venant transfer matrix for canal 1 for nonuniform flow conditions

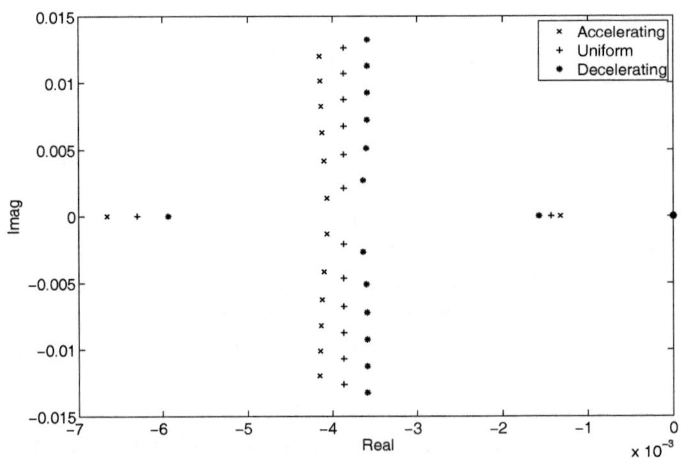

Fig. 3.20 Location of the poles of Saint-Venant transfer matrix for canal 2 for nonuniform flow conditions

3.4.1.4 Input–Output Transfer Matrix

Once the state-transition matrix $\Phi(x,s)$ is computed, the input–output transfer matrix can easily be obtained as follows:

$$\begin{pmatrix} \hat{y}(0,s) \\ \hat{y}(L,s) \end{pmatrix} = \begin{pmatrix} p_{11}(s) & p_{12}(s) \\ p_{21}(s) & p_{22}(s) \end{pmatrix} \begin{pmatrix} \hat{q}(0,s) \\ \hat{q}(L,s) \end{pmatrix},$$

with

$$p_{11}(s) = \frac{1}{T_0(0)}\left(\phi_{12}(0,s) - \phi_{11}(0,s)\frac{\phi_{22}(L,s)}{\phi_{21}(L,s)}\right), \tag{3.83a}$$

$$p_{12}(s) = \frac{1}{T_0(0)}\frac{\phi_{11}(0,s)}{\phi_{21}(L,s)}, \tag{3.83b}$$

$$p_{21}(s) = \frac{1}{T_0(L)}\left(\phi_{12}(L,s) - \phi_{11}(L,s)\frac{\phi_{22}(L,s)}{\phi_{21}(L,s)}\right), \tag{3.83c}$$

$$p_{22}(s) = \frac{1}{T_0(L)}\frac{\phi_{11}(L,s)}{\phi_{21}(L,s)}. \tag{3.83d}$$

Example 3.12 (Bode plots for nonuniform flow). We compute the Bode plots for the three situations: uniform flow, decelerating flow (M1 backwater curve), and accelerating flow (M2 backwater curve). The results are depicted in Figs. 3.21 and 3.22.

The integrator gain increases for accelerating flow, while it decreases for decelerating flow. This also is consistent with physical intuition: in a first approximation, the integrator gain is inversely proportional to the area of the pool. In the case of a trapezoidal geometry, the area increases when the water level increases. For a constant discharge, this corresponds to a decelerating flow. This explains why the integrator gain tends to decrease for decelerating flow. The reverse occurs for accelerating flow.

For the oscillating modes, we see that their damping decreases for decelerating flow, while it increases for accelerating flow. This will be confirmed by a direct study of the poles.

3.4.1.5 Distributed Transfer Matrix

Using the proposed numerical solution, we can also compute the distributed transfer matrix $G(x,s)$. Indeed, the proposed numerical solution enables us to compute the state-transition matrix $\Phi(x,s)$ for all $x \in [0,L]$. Then, the distributed transfer matrix $G(x,s)$ is directly obtained as follows:

$$\begin{pmatrix} \hat{y}(x,s) \\ \hat{q}(x,s) \end{pmatrix} = G(x,s)\begin{pmatrix} \hat{q}(0,s) \\ \hat{q}(L,s) \end{pmatrix}, \tag{3.84}$$

with

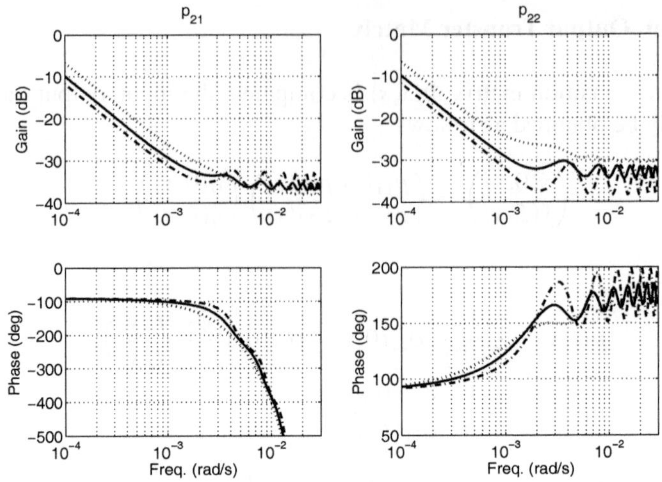

Fig. 3.21 Bode plots of transfer functions $p_{21}(s)$ and $p_{22}(s)$ for canal 1 in accelerating ($\cdots$), uniform (—), and decelerating ($- \cdot -$) flow

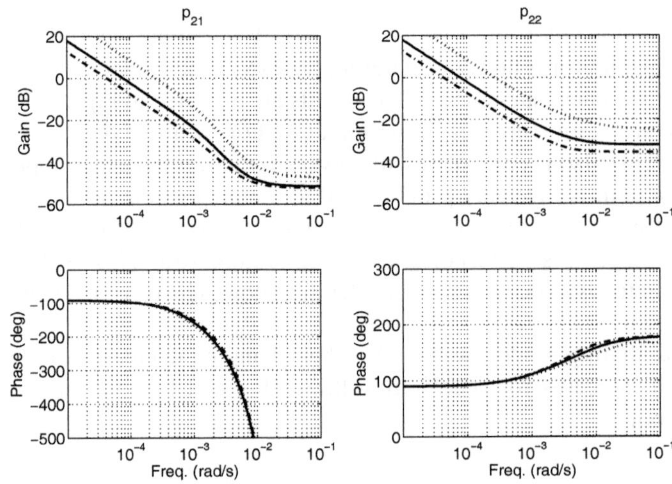

Fig. 3.22 Bode plots of transfer functions $p_{21}(s)$ and $p_{22}(s)$ for canal 2 in accelerating ($\cdots$), uniform (—), and decelerating ($- \cdot -$) flow

$$g_{11}(x,s) = \frac{1}{T_0(x)}\left(\phi_{12}(x,s) - \phi_{11}(x,s)\frac{\phi_{22}(L,s)}{\phi_{21}(L,s)}\right), \tag{3.85a}$$

$$g_{12}(x,s) = \frac{1}{T_0(x)}\frac{\phi_{11}(x,s)}{\phi_{21}(L,s)}, \tag{3.85b}$$

$$g_{21}(x,s) = \phi_{22}(x,s) - \phi_{21}(x,s)\frac{\phi_{22}(L,s)}{\phi_{21}(L,s)}, \tag{3.85c}$$

$$g_{22}(x,s) = \frac{\phi_{21}(x,s)}{\phi_{21}(L,s)}. \tag{3.85d}$$

These expressions generalize to the nonuniform flow case results already obtained in Sect. 3.2.1.5 for the horizontal frictionless case and in Sect. 3.3.1.5 for the uniform flow case.

3.4.2 Properties of the Transfer Matrix

3.4.2.1 Delays

In the nonuniform flow case, we use the proposed numerical scheme to show that the delays of the Saint-Venant transfer functions can be computed by integrating the characteristics lines along the channel.

Computations of the Delays for the Interconnection of Two Pools

Let us first compute the relations between transfer functions $p_{ij}(s)$ for the interconnection of two pools. We denote by $P^{(1)}(s)$ the transfer matrix for the upstream pool, by $P^{(2)}(s)$ the one for the downstream pool, and by $P(s)$ the transfer matrix resulting from the interconnection.

In the following, $p_{ij}^{(1)}(s)$ denotes the terms of transfer matrix corresponding to the upstream part, $p_{ij}^{(2)}(s)$ those corresponding to the downstream part, and $p_{ij}(s)$ those corresponding to the whole pool.

The whole transfer matrix $P(s)$ is therefore obtained by multiplication of the corresponding state-transition matrices and rearranging in order to obtain the desired inputs and outputs, which leads to:

$$p_{11}(s) = p_{11}^{(1)}(s) + \frac{p_{12}^{(1)}(s)p_{21}^{(1)}(s)}{p_{11}^{(2)}(s) - p_{22}^{(1)}(s)}, \tag{3.86a}$$

$$p_{12}(s) = -\frac{p_{12}^{(1)}(s)p_{12}^{(2)}(s)}{p_{11}^{(2)}(s) - p_{22}^{(1)}(s)}, \tag{3.86b}$$

$$p_{21}(s) = \frac{p_{21}^{(1)}(s)p_{21}^{(2)}(s)}{p_{11}^{(2)}(s) - p_{22}^{(1)}(s)}, \tag{3.86c}$$

$$p_{22}(s) = p_{22}^{(2)}(s) - \frac{p_{12}^{(2)}(s)p_{21}^{(2)}(s)}{p_{11}^{(2)}(s) - p_{22}^{(1)}(s)}. \tag{3.86d}$$

We can easily show that the transfer function $\frac{1}{p_{11}^{(2)}(s)-p_{22}^{(1)}(s)}$ has no delay. Therefore, the delay part of transfer function $p_{21}(s)$ is obtained as the sum of the delays $\tau_1^{(1)} + \tau_1^{(2)}$. Similar results hold for transfer function $p_{12}(s)$, with the delay being given by $\tau_2^{(1)} + \tau_2^{(2)}$. Finally, transfer functions $p_{11}(s)$ and $p_{22}(s)$ have no delay.

Computation of the Delays for the General Case

Using the approximation (3.82) for the state-transition matrix, and the above results on the interconnection of two pools, we conclude that the delays of $p_{21}(s)$ and of $p_{12}(s)$ can be obtained by integrating the characteristics lines along the channel.

To this end, let us consider a spatial discretization of the interval $[0,L]$ with a constant space step Δx, and prove the desired result by induction. Let us denote by $p_{ij}^{[0,n]}$ the transfer functions for the interval $[0,n\Delta x]$. For the first section of channel between 0 and Δx, we know from the results of Sect. 3.3.2.1 that the transfer functions $p_{21}^{[0,1]}(s)$ and $p_{12}^{[0,1]}(s)$ include a delay equal to $\Delta x/\alpha(0)$ and $\Delta x/\beta(0)$, respectively.

Let us now assume that the transfer functions for the interval $[0,n\Delta x]$ are given by:

$$
\begin{aligned}
p_{12}^{[0,n]}(s) &= \tilde{p}_{12}^{[0,n]}(s)e^{-\sum_{k=0}^{n-1}\frac{\Delta x}{\beta(x_k)}s},\\
p_{21}^{[0,n]}(s) &= \tilde{p}_{21}^{[0,n]}(s)e^{-\sum_{k=0}^{n-1}\frac{\Delta x}{\alpha(x_k)}s},
\end{aligned}
$$

where $\tilde{p}_{12}^{[0,n]}(s)$ and $\tilde{p}_{21}^{[0,n]}(s)$ are outer transfer functions.

We also know from Sect. 3.3.2.1 that the transfer functions for the interval $[n\Delta x,(n+1)\Delta x]$ are given by:

$$
\begin{aligned}
p_{12}^{[n,n+1]}(s) &= \tilde{p}_{12}^{[n,n+1]}(s)e^{-\frac{\Delta x}{\beta(x_n)}s},\\
p_{21}^{[n,n+1]}(s) &= \tilde{p}_{21}^{[n,n+1]}(s)e^{-\frac{\Delta x}{\alpha(x_n)}s},
\end{aligned}
$$

where $\tilde{p}_{12}^{[n,n+1]}(s)$ and $\tilde{p}_{21}^{[n,n+1]}(s)$ are outer transfer functions.

Finally, the transfer function for the interval $[0,(n+1)\Delta x]$ can be obtained using (3.86), leading to:

$$
\begin{aligned}
p_{12}^{[0,n+1]}(s) &= -\frac{\tilde{p}_{12}^{[0,n]}(s)\tilde{p}_{12}^{[n,n+1]}(s)}{p_{11}^{[n,n+1]}(s)-p_{22}^{[0,n]}(s)}e^{-\sum_{k=0}^{n}\frac{\Delta x}{\beta(x_k)}s},\\
p_{21}^{[0,n+1]}(s) &= \frac{\tilde{p}_{21}^{[0,n]}(s)\tilde{p}_{21}^{[n,n+1]}(s)}{p_{11}^{[n,n+1]}(s)-p_{22}^{[0,n]}(s)}e^{-\sum_{k=0}^{n}\frac{\Delta x}{\alpha(x_k)}s}.
\end{aligned}
$$

To conclude, we use the fact that the transfer function $\frac{1}{p_{11}^{[n,n+1]}(s)-p_{22}^{[0,n]}(s)}$ has no delay.

As $\Delta x \to 0$, the sum $\sum_{k=0}^{n}\frac{\Delta x}{\alpha(x_k)}$ tends towards $\int_0^L \frac{dx}{\alpha(x)}$, and the same occurs for $\sum_{k=0}^{n}\frac{\Delta x}{\beta(x_k)}$, which tends towards $\int_0^L \frac{dx}{\beta(x)}$.

We finally obtain the delays for the transfer functions $p_{21}(s)$ and $p_{12}(s)$ in the nonuniform case as follows:

$$
\tau_1 = \int_0^L \frac{dx}{V_0(x)+C_0(x)} \tag{3.87}
$$

for $p_{21}(s)$ and

$$\tau_2 = \int_0^L \frac{dx}{C_0(x) - V_0(x)} \tag{3.88}$$

for $p_{12}(s)$.

This result is consistent with our physical intuition, and confirms that an open channel in nonuniform flow behaves very similarly to an open channel in uniform flow: the main difference lies in the fact that the results can no longer be expressed in a closed form, but the proposed numerical scheme enables us to compute the transfer function for any realistic flow configuration.

Example 3.13 (Delays for nonuniform flow). Let us define the dimensionless time $t^* = \frac{\alpha_n t}{L}$, where α_n stands for the uniform value of $\alpha = V_0 + C_0$. This enables us to compare the delays τ_1 and τ_2 for various nonuniform flow conditions: for the uniform flow, we have $\tau_1^* = 1$, and $\tau_2^* = \frac{\alpha_n}{\beta_n}$, where the subscript n denotes the uniform flow values.

Fig. 3.23 depicts the way the dimensionless delays vary with the flow in both canals. The two canals have a very different behavior: the delay τ_1 varies much more with the downstream water level in the case of canal 1 than in the case of canal 2. Indeed, the backwater curve affects only a small portion of canal 2, while it affects the whole canal 1. The delay τ_2 is much larger than τ_1 in the case of canal 2, because β is relatively much lower than α for canal 2 than for canal 1.

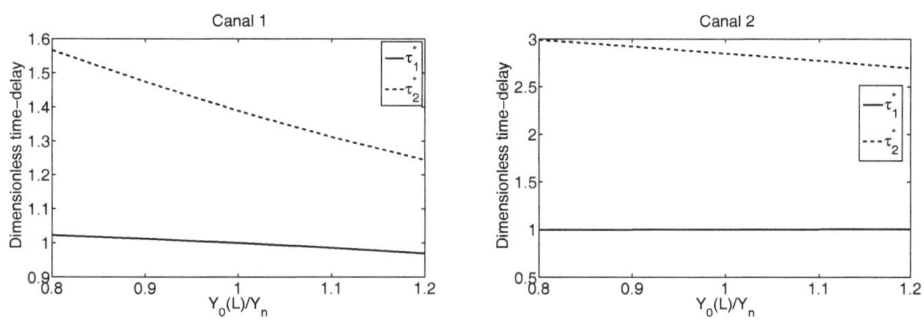

Fig. 3.23 Variation of the dimensionless delays τ_1^* and τ_2^* as a function of $Y_0(L)/Y_n$

3.4.2.2 Asymptotic Estimates of the Poles

It is also possible to derive a closed-form expression for the high frequency behavior of the system.

We start from the characteristic form derived in Sect. 2.4.2:

$$\frac{\partial \chi}{\partial t} + \mathbf{D}(x)\frac{\partial \chi}{\partial x} + \mathbf{E}(x)\chi = 0, \tag{3.89}$$

where $\chi(x,t) = \mathbf{X}(x)\xi(x,t)$ with $\xi(x,t) = \begin{pmatrix} T_0(x)y(x,t) \\ q(x,t) \end{pmatrix}$, $\mathbf{X}(x) = \begin{pmatrix} \beta(x) & 1 \\ -\alpha(x) & 1 \end{pmatrix}$, $\mathbf{D}(x) = \begin{pmatrix} \alpha(x) & 0 \\ 0 & -\beta(x) \end{pmatrix}$, and $\mathbf{E}(x) = [\mathbf{X}(x)\mathbf{B}(x) - \mathbf{D}(x)\mathbf{X}'(x)]\mathbf{X}(x)^{-1}$, with $\mathbf{B}(x) = \begin{pmatrix} 0 & 0 \\ -\gamma(x) & \delta(x) \end{pmatrix}$. The elements of matrix $\mathbf{E}(x)$ are given by (2.33).

Applying the Laplace transform to (3.89) leads to the ODE:

$$\frac{\partial \hat{\chi}}{\partial x}(x,s) = \mathbf{C}(x,s)\hat{\chi}(x,s), \tag{3.90}$$

with $\mathbf{C}(x,s) = -\mathbf{D}(x)^{-1}(sI + \mathbf{E}(x))$.

For high frequencies, the diagonal terms in $\mathbf{C}(x,s)$ dominate the anti-diagonal terms. Then, the high frequency approximate solution of (3.90) is given by:

$$\hat{\chi}(x,s) \approx \exp\Big(\int_0^x \mathbf{C}_{\text{diag}}(v,s)\text{d}v\Big)\hat{\chi}(0,s),$$

where $\mathbf{C}_{\text{diag}}(x,s) = \begin{pmatrix} -\frac{s}{\alpha(x)} - r_1(x) & 0 \\ 0 & \frac{s}{\beta(x)} + r_2(x) \end{pmatrix}$, with

$$r_1(x) = \frac{\alpha(x)\delta(x) - \gamma(x) - \alpha(x)\beta'(x)}{\alpha(x)(\alpha(x)+\beta(x))},$$

$$r_2(x) = \frac{\beta(x)\delta(x) + \gamma(x) + \alpha'(x)\beta(x)}{\beta(x)(\alpha(x)+\beta(x))}.$$

Therefore, using the previous developments and after straightforward manipulations, a high frequency approximation of the poles is obtained by solving:

$$\exp\left(\int_0^L \left[\Big(\frac{1}{\alpha(x)} + \frac{1}{\beta(x)}\Big)s + r_1(x) + r_2(x)\right]\text{d}x\right) = 1,$$

which leads to:

$$p_{\pm k} \approx -\frac{\int_0^L [r_1(x) + r_2(x)]\text{d}x}{\tau_1 + \tau_2} \pm \frac{2jk\pi}{\tau_1 + \tau_2}, \tag{3.91}$$

with $\tau_1 = \int_0^L \frac{\text{d}x}{\alpha(x)}$ and $\tau_2 = \int_0^L \frac{\text{d}x}{\beta(x)}$.

We recover an expression similar to that already obtained in the uniform case. The high frequency poles have an imaginary part, which is linked to the forward and backward delays, and a real part, which is linked to the forward and backward damping. Let us note that:

$$r_1 + r_2 = \frac{2\alpha\beta\delta + (\alpha - \beta)\gamma}{\alpha\beta(\alpha+\beta)} + \frac{\alpha' - \beta'}{\alpha+\beta},$$

where the argument x has been dropped for readability.

Therefore, the damping of the high frequency poles is the sum of a positive term plus a term whose sign depends on the derivative of $\alpha(x) - \beta(x) = 2V_0(x)$ with respect to x. This means that the system will tend to be more damped if the velocity tends to increase along the canal. On the contrary, it shows that the system will tend to be more oscillating if the velocity tends to decrease along the canal. This is consistent with the results obtained previously: an accelerating flow tends to dampen the system, while a decelerating flow will be more prone to possible oscillations.

As in the uniform case, we may show that for high frequencies, the Saint-Venant equations are close to the following damped wave equation, written here with an abuse of notation (the operator $\partial/\partial x$ does not apply to $r_1(x), r_2(x), \alpha(x)$ and $\beta(x)$):

$$\left(\frac{\partial}{\partial x} + r_1(x) + \frac{1}{\alpha(x)}\frac{\partial}{\partial t}\right)\left(\frac{\partial}{\partial x} - r_2(x) - \frac{1}{\beta(x)}\frac{\partial}{\partial t}\right)q = 0,$$

with boundary conditions $q(0,t)$ and $q(L,t)$.

The high frequency modes are therefore the result of the interaction of two gravity waves, one traveling downstream at speed $\alpha(x) = V_0(x) + C_0(x)$ with attenuation factor $r_1(x)$, and one traveling upstream at speed $\beta(x) = C_0(x) - V_0(x)$ with attenuation factor $r_2(x)$. The high frequency estimate of the poles is the direct extension of the result obtained in the uniform case.

Example 3.14 (High frequency estimate of the poles). Tables 3.3 and 3.4 give the high frequency poles approximation and the exact value for both canals, for accelerating and decelerating flows, respectively. The poles corresponding to $k = 1,\ldots,5$ and $k = 10$ are computed with the numerical method described in Sect. 3.4.1.3, and compared to the high frequency approximation obtained with (3.91). We see that the high frequency approximation gives good results when k is large enough. This is consistent with the remarks done in uniform flow, where the high frequency approximation is also valid for $k \gg k_h$.

Table 3.3 Exact value and high frequency approximation for the poles of the two canals, accelerating flow

k	Canal 1		Canal 2	
	Exact	Approx.	Exact	Approx.
±1	$-0.0022 \pm 0.0031j$	$-0.0023 \pm 0.0037j$	-0.0013, -0.0067	$-0.0044 \pm 0.0018j$
±2	$-0.0022 \pm 0.0072j$	$-0.0023 \pm 0.0075j$	$-0.0041 \pm 0.0013j$	$-0.0044 \pm 0.0036j$
±3	$-0.0022 \pm 0.0110j$	$-0.0023 \pm 0.0112j$	$-0.0041 \pm 0.0042j$	$-0.0044 \pm 0.0054j$
±4	$-0.0022 \pm 0.0148j$	$-0.0023 \pm 0.0150j$	$-0.0041 \pm 0.0063j$	$-0.0044 \pm 0.0072j$
±5	$-0.0022 \pm 0.0186j$	$-0.0023 \pm 0.0187j$	$-0.0041 \pm 0.0082j$	$-0.0044 \pm 0.0089j$
±10	$-0.0022 \pm 0.0374j$	$-0.0023 \pm 0.0375j$	$-0.0042 \pm 0.0175j$	$-0.0044 \pm 0.0179j$

Table 3.4 Exact value and high frequency approximation for the poles of the two canals, decelerating flow

k	Canal 1		Canal 2	
	Exact	Approx.	Exact	Approx.
± 1	$-0.0011 \pm 0.0043j$	$-0.0010 \pm 0.0044j$	$-0.0016, -0.0059$	$-0.0035 \pm 0.0019j$
± 2	$-0.0011 \pm 0.0087j$	$-0.0010 \pm 0.0088j$	$-0.0036 \pm 0.0027j$	$-0.0035 \pm 0.0039j$
± 3	$-0.0011 \pm 0.0131j$	$-0.0010 \pm 0.0131j$	$-0.0036 \pm 0.0051j$	$-0.0035 \pm 0.0058j$
± 4	$-0.0011 \pm 0.0175j$	$-0.0010 \pm 0.0175j$	$-0.0036 \pm 0.0072j$	$-0.0035 \pm 0.0077j$
± 5	$-0.0011 \pm 0.0219j$	$-0.0010 \pm 0.0219j$	$-0.0036 \pm 0.0093j$	$-0.0035 \pm 0.0096j$
± 10	$-0.0011 \pm 0.0438j$	$-0.0010 \pm 0.0438j$	$-0.0036 \pm 0.0191j$	$-0.0035 \pm 0.0193j$

3.4.2.3 Distributed Modal Decomposition

We extend here the results obtained in Sects. 3.2.2.2 and 3.3.2.3 to the nonuniform flow case, where the distributed transfer matrix is given by (3.84).

We know using (3.91) that the high frequency poles are complex conjugate, and close to a vertical line in the left half plane. Using similar arguments as in the uniform case, the Cauchy residues theorem implies that each transfer function $g_{ij}(x,s)$ can be decomposed as an infinite sum:

$$g_{ij}(x,s) = b_{ij}(x) + \frac{a_{ij}^{(0)}(x)}{s} + \sum_{k=-\infty, k\neq 0}^{\infty} \frac{a_{ij}^{(k)}(x)s}{p_k(s-p_k)}, \tag{3.92}$$

where the coefficient $a_{ij}^{(k)}(x) = \lim_{s\to p_k}(s-p_k)g_{ij}(x,s)$ is the residue of transfer function $g_{ij}(x,s)$ at the pole p_k and $b_{ij}(x) = \frac{\partial}{\partial s}[sg_{ij}(x,s)]|_{s=0}$.

Contrarily to the uniform case, we have no analytical expression for the residues $a_{ij}^{(k)}(x)$. However, the proposed numerical method also enables us to compute the terms numerically.

The details of the computation are given in Appendix D.

Example 3.15 (Coefficients $a_{ij}^{(k)}(x)$ for nonuniform flow). The absolute value of functions $a_{11}^{(k)}(x)$ of the distributed modal decomposition of $g_{11}(x,s)$ of the example canals 1 and 2 are depicted in Figs. 3.24 and 3.25. We can see that in both canals the modes in nonuniform flow are similar to those in uniform flow. For canal 1, the spatial functions $a_{11}^{(k)}(x)$ are only slightly modified in nonuniform flow conditions. The change in the integrator gain $a_{11}^{(0)}(x)$ amounts to about 10% in decelerating flow, and about 20% in accelerating flow. We see that the change in flow speed modifies the location of the nodes of oscillation: the nodes are slightly moved downstream for accelerating flow and upstream for decelerating flow.

For canal 2, the nonuniformity has an important effect on the modal decomposition. The nodes are modified following the same pattern observed for canal 1, but the change in amplitude is much more important, especially in the case of accelerating flow.

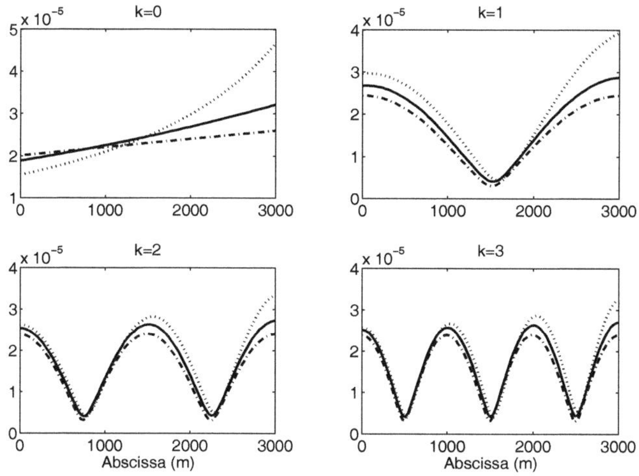

Fig. 3.24 Coefficients $|a_{11}^{(k)}(x)|$ of the distributed modal decomposition for canal 1 in accelerating $(\cdots)$, uniform (—), and decelerating $(- \cdot -)$ flow

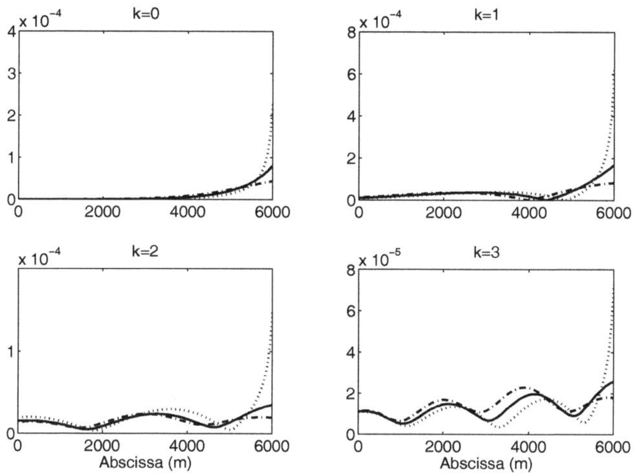

Fig. 3.25 Coefficients $|a_{11}^{(k)}(x)|$ of the distributed modal decomposition for canal 2 in accelerating $(\cdots)$, uniform (—), and decelerating $(- \cdot -)$ flow

Example 3.16 (Coefficients $b_{ij}(x)$ for nonuniform flow). Figure 3.26 depicts the value of coefficients $b_{11}(x)$ and $b_{12}(x)$ of the distributed modal decomposition of the transfer function $g_{11}(x,s)$ and $g_{12}(x,s)$ for canals 1 and 2. We observe the same trends as already discussed for coefficients $a_{11}^{(k)}(x)$.

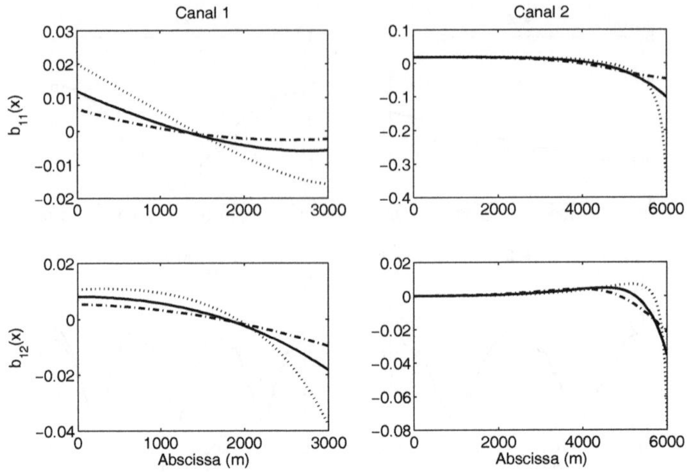

Fig. 3.26 Functions $b_{11}(x)$ and $b_{12}(x)$ of the distributed modal decomposition for canals 1 and 2 in accelerating ($\cdots$), uniform (—) and decelerating ($-\cdot-$) flow

3.4.2.4 Modal Decomposition of the Delay-free Part

The modal decomposition of $\tilde{g}_{ij}(x,s)$ can also be obtained as

$$\tilde{g}_{ij}(x,s) = \tilde{b}_{ij}(x) + \frac{\tilde{a}_{ij}^{(0)}(x)}{s} + \sum_{k=-\infty,k\neq 0}^{\infty} \frac{\tilde{a}_{ij}^{(k)}(x)s}{p_k(s-p_k)}, \tag{3.93}$$

where the coefficient $\tilde{a}_{ij}^{(k)}(x) = \lim_{s\to p_k}(s-p_k)\tilde{g}_{ij}(x,s)$ is the residue of transfer function $\tilde{g}_{ij}(x,s)$ at the pole p_k.

Details of the computation are given in Appendix D.

3.4.3 Time Domain Expressions

The time domain expressions obtained in the uniform case can only be partially extended to the nonuniform case. Indeed, the explicit expressions using the characteristic form cannot be extended, since the change of variable no longer diagonalizes the system. Similarly, the infinite series expansion was based on the explicit transfer function expression, and cannot as such be used in nonuniform flow.

3.4.3.1 State-space Expressions

The modal decomposition can be used in the nonuniform flow in a similar way as in the uniform flow case to obtain a numerical means to solve the linearized Saint-Venant equations. This point is not developed here, but will be investigated in more detail in Chap. 4.

3.4.4 Characterization of Nonuniform Flow Regimes

3.4.4.1 Backwater Curves

We focus on the two example canals already introduced, and consider the hydraulic conditions where the reference downstream water level is fixed to the normal depth corresponding to the maximum discharge defined in Table 2.1 and where the discharge can vary from $Q_{\max}/8$ to $Q_{\max}$. The idea here is to investigate a large set of linearization points, which correspond to different functioning points of the canal.

The corresponding backwater curves are depicted in Fig. 3.27.

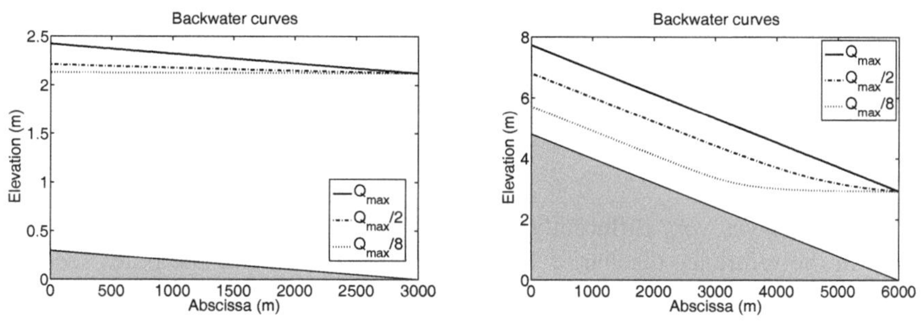

Fig. 3.27 Backwater curves for canals 1 and 2 for various discharges and a constant reference downstream boundary condition $Y_0(L)$

3.4.4.2 Poles

We used the numerical procedure proposed in Sect. 3.4.1.3 to compute the poles of the open channel for the different reference discharges. The poles are depicted in Fig. 3.28 for canal 1 and in Fig. 3.29 for canal 2. The canals have a very different behavior when the discharge changes. Canal 1 mainly behaves as in the previous case, when the flow was decelerating. Indeed, decreasing the discharge leads to a decelerating flow. However, since the discharge and the water level change, the velocity and the celerity change and modify the poles. In the present case, only the real part of the poles change, their imaginary part remains almost constant. We note that when the discharge decreases, the poles migrate towards the imaginary axis, therefore their damping diminishes.

The fact that the imaginary part of the poles remains constant is due to the fact that the sum of the delays $\tau_1 + \tau_2$ remains constant when the discharge changes (see Fig. 3.32).

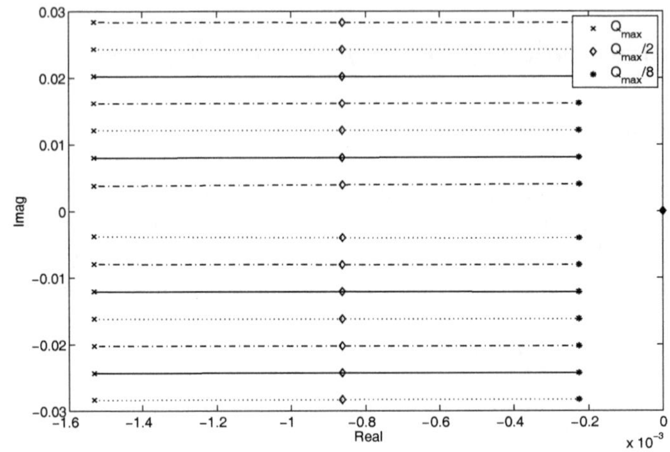

Fig. 3.28 Location of the poles of Saint-Venant transfer matrix for canal 1 around nonuniform flow conditions

Canal 2 behaves very differently. Indeed, the poles are highly modified by the change in the reference discharge. First of all, the damping of the poles seems to increase as the discharge decreases. This is due to the fact that the discharge and the water depth vary at the same time, contrarily to the case studied earlier in this chapter, where only the water depth could change. Second, the oscillating frequency changes. This can be explained by the fat that the sum $\tau_1 + \tau_2$ changes when Q decreases (see Fig. 3.32).

We observe that one pair of poles (corresponding to $k = 4$) changes direction in the complex plane when Q becomes lower than $Q_{\max}/2$. This is also the case for the pair of poles corresponding to $k = 6$, but to a lesser extent. This behavior may be explained by analyzing the backwater curve (see Fig. 3.27) corresponding to $Q_{\max}/8$. The flow can be approximated by the interconnection of two parts, the upstream part with a constant depth corresponding to the uniform depth for $Q_{\max}/8$, and the downstream part with a variable depth with more oscillating poles. The poles are therefore the result of the interconnection between a rather damped upstream part and the more oscillating downstream part.

A more detailed explanation of this phenomenon can be inferred from the high frequency poles' approximation (3.91). Indeed, let us separate the pool into two parts, one from 0 to x_1, and the other one from x_1 to L. We denote by $p_k^{(1)}$ the

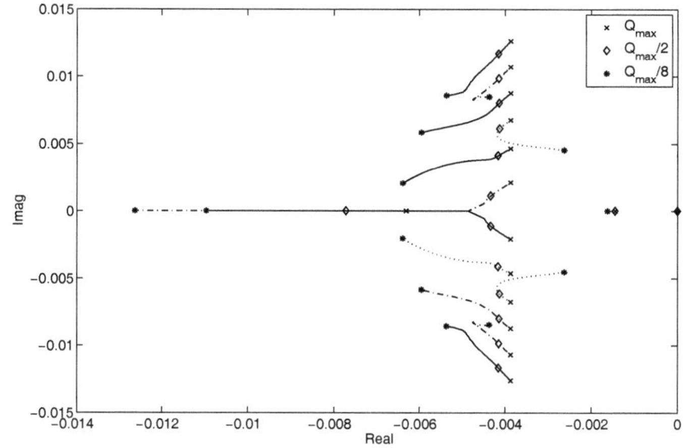

Fig. 3.29 Location of the poles of Saint-Venant transfer matrix for canal 2 around nonuniform flow conditions

poles of the upstream part and $p_k^{(2)}$ the poles of the downstream part. Using (3.91), we can derive a relation between the high frequency approximation of the poles corresponding to the total length and those corresponding to a part of the channel. The relation is as follows:

$$\begin{aligned} p_k &= -\frac{\int_0^{x_1}[r_1(x)+r_2(x)]\mathrm{d}x + \int_{x_1}^{L}[r_1(x)+r_2(x)]\mathrm{d}x}{\tau_1+\tau_2} \pm \frac{2jk\pi}{\tau_1+\tau_2} \\ &= \frac{\tau_1^{(1)}+\tau_2^{(1)}}{\tau_1+\tau_2} p_k^{(1)} + \frac{\tau_1^{(2)}+\tau_2^{(2)}}{\tau_1+\tau_2} p_k^{(2)}. \end{aligned}$$

Therefore the high frequency poles corresponding to the total length are the linear combination of the high frequency poles obtained with both parts, weighted according to the ratio of the sum of their delays over the sum of the total delays. This relation only holds for high frequencies. The behavior of low frequency poles is more difficult to assess with closed-form expressions.

3.4.4.3 Bode Plots

The frequency response of both canals are plotted in Figs. 3.30 and 3.31, respectively. For canal 1, this confirms that the modes remain at the same frequency, but their amplitude changes, it increases as the discharge decreases.

For canal 2, the change in the integrator gain is clearly visible: the integrator gain decreases as the discharge Q decreases. The delay τ_1 is also strongly affected (see also Fig. 3.32).

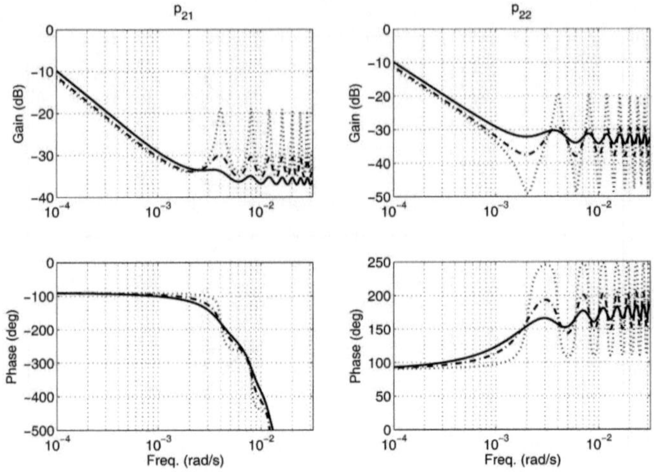

Fig. 3.30 Bode plots of transfer functions $p_{21}(s)$ and $p_{22}(s)$ for canal 1 around $Q_{\max}/8$ ($\cdots$), $Q_{\max}/2$ ($- \cdot -$), and $Q_{\max}$ (—)

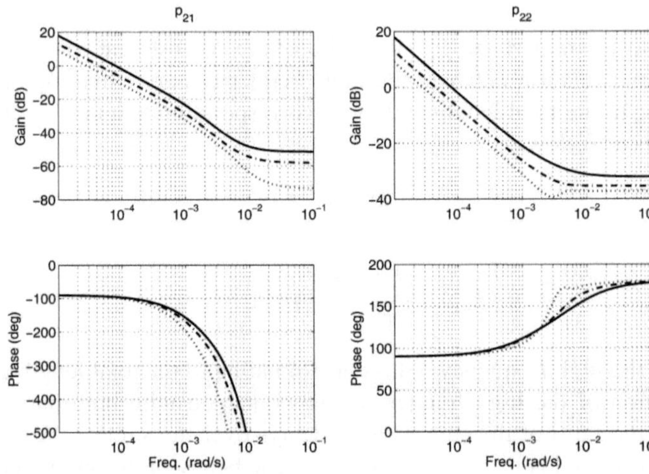

Fig. 3.31 Bode plots of transfer functions $p_{21}(s)$ and $p_{22}(s)$ for canal 2 around $Q_{\max}/8$ ($\cdots$), $Q_{\max}/2$ ($- \cdot -$), and $Q_{\max}$ (—)

3.4.4.4 Delays

The dimensionless delays $\tau_1/\tau_1(Q_{\max})$ and $\tau_2/\tau_1(Q_{\max})$ are depicted in Fig. 3.32 for both canals and various discharges. As already noted above, for canal 1, the sum of the delays is almost constant, even if τ_1 decreases and τ_2 increases as Q decreases.

For canal 2, only τ_1 appears to be affected by the reference discharge, τ_2 remains almost constant.

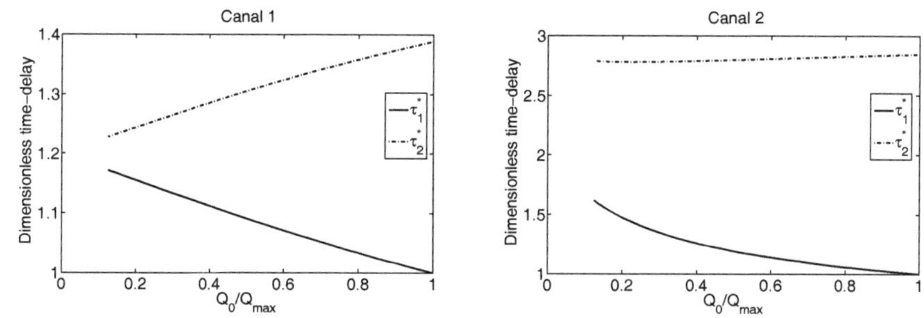

Fig. 3.32 Dimensionless delays for canal 1 and 2, for various discharges and a constant reference downstream boundary condition $Y_0(L)$

3.5 Summary

In this chapter, we obtained the Saint-Venant transfer matrix, first for the horizontal frictionless case, then for the uniform flow case, and finally for the general nonuniform flow case. We analyzed the transfer matrix for boundary conditions in terms of discharges, and computed its poles, first analytically for the uniform case, then numerically for the backwater case. We also characterized the model in terms of delays, which gives an a priori evaluation of the available bandwidth for the controlled system. These results enable us to use classical frequency domain tools such as the Nyquist criteria for closed-loop stability, since as we show in Appendix E, the transfer matrix belongs to the Callier–Desoer class of transfer matrices. Finally, we used a numerical scheme to study the effect of the flow nonuniformity on the linearized Saint-Venant transfer matrix.

The transfer matrix expressions in uniform flow have been obtained in many research papers [6, 7, 18, 8, 10, 21, 3, 5]. Inverse Laplace transforms have already been obtained for lateral flow in [17], and the Green function of linearized Saint-Venant equations has been obtained in [19]. A similar approach was developed independently in the thesis [5], where linearized open channel flow is studied in the frequency domain. The poles' expression are obtained using discharges as boundary conditions, and the corresponding residues are computed. However, the computations are done in uniform flow, and the static term is not computed in the modal decomposition.

Only few results have been published on the characterization of nonuniform open channel flow [20, 22]. The results in this chapter use a numerical method first proposed in [15] and also contain new results, such as the asymptotic approximation of the poles and the numerical method to compute the modal factors [16].

In summary, the dynamic behavior of an open channel in nonuniform flow is qualitatively very similar to its behavior in uniform flow. The only difference is that in nonuniform flow, the nice analytical expressions can no longer be used, but efficient numerical methods are available to compute poles, delays, and modal factors.

References

[1] Abramowitz M, Stegun IA (1972) Handbook of mathematical functions, Applied Mathematics Series, vol 55, 10th edn. National Bureau of Standards
[2] Aström K (2000) Limitations on control system performance. European J of Control 6:1–19
[3] Baume JP, Sau J (1997) Study of irrigation canal dynamics for control purposes. In: Int. Workshop on the Regulation of Irrigation Canals, RIC'97, Marrakech, Morroco, pp 3–12
[4] Blanes S, Casa F, Ros J (2000) Improved high order integrators based on the magnus expansion. Tech. rep., Department of Applied Mathematics and Theoretical Physics, University of Cambridge
[5] Chapuis J (1998) Modelierung und Neues Konzept fr die Regelung von Laufwasserkraftwerken. PhD thesis, ETH Zrich
[6] Constantinides C, Kia-Koojoori T, Jacquot R (1976) Stability of a feedback controlled distributed system by model representation. International J of Systems Science 7(9):989–1003
[7] Corriga G, Patta F, Sanna S, Usai G (1979) A mathematical model for open-channel networks. Appl Math Mod 3:51–54
[8] Corriga G, Sanna S, Usai G (1989) Estimation of uncertainty in an open-channel network mathematical model. Appl Math Mod 13:651–657
[9] Courant R, Hilbert D (1962) Methods of mathematical physics, vol II, third printing (66) edn. John Wiley
[10] Ermolin Y (1992) Study of open-channel dynamics as controlled process. J Hydraul Eng 118(1):59–71
[11] Hoffman K (1962) Banach spaces of analytic functions. Prentice Hall, London
[12] Iserles A (2000) On the global error of discretization methods for highly oscillatory ordinary differential equations. Tech. Rep. NA2000/11, Dept. of Applied Math. and Theoretical Physics, Univ. of Cambridge
[13] Kailath T (1980) Linear systems. Information and system sciences series, Prentice Hall, London
[14] Litrico X, Fromion V (2001) About optimal performance and approximation of open-channel hydraulic systems. In: Proc. 40th IEEE Conf. on Decision and Control, Orlando, pp 4511–4516
[15] Litrico X, Fromion V (2004) Frequency modeling of open channel flow. J Hydraul Eng 130(8):806–815
[16] Litrico X, Fromion V (2009) Modal decomposition of linearized open channel flow model. Networks Heterogeneous Media 4(2):325–357
[17] Moramarco T, Fan Y, Bras RL (1999) Analytical solution for channel routing with uniform lateral inflow. J Hydraul Eng 125(7):707–713

[18] Napiórkowski J, Dooge J (1988) Analytical solution of channel flow model with downstream control. Hydrological Sciences J 33(3-6):269–287
[19] Ridolfi L, Porporato A, Revelli R (2006) Green's function of the linearized de Saint-Venant equations. J of Eng Mechanics 132(2):125–132
[20] Schuurmans J (1997) Control of water levels in open-channels. PhD thesis, ISBN 90-9010995-1, Delft University of Technology
[21] Schuurmans J, Bosgra O, Brouwer R (1995) Open-channel flow model approximation for controller design. Appl Math Mod 19:525–530
[22] Tsai CW (2005) Flood routing in mild-sloped rivers–wave characteristics and downstream backwater effect. J of Hydrology 308(1-4):151–167

Chapter 4
Finite Dimensional Models of Open Channel Flow

Abstract This chapter examines various ways to obtain finite dimensional models for linearized open channel flow equations. We first study rational models obtained based on the modal decomposition derived in Chap. 3. Then we put the problem as a convex optimization one. Finally, we analyze the properties of a finite dimensional discrete linear model obtained based on the classical finite difference Preissmann scheme. These models are useful for simulation purposes and for controller design.

4.1 Introduction

A classical way to obtain models for systems described by partial differential equations is to use a finite difference scheme. This approach is inspired from the methods used to solve the nonlinear equations for simulation purposes. The model obtained can then be linearized around a stationary regime in order to obtain a linear model suited for controller design. Such models have been obtained using either an implicit finite difference scheme [9] or an explicit one [2]. The main advantage of these models is that they can be obtained for any flow configuration by linearizing the discrete equations. However, the main drawback of this method is that the state dimension can become very large for real systems (with a space step of 100 m, a channel of 10 km leads to a state-space representation of dimension 200). This is why it is interesting to obtain accurate models of smaller dimensions. These finite difference-based models are also not valid for high frequencies. The numerical schemes, used to solve the equations in the time domain, are not always suited for controller design, since they do not capture the whole frequency domain behavior.

Other methods include the collocation method, which ends up with a discretized model in space, but continuous in time (see [5, 6]). This method leads to results similar to those based on the finite difference schemes, since one needs to consider a large number of states to have a good representation of the system's dynamics.

It is also possible to directly identify a model on the real data [15, 11]. This method provides an input–output model of dimension chosen by the user. This pragmatic approach is efficient for designing controllers, but it is not able to provide insight into the physics of the system. More importantly, it is not able to consider deviations from the present hydraulic conditions, or to simulate a hypothetical channel.

In Chap. 3, we developed a method to compute the linearized Saint-Venant transfer matrix around any type of steady flow. We finally ended up with an input–output transfer matrix linking the water levels to the boundary discharges

$$\begin{pmatrix} \hat{y}(0,s) \\ \hat{y}(L,s) \end{pmatrix} = \begin{pmatrix} p_{11}(s) & p_{12}(s) \\ p_{21}(s) & p_{22}(s) \end{pmatrix} \begin{pmatrix} \hat{q}(0,s) \\ \hat{q}(L,s) \end{pmatrix}. \tag{4.1}$$

The difficulties to use the linearized Saint-Venant equations associated to (4.1) is mainly linked to the fact that $p_{ij}(s)$ are irrational transfer functions and that system (4.1) in general does not have a closed-form solution. These two difficulties can be nevertheless bypassed. Indeed, on the one hand, it is possible for any value of $s \in \mathbb{C}$ to obtain the numerical value associated to $p_{ij}(s)$. On the other hand, an accurate rational transfer representation of the system in the finite band of frequency is sufficient to control the system. This band of frequency is finite since it is at least constrained by the actuators' bandwidth.

We present below three different methods to obtain a rational approximation of the linearized Saint-Venant equations, two leading to continuous-time rational models, and one leading to a discrete-time rational model:

- First, a method based on the modal decomposition of the transfer matrix. This enables us to obtain a distributed rational continuous-time approximation of the system.
- Second, we assume that the poles are known, and turn the continuous-time rational approximation problem into a convex optimization problem, which is easy to solve.
- Finally we analyze a discrete-time model obtained by discretizing the equations according to the Preissmann discretization scheme.

4.2 Distributed Rational Approximations

The modal decompositions obtained in Chap. 3 can be used to obtain a rational approximate model of open channel flow. We first examine the horizontal frictionless case, then the uniform case, and finally extend the results to the nonuniform case.

4.2.1 Horizontal Frictionless Case

4.2.1.1 Rational Approximation

The distributed modal decomposition obtained in Sect. 3.2.2.2 provides a way to get a rational approximation of the distributed Saint-Venant transfer matrix $G(x,s)$ given by (3.13). Indeed, if we truncate the infinite sum to a given number of poles n_p, (3.19) leads to

$$g_{ij}^{[n_p]}(x,s) = b_{ij}(x) + \frac{a_{ij}^{(0)}(x)}{s} + \sum_{k=-n_p,k\neq 0}^{n_p} \frac{s a_{ij}^{(k)}(x)}{p_k(s-p_k)}, \tag{4.2}$$

where $g_{ij}^{[n_p]}(x,s)$ are the finite dimensional approximations of the distributed transfer functions $g_{ij}(x,s)$. In the horizontal frictionless case, $a_{ij}^{(k)}(x)$ are given by (3.22) and $b_{ij}(x)$ are given by (3.23).

4.2.1.2 Rational Approximation with Delay

It is also interesting to have an approximation including the propagation delay. Using the distributed modal decomposition of the delay-free part, we get a rational approximation of Saint-Venant transfer matrix with delay. Once again, truncating the infinite sum to a given number of poles n_p, (3.19) leads to

$$g_{11}^{[n_p]}(x,s) = e^{-\frac{x}{\alpha}s}\left(\tilde{b}_{11}(x) + \frac{\tilde{a}_{11}^{(0)}(x)}{s} + \sum_{k=-n_p,k\neq 0}^{n_p} \frac{s\tilde{a}_{11}^{(k)}(x)}{p_k(s-p_k)}\right), \tag{4.3a}$$

$$g_{12}^{[n_p]}(x,s) = e^{-\frac{L-x}{\beta}s}\left(\tilde{b}_{12}(x) + \frac{\tilde{a}_{12}^{(0)}(x)}{s} + \sum_{k=-n_p,k\neq 0}^{n_p} \frac{\tilde{a}_{12}^{(k)}(x)}{p_k(s-p_k)}\right), \tag{4.3b}$$

$$g_{21}^{[n_p]}(x,s) = e^{-\frac{x}{\alpha}s}\left(\tilde{b}_{21}(x) + \frac{\tilde{a}_{21}^{(0)}(x)}{s} + \sum_{k=-n_p,k\neq 0}^{n_p} \frac{\tilde{a}_{21}^{(k)}(x)}{p_k(s-p_k)}\right), \tag{4.3c}$$

$$g_{22}^{[n_p]}(x,s) = e^{-\frac{L-x}{\beta}s}\left(\tilde{b}_{22}(x) + \frac{\tilde{a}_{22}^{(0)}(x)}{s} + \sum_{k=-n_p,k\neq 0}^{n_p} \frac{\tilde{a}_{22}^{(k)}(x)}{p_k(s-p_k)}\right), \tag{4.3d}$$

where $\tilde{a}_{ij}^{(k)}(x)$ and $\tilde{b}_{ij}(x)$ are given by (3.27) and (3.28).

4.2.1.3 Frequency Plots

We first compare the Bode plots of the approximate input–output transfer functions and the exact transfer functions $p_{ij}(s)$ for various values of n_p:

- $n_p = 0$: this corresponds to considering only the integrator and the static gain in the approximation;
- $n_p = 10$: this corresponds to considering the integrator, the static gain and 10 pairs of complex conjugate poles in the approximation.

Figure 4.1 compares the Bode plots of $\tilde{p}_{21}(s)$, $\tilde{p}_{22}(s)$ and their rational approximations for $n_p = 0$ and $n_p = 10$.

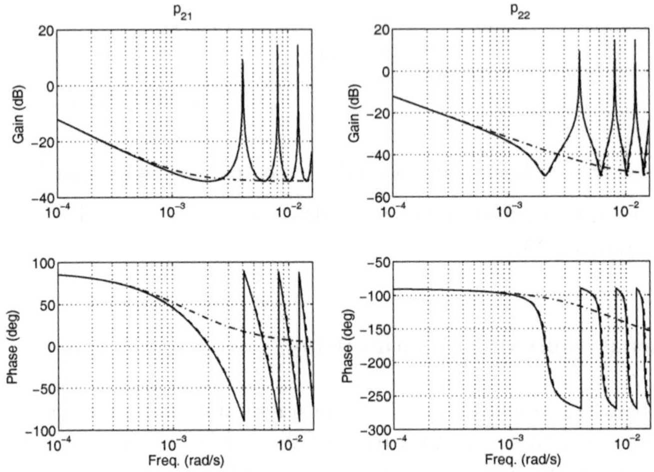

Fig. 4.1 Bode plots of transfer functions $\tilde{p}_{21}(s)$ and $\tilde{p}_{22}(s)$ for a horizontal frictionless channel and their finite dimensional approximations with 1 and 21 poles

The very simple approximation where only the integrator and the static gain, and zero poles are used provides a low frequency approximation of the system. The oscillating poles are not taken into account in the approximation.

We observe that the rational approximation with 21 poles is very close to the exact one, at least for frequencies lower than 3×10^{-1} rad/s. For higher frequencies, there is a visible mismatch, which corresponds to the fact that the higher order poles are not considered in the rational approximation.

This is also clear from Fig. 4.2, which depicts the approximation error for various numbers of poles n_p. We see that the error decreases as the number of poles increases. The low frequency behavior is already captured with a low order model, using only the integrator and the static gain. The approximation improves as one takes into account an increasing number of oscillating poles.

To see the effect of this approximation in the time domain, we compute the step response of the system.

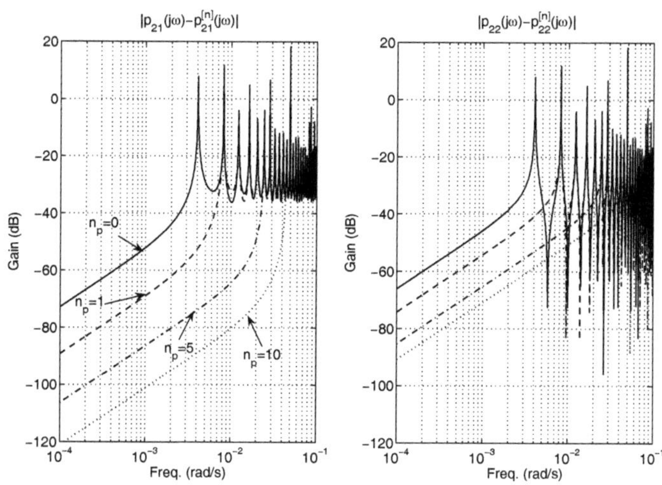

Fig. 4.2 Error plot for the rational approximation of transfer functions $p_{21}(s)$ and $p_{22}(s)$ for a horizontal frictionless channel

4.2.1.4 Step Responses

The rational decomposition enables us to compute the system's response to a given input $u(t)$. For simplicity, we consider the step response corresponding to $u(s) = 1/s$.

We know that the poles of the open channel are given by (3.9), leading to a pole in zero, possibly negative real poles, and then an infinite series of complex conjugate poles. Let us first compute the step response of an integrator, of a negative real pole, and of a pair of complex conjugate poles.

Integrator

An integrator corresponds to a pole equal to zero

$$f(s) = \frac{a_0}{s}.$$

The corresponding differential equation is the classical equation of a reservoir

$$\frac{\mathrm{d}y}{\mathrm{d}t} = a_0.$$

In this case, the step response is a ramp, with the equation

$$y(t) = a_0 t + y(0).$$

Real Pole

Let $f(s)$ be a transfer function with a real pole $p < 0$

$$f(s) = \frac{sa_1}{p(s-p)},$$

with $a_1 \in \mathbb{R}$.

Then the step response is given by

$$y(t) = \frac{a_1}{p} e^{pt}. \tag{4.4}$$

Complex Conjugate Poles

Let us now consider a system with two complex conjugate poles p and $\overline{p}$ (complex poles are always complex conjugate for a real system)

$$f(s) = \frac{sa_1}{p(s-p)} + \frac{s\overline{a}_1}{\overline{p}(s-\overline{p})}. \tag{4.5}$$

Let us set $\frac{a_1}{p} = \rho e^{j\phi}$ and $p = r + j\omega$, then the step response is given by

$$y(t) = \frac{a_1}{p} e^{pt} + \frac{\overline{a}_1}{\overline{p}} e^{\overline{p}t} \tag{4.6a}$$

$$= 2\rho e^{rt} \cos(\omega t + \phi). \tag{4.6b}$$

We see that a second-order system with complex conjugate poles has an oscillating step response, with an oscillation period ω equal to the imaginary part of the poles and a damping ratio r equal to their real part.

Superposition

Now, using the superposition theorem, which applies for linear systems, we obtain the step response of each transfer function as the sum of the step responses linked to the poles

$$g_{ij}^{[n_p]}(x,t) = b_{ij}(x) + a_{ij}^{(0)}(x)t + \sum_{k=-n_p,\ k\neq 0}^{n_p} \frac{a_{ij}^{(k)}(x)}{p_k}(1 - e^{p_k t}). \tag{4.7}$$

Example 4.1 (Step response of a horizontal frictionless channel). The step responses of each transfer function $p_{ij}(s)$ are computed with two different methods: first the analytical method given by (3.34), which is the exact time response, second the rational approximation with delay based on the delay-free modal decomposition.

We compare here two cases, where $n_p = 0$, i.e., where only the integrator and the static term $\tilde{b}_{ij}(x)$ are used, and where $n_p = 10$, i.e., where the approximation uses the integrator, the static term, and 10 pairs of poles. The results are depicted in Fig. 4.3. The approximate solution with 21 poles is very close to the analytical solution. The low approximate solution with only the integrator and the static term leads to a low frequency model of the channel, which predicts correctly the average water level. However, the peaks are not reproduced by this low order model.

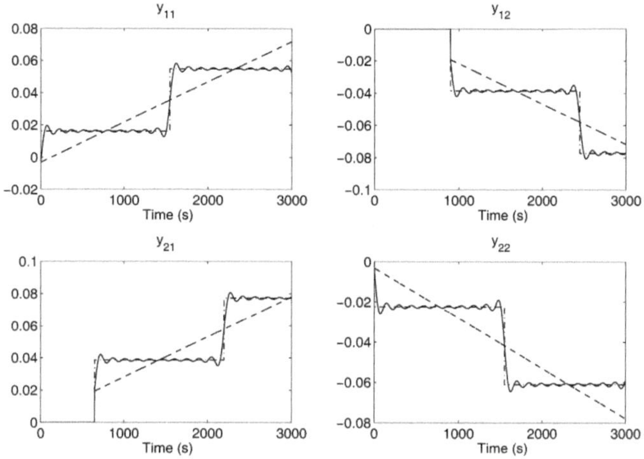

Fig. 4.3 Step responses of Saint-Venant transfer functions $p_{11}(s)$, $p_{12}(s)$, $p_{21}(s)$, and $p_{22}(s)$ for a horizontal frictionless channel. Comparison between the exact solution (– · –), that obtained with one pole (– –), and with 21 poles (—)

4.2.2 *Uniform Flow Case*

In the uniform flow case, the same approach can be used. We use the modal decompositions obtained in Sect. 3.3.2.3 to get a rational approximation of Saint-Venant transfer matrix.

4.2.2.1 Rational Approximation

The distributed modal decomposition obtained in Sect. 3.3.2.3 provides a way to get a rational approximation of Saint-Venant transfer matrix. Indeed, if we truncate the infinite sum to a given number of poles n, (3.56) leads to

$$g_{ij}^{[np]}(x,s) = b_{ij}(x) + \frac{a_{ij}^{(0)}(x)}{s} + \sum_{k=-n_p,k\neq 0}^{n_p} \frac{a_{ij}^{(k)}(x)}{s-p_k}, \tag{4.8}$$

where $g_{ij}^{[np]}(x,s)$ are the finite dimensional approximations of the distributed transfer functions $g_{ij}(x,s)$ given by (3.52) and $a_{ij}^{(k)}(x)$ are the residues, given by (3.59–3.60).

4.2.2.2 Rational Approximation with Delay

We use the modal decomposition of $\tilde{g}_{ij}(x,s)$ to obtain a rational approximation plus delay of the distributed transfer functions

$$\tilde{g}_{ij}^{[np]}(x,s) = \tilde{b}_{ij}(x) + \frac{\tilde{a}_{ij}^{(0)}(x)}{s} + \sum_{k=-n_p,k\neq 0}^{n_p} \frac{s\tilde{a}_{ij}^{(k)}(x)}{p_k(s-p_k)}. \tag{4.9}$$

The final rational approximation with delay is given by

$$g_{11}^{[np]}(x,s) = e^{-\frac{x}{\alpha}s}\left(\tilde{b}_{11}(x) + \frac{\tilde{a}_{11}^{(0)}(x)}{s} + \sum_{k=-n_p,k\neq 0}^{n_p} \frac{s\tilde{a}_{11}^{(k)}(x)}{p_k(s-p_k)}\right) \tag{4.10a}$$

$$g_{12}^{[np]}(x,s) = e^{-\frac{L-x}{\beta}s}\left(\tilde{b}_{12}(x) + \frac{\tilde{a}_{12}^{(0)}(x)}{s} + \sum_{k=-n_p,k\neq 0}^{n_p} \frac{\tilde{a}_{12}^{(k)}(x)}{p_k(s-p_k)}\right) \tag{4.10b}$$

$$g_{21}^{[np]}(x,s) = e^{-\frac{x}{\alpha}s}\left(\tilde{b}_{21}(x) + \frac{\tilde{a}_{21}^{(0)}(x)}{s} + \sum_{k=-n_p,k\neq 0}^{n_p} \frac{\tilde{a}_{21}^{(k)}(x)}{p_k(s-p_k)}\right) \tag{4.10c}$$

$$g_{22}^{[np]}(x,s) = e^{-\frac{L-x}{\beta}s}\left(\tilde{b}_{22}(x) + \frac{\tilde{a}_{22}^{(0)}(x)}{s} + \sum_{k=-n_p,k\neq 0}^{n_p} \frac{\tilde{a}_{22}^{(k)}(x)}{p_k(s-p_k)}\right). \tag{4.10d}$$

4.2.2.3 Frequency Plots

Figures 4.4 and 4.5 depict the Bode plots of transfer functions $\tilde{p}_{21}(s)$ and $\tilde{p}_{22}(s)$ and their rational approximations for canal 1 and 2, respectively. We compare the exact frequency response to that obtained with 1 pole and that with 21 poles. The low order approximation with only one pole appears to efficiently capture the average frequency behavior of the open channel. In Fig. 4.4, we see that the oscillating modes are not captured by this approximation, but that the integrator and the high frequency gain are efficiently modeled with this low order model. For canal 2, Fig. 4.5 shows that the real negative poles corresponding to the first mode ($k = 1$) modify the frequency response in a way that cannot be accurately captured by a low order model.

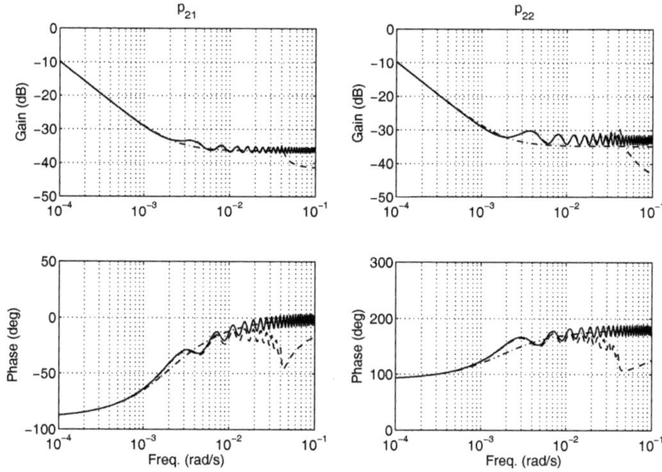

Fig. 4.4 Bode plots of transfer functions $\tilde{p}_{21}(s)$ and $\tilde{p}_{22}(s)$ for canal 1 and rational approximations based on the modal decomposition. Comparison between the exact solution (—), the approximation with 1 pole (– · –), and that with 21 poles (– –)

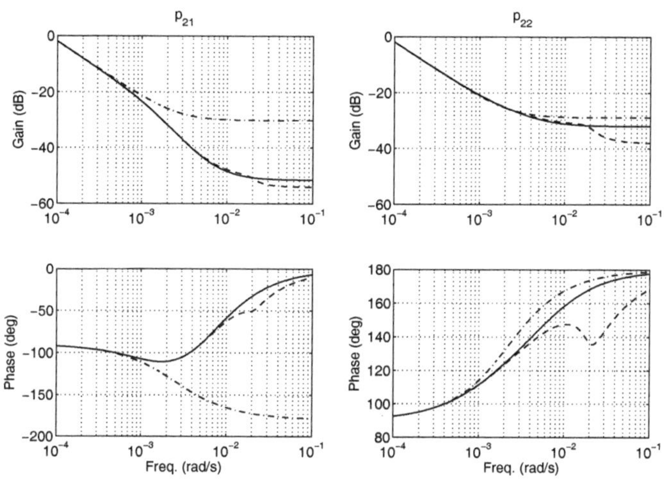

Fig. 4.5 Bode plots of transfer functions $\tilde{p}_{21}(s)$ and $\tilde{p}_{22}(s)$ for canal 2 and rational approximations based on modal decomposition. Comparison between the exact solution (—), the approximation with 1 pole (– · –), and that with 21 poles (– –)

Figures 4.6 and 4.7 depict the approximation error for transfer functions $p_{21}(s)$ and $p_{22}(s)$ for various approximations obtained with different numbers of poles n_p. We observe that the higher the number of poles considered in the approximation, the larger the frequency band with a small error. However, the error may be larger in high frequencies.

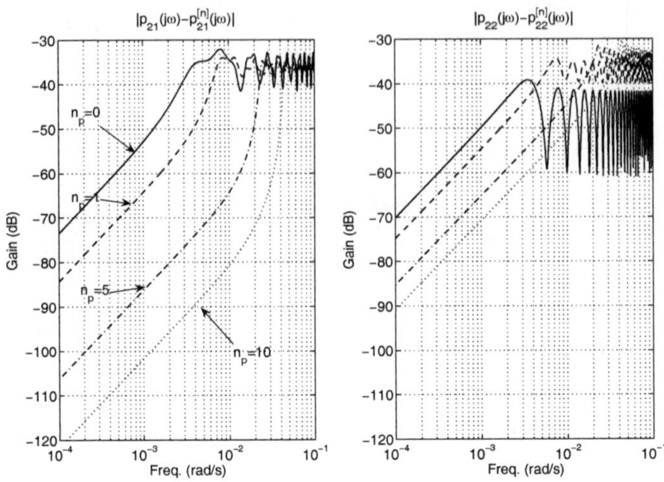

Fig. 4.6 Error plot for the rational approximation of transfer functions $p_{21}(s)$ and $p_{22}(s)$ based on modal decomposition for canal 1. Comparison of different orders of approximation

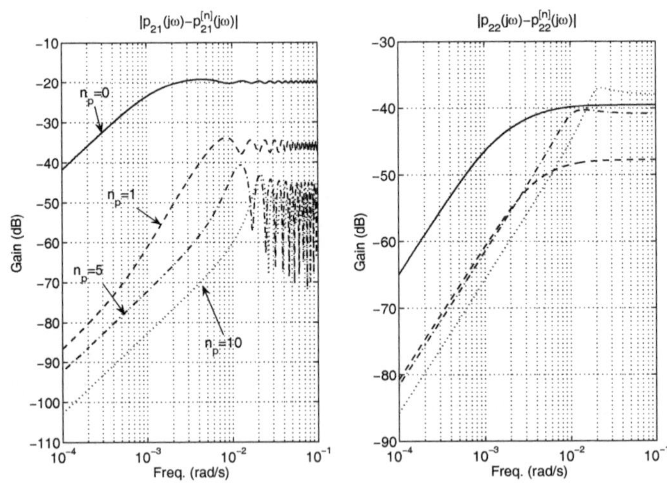

Fig. 4.7 Error plot for the rational approximation of transfer functions $p_{21}(s)$ and $p_{22}(s)$ based on modal decomposition for canal 2. Comparison of different orders of approximation

4.2.2.4 Step Responses

We display below the step responses computed using the approximate rational models for the example canals 1 and 2.

The distributed step response of transfer function $g_{11}(x,s)$ is depicted in Fig. 4.8 for canal 1. This distributed step response was computed using a rational model with delay, based on 21 poles. One of the interesting facts about the distributed rational approximation is that it enables us to easily compute the time domain response of the whole open channel, not only in its boundaries. We observe that the water surface is modified is the same way as the integrator gain.

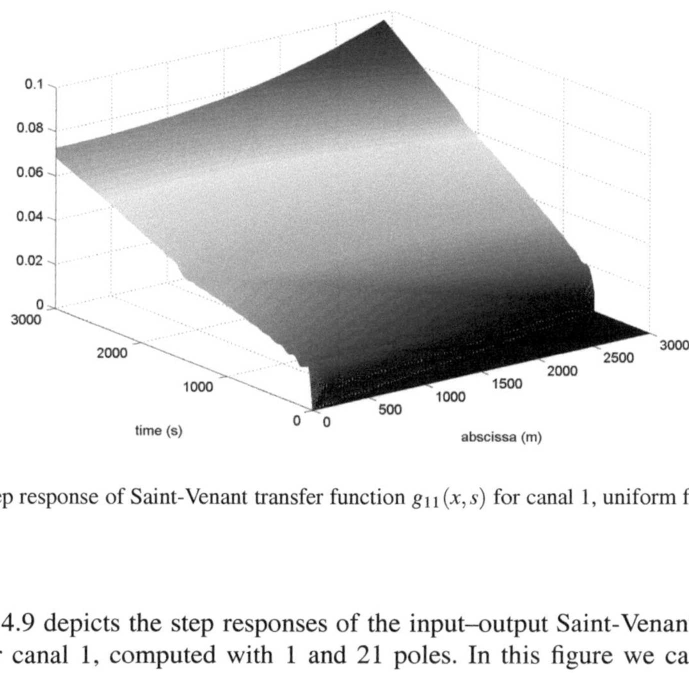

Fig. 4.8 Step response of Saint-Venant transfer function $g_{11}(x,s)$ for canal 1, uniform flow conditions

Figure 4.9 depicts the step responses of the input–output Saint-Venant transfer matrix for canal 1, computed with 1 and 21 poles. In this figure we can clearly see the various elements pointed out in the frequency domain analysis. First of all, step responses of transfer functions $p_{12}(s)$ and $p_{21}(s)$ are delayed, one with a delay equal to $\tau_2 = 900$ s and the other one with a delay equal to $\tau_1 = 648$ s. The transfer functions $p_{11}(s)$ and $p_{22}(s)$ have no delay, and the effect of the step input directly affects the water level. Transfer functions $p_{11}(s)$ and $p_{12}(s)$ have the same integrator coefficient (only the sign is opposite); this also true of transfer functions $p_{21}(s)$ and $p_{22}(s)$. In the step response of transfer function $p_{11}(s)$, we see a small bump at time $\tau_1 + \tau_2 = 1548$ s, corresponding to the arrival of the wave reflected on the downstream boundary condition.

The distributed step response of transfer function $g_{11}(x,s)$ is depicted in Fig. 4.10 for canal 2.

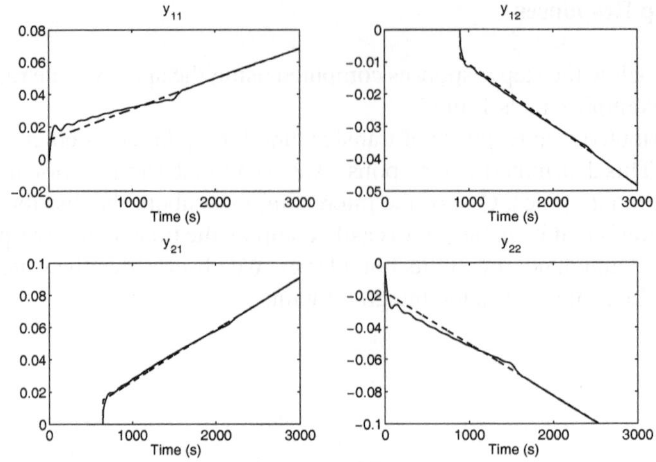

Fig. 4.9 Step responses of Saint-Venant transfer functions $p_{11}(s)$, $p_{12}(s)$, $p_{21}(s)$, and $p_{22}(s)$ for canal 1 for uniform flow conditions. Comparison between the approximation with 1 pole $(- \cdot -)$ and that with 21 poles (—)

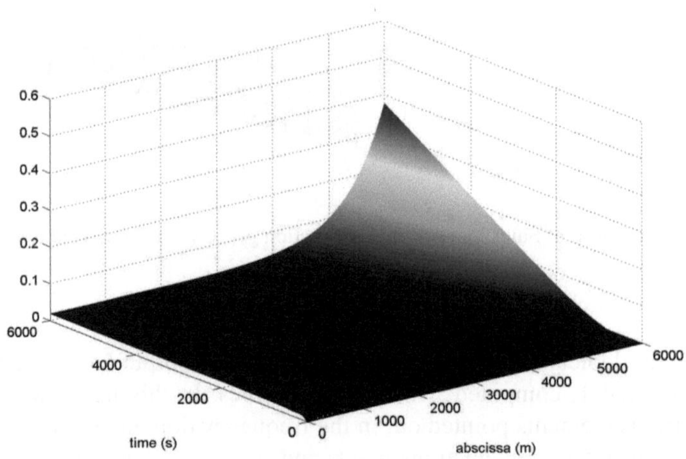

Fig. 4.10 Step response of Saint-Venant transfer function $g_{11}(x,s)$ for canal 2, uniform flow conditions

Figure 4.11 depicts the step responses of Saint-Venant transfer matrices for canal 2, computed with 1 and 21 poles. The same features already mentioned for canal 1 are also present. This time the delays are $\tau_1 = 881$ s and $\tau_2 = 2508$ s. It is also visible that the low order approximation with only one pole leads to a fairly good approximation of the low frequency dynamics of the open channel.

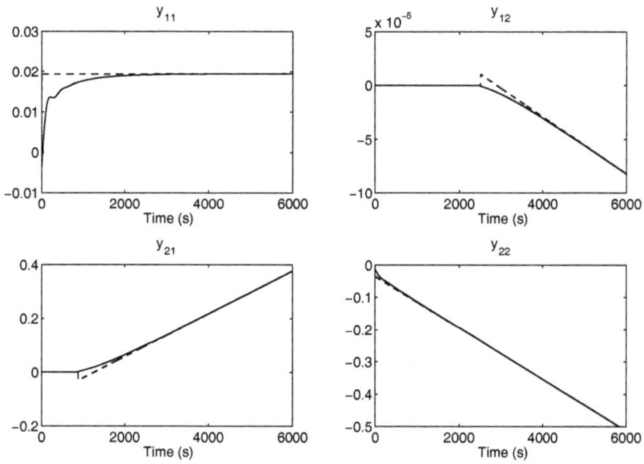

Fig. 4.11 Step responses of Saint-Venant transfer functions $p_{11}(s)$, $p_{12}(s)$, $p_{21}(s)$ and $p_{22}(s)$ for canal 2 for uniform flow conditions. Comparison between the approximation with 1 pole (– · –) and that with 11 poles (—)

4.2.3 Nonuniform Flow Case

An interesting feature of this approach is that it can be directly extended to the general case of nonuniform flow. Indeed, using the numerical approach developed in Sect. 3.4.2.3, we can compute the poles p_k and the corresponding coefficients $a_{ij}^{(k)}(x)$ and $b_{ij}(x)$ (see (D.3) and (D.6)). Then, we can obtain a distributed rational approximation of the open channel for any flow configuration.

We can also use this approach to compute a rational approximation with delay, using the delay-free decomposition (3.93) and the numerical method to compute the coefficients $\tilde{a}_{ij}^{(k)}(x)$ and $\tilde{b}_{ij}(x)$ (see (D.15) and (D.16)).

4.2.4 Summary

The proposed method is based on the numerical method developed in Chap. 3 to compute the poles of the system in nonuniform flow conditions. Once the poles are computed, the developments of Chap. 3 enable us to compute the modal factors corresponding to the poles, leading to an accurate numerical approximation of the open channel.

One of the main advantages of this method is that it provides a distributed approximation of the flow dynamics, with or without delays. The method works for any type of flow configuration.

A limitation of the method is that the approximation is valid only for frequencies lower than that corresponding to the higher order pole. We cannot control the error for higher frequencies. A remedy to this problem can be found by using an optimization framework [7].

4.3 Rational Model Using the Preissmann Implicit Scheme

A common way to obtain finite dimensional models for partial differential equations is to first use a finite difference scheme to discretize the equation (in space, or in time and in space), then to linearize this model to obtain a linear rational model. This approach has been used by various authors to get a finite dimensional linear model from the Saint-Venant equations [2, 10].

The finite difference Preissmann scheme is defined by the following equations:

$$f(x,t) = \theta\left[\phi f_{i+1}^{k+1} + (1-\phi) f_i^{k+1}\right] + (1-\theta)\left[\phi f_{i+1}^{k} + (1-\phi) f_i^{k}\right], \quad (4.11a)$$

$$\frac{\partial f}{\partial x}(x,t) = \theta \frac{f_{i+1}^{k+1} - f_i^{k+1}}{\Delta x} + (1-\theta)\frac{f_{i+1}^{k} - f_i^{k}}{\Delta x}, \quad (4.11b)$$

$$\frac{\partial f}{\partial t}(x,t) = \phi \frac{f_{i+1}^{k+1} - f_{i+1}^{k}}{\Delta t} + (1-\phi)\frac{f_{i}^{k+1} - f_i^{k}}{\Delta t}, \quad (4.11c)$$

where i is the space index, k the time index, and $\theta \in [0,1]$, $\phi \in [0,1]$ are weighting coefficients.

This numerical scheme has already been studied in the literature. Lyn and Goodwin [8] studied the stability of this generalized Preissmann scheme using a Fourier series method. In the homogeneous case (corresponding to the horizontal frictionless case), a necessary condition for numerical stability is given by:

$$\frac{\phi - \frac{1}{2}}{C_r} + (\theta - \frac{1}{2}) \geq 0,$$

with C_r the Courant number, $C_r = \alpha \Delta t / \Delta x$.

Here, instead of linearizing the discretized equations, as is usually done, we apply the Preissmann scheme to the linearized Saint-Venant equations. The approaches are not equivalent, since the first one depends on the way the nonlinear terms are discretized. We focus here on a way to obtain a rational model for the linearized equations.

4.3.1 Study of the Discretized System

We now use the frequency domain analysis tools for the discretized equations in the uniform flow case. The discretized solution ξ_i^k is defined as the solution $\xi(x,t)$ evaluated at the discretization point $(i\Delta x, k\Delta t)$. The discretization of the linearized Saint-Venant equations around a uniform flow according to the generalized Preissmann scheme given by (4.11) leads to:

$$A_1 \xi_{i+1}^{k+1} + A_2 \xi_i^{k+1} + B_1 \xi_{i+1}^k + B_2 \xi_i^k = 0, \tag{4.12}$$

with

$$\begin{aligned}
A_1 &= \frac{\phi}{\Delta t} I + \frac{\theta}{\Delta x} \mathbf{A} + \theta\phi \mathbf{B}, \\
A_2 &= \frac{(1-\phi)}{\Delta t} I - \frac{\theta}{\Delta x} \mathbf{A} + \theta(1-\phi)\mathbf{B}, \\
B_1 &= -\frac{\phi}{\Delta t} I + \frac{(1-\theta)}{\Delta x} \mathbf{A} + (1-\theta)\phi \mathbf{B}, \\
B_2 &= -\frac{(1-\phi)}{\Delta t} I - \frac{(1-\theta)}{\Delta x} \mathbf{A} + (1-\theta)(1-\phi)\mathbf{B},
\end{aligned}$$

where Δt, Δx are, respectively, the time and space discretization steps, $\theta \in [0,1]$ and $\phi \in [0,1]$ are weighting factors, and where $\mathbf{A}$ and $\mathbf{B}$ are matrices of the linearized Saint-Venant equations given by:

$$\mathbf{A} = \begin{pmatrix} 0 & 1 \\ \alpha\beta & \alpha - \beta \end{pmatrix} \text{ and } \mathbf{B} = \begin{pmatrix} 0 & 0 \\ -\gamma & \delta \end{pmatrix}.$$

Similarly to the z-transform, let us introduce the space transform[1] such that $\xi_{i+1}^k = l\xi_i^k$. Then, we have the following relations:

$$\begin{aligned}
\xi_{i+1}^{k+1} &= lz\xi_i^k, \\
\xi_i^{k+1} &= z\xi_i^k, \\
\xi_{i+1}^k &= l\xi_i^k.
\end{aligned}$$

Putting these expressions back into (4.12), the condition for the existence of a nontrivial solution is:

$$\det(lzA_1 + zA_2 + lB_1 + B_2) = 0.$$

After elementary manipulations, this condition can be rewritten as follows:

[1] This is directly related to the classical approach for the study of stability and accuracy of numerical schemes, where one considers periodic solutions $\xi(x,t) = \xi_0 e^{\lambda x + st}$, with λ the spatial frequency and s the time frequency (see, e.g., [12, 3, 4, 13, 14]). In that case, we have $z = e^{s\Delta t}$ and $l = e^{\lambda \Delta x}$.

$$\det(\sigma I + \rho \mathbf{A} + \mathbf{B}) = 0, \tag{4.13}$$

with

$$\sigma = \frac{1}{\Delta t}\frac{z-1}{1+\theta(z-1)},$$
$$\rho = \frac{1}{\Delta x}\frac{l-1}{1+\phi(l-1)}.$$

Condition (4.13) then becomes:

$$\sigma^2 + (\alpha-\beta)\rho\sigma - \alpha\beta\rho^2 + \delta\sigma + \gamma\rho = 0. \tag{4.14}$$

Let us note that this equation is exactly the one obtained in the continuous-time case (see (3.41)), with $\sigma = s$ and $\rho = \lambda$. We have solved this equation for λ, to get the expressions $\lambda(s)$, and the boundary conditions have given the poles, i.e., the values of s such that the boundary conditions hold. We will proceed similarly in the discrete-time case, by first solving for ρ (and l), and then finding the poles σ, leading to the discrete-time poles z.

This equation can be solved for ρ, giving two solutions $\rho_1(\sigma)$ and $\rho_2(\sigma)$:

$$\rho_1(\sigma) = \frac{(\alpha-\beta)\sigma + \gamma - \sqrt{d(\sigma)}}{2\alpha\beta} \tag{4.15a}$$

$$\rho_2(\sigma) = \frac{(\alpha-\beta)\sigma + \gamma + \sqrt{d(\sigma)}}{2\alpha\beta}, \tag{4.15b}$$

with $d(\sigma) = (\alpha+\beta)^2\sigma^2 + 2\sigma[(\alpha-\beta)\gamma + 2\alpha\beta\delta] + \gamma^2$. Let us note that these two solutions are formally identical to the continuous-time eigenvalues $\lambda_1(s)$ and $\lambda_2(s)$ given by (3.42a–3.42b).

4.3.2 Poles of the Discretized System

Let us denote by $l_1(z)$ and $l_2(z)$ the values of l corresponding to the two solutions $\rho_1(\sigma)$ and $\rho_2(\sigma)$ given by (4.15). The solution ξ_i^k is then obtained as the linear combination of two vectors:

$$\xi_i^k = \xi_1 l_1(z)^i z^k + \xi_2 l_2(z)^i z^k,$$

where ξ_1 and ξ_2 are determined by the boundary conditions. The upstream boundary condition in terms of discharge implies that the sum of the second components of ξ_1 and ξ_2 is equal to zero, or

$$\xi_1(2) = -\xi_2(2).$$

Combined with the downstream boundary condition, this equality implies that:

$$l_1(z)^N = l_2(z)^N,$$

where N is the number of space steps in the domain $N = \frac{L}{\Delta x}$. Therefore, the discrete-time poles of the discretized system are given by the complex values of z such that

$$l_1(z) = l_2(z)e^{\frac{2jn\pi\Delta x}{L}}, \tag{4.16}$$

with $n \in (-N, N)$.

We observe that the discretized system has a finite number of poles, equal to $2N-1$. This is in contrast with the continuous-time case, where the system has an infinite number of poles. The discretization implies a loss in the representation of the infinite-dimensional system's dynamics. This can be quantified in terms of the number of poles correctly reproduced by the discretized system.

Let us define $\eta_n = e^{\frac{2jn\pi\Delta x}{L}}$. Then, using the relation $l = \frac{1+(1-\phi)\rho\Delta x}{1-\phi\rho\Delta x}$, (4.16) becomes:

$$(1-\eta_n)[1-\phi(1-\phi)\Delta x^2\rho_1\rho_2] + (1-\phi)\Delta x(\rho_1 - \eta_n\rho_2) - \phi\Delta x(\rho_2 - \eta_n\rho_1) = 0.$$

Furthermore, using the fact that $\rho_1(\sigma)\rho_2(\sigma) = -\frac{\sigma(\sigma+\delta)}{\alpha\beta}$ and the expressions of $\rho_1(\sigma)$ and $\rho_2(\sigma)$ given by (4.15), we get:

$$\varepsilon_n\left[1 + (1-2\phi)\Delta x\frac{(\alpha-\beta)\sigma+\gamma}{2\alpha\beta} + \phi(1-\phi)\Delta x^2\frac{\sigma(\sigma+\delta)}{\alpha\beta}\right] = \Delta x\frac{\sqrt{d(\sigma)}}{2\alpha\beta} \tag{4.17}$$

with $\varepsilon_n = \frac{1-\eta_n}{1+\eta_n}$.

In the general case, (4.17) leads to a fourth order polynomial equation in σ, which can be solved analytically using the Cardan formulas (see [1], p. 17). To simplify the expressions, we only consider here the case $\phi = 0.5$. The fourth order polynomial is then given by:

$$\sigma^4 + a_3\sigma^3 + a_2\sigma^2 + a_1\sigma + a_0 = 0, \tag{4.18}$$

with

$$a_3 = 2\delta, \tag{4.19a}$$

$$a_2 = \delta^2 + \frac{4}{\Delta x^2}\left(2\alpha\beta - \frac{(\alpha+\beta)^2}{\varepsilon_n^2}\right), \tag{4.19b}$$

$$a_1 = \frac{8\alpha\beta}{\varepsilon_n^2\Delta x^2}\left(\delta(\varepsilon_n^2 - 2) - \frac{(\alpha-\beta)\gamma}{\alpha\beta}\right), \tag{4.19c}$$

$$a_0 = \frac{4}{\Delta x^2}\left(4\frac{\alpha^2\beta^2}{\Delta x^2} - \frac{\gamma^2}{\varepsilon_n^2}\right). \tag{4.19d}$$

Let us note that since $\varepsilon_n = -\varepsilon_{N-n}$, we only need to compute the solutions for $k \in [1, \lfloor N/2 \rfloor]$, where $\lfloor x \rfloor$ denotes the integer part of x.

The poles of the linearized open channel flow discretized with the Preissmann scheme can therefore be computed by first solving (4.18) for σ, and then computing the discrete-time poles with the following equation:

$$z_n = \frac{1+(1-\theta)\sigma_n \Delta t}{1-\theta\sigma_n \Delta t}. \tag{4.20}$$

This result shows that the effects of the spatial and temporal discretization can be analyzed separately. The spatial discretization first leads to a finite dimensional continuous-time system, with poles given by the values of σ solutions of (4.17). Then, the temporal discretization adds another approximation, leading to a discrete-time system.

Let us note that (4.20) is a Möbius transformation. When $\theta = 0.5$, it is equivalent to the bilinear transform, which transforms the left half plane into the unit circle. The bilinear transform is widely used since it ensures that a stable continuous system is transformed into a stable discrete one.

4.3.2.1 Horizontal Frictionless Case

The solution is easy to obtain in the horizontal frictionless case, where $\gamma = \delta = 0$. In that case, the two solutions $\rho_1(\sigma)$ and $\rho_2(\sigma)$ of (4.14) are given by:

$$\rho_1(\sigma) = -\frac{\sigma}{\alpha},$$
$$\rho_2(\sigma) = \frac{\sigma}{\beta},$$

and (4.17) simplifies to:

$$\varepsilon_n \left[1+(1-2\phi)\Delta x \frac{(\alpha-\beta)\sigma}{2\alpha\beta} + \phi(1-\phi)\Delta x^2 \frac{\sigma^2}{\alpha\beta}\right] = \Delta x \frac{(\alpha+\beta)\sigma}{2\alpha\beta}, \tag{4.21}$$

which leads to the second-order polynomial equation in σ:

$$2\varepsilon_n\phi(1-\phi)\Delta x^2\sigma^2 - [\alpha+\beta-\varepsilon_n(1-2\phi)(\alpha-\beta)]\Delta x\sigma + 2\alpha\beta\varepsilon_n = 0.$$

If $\varepsilon_n \neq 0$ and $\phi \neq 1$, this equation has two solutions given by:

$$\sigma_1 = \frac{b_2 - \sqrt{b_2^2 - 4b_1b_3}}{2b_1\Delta x},$$
$$\sigma_2 = \frac{b_2 + \sqrt{b_2^2 - 4b_1b_3}}{2b_1\Delta x},$$

with $b_1 = 2\varepsilon_n\phi(1-\phi)$, $b_2 = \alpha+\beta-\varepsilon_n(1-2\phi)(\alpha-\beta)$, and $b_3 = 2\alpha\beta\varepsilon_n$.

The discrete-time poles are then obtained using (4.20).

The discrete-time poles can be compared to the continuous-time poles by using the following relation:

$$p = \frac{1}{\Delta t}\log(z),$$

which gives the equivalent continuous-time poles from the discrete-time ones.

Example 4.2 (Discrete poles of a horizontal frictionless channel). Let us consider the horizontal frictionless canal of Example 3.1. We choose a fixed space step $\Delta x =$ 300 m, which corresponds to one tenth of the canal length, and a fixed time step $\Delta t = 64.8$ s, which corresponds to one tenth of the downstream propagation delay. This leads to a Courant number equal to 1. In this case, the poles of the discrete-time model consist in an integrator and nine pairs of complex poles. Figure 4.12 compares the continuous poles to the values of σ_n. Only the imaginary part of the nine poles in the upper half plane is plotted since the real part is zero and the poles are complex conjugate. We observe that the first three modes match the continuous one, but the frequency of modes 4 to 6 is not correctly reproduced. Then, the error on the last three modes increases in an exponential way with the mode number.

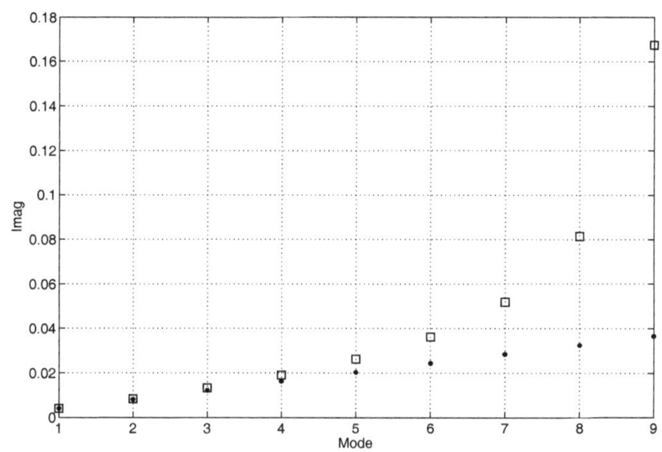

Fig. 4.12 Plot of σ_n ($\square$) for a horizontal frictionless channel. Comparison with the continuous poles ($\cdot$)

One may therefore think that the error due to the time sampling will increase the global error. However, the choice of the sampling period Δt may, in fact, decrease the error on the discretized poles. This will be studied in the following.

We first study the effect of the weighting coefficient θ. Figure 4.13 depicts the equivalent continuous poles obtained with the Preissmann discretization scheme for various values of $\theta \in [0.1, 0.9]$ and $\phi = 0.5$. When $\theta < 0.5$, the poles are unstable, in the right half plane. When $\theta = 0.5$, the poles obtained with the discrete-time model are on the imaginary axis, close to the continuous poles for low frequencies. When $\theta > 0.5$, the poles move towards the left in the complex plane, which means that their damping increases.

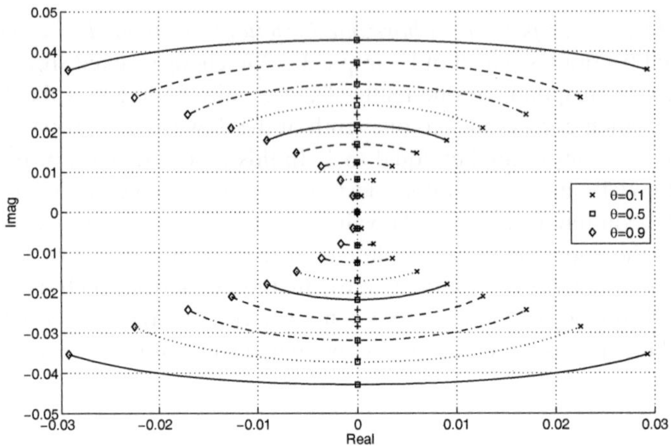

Fig. 4.13 Poles of the discretized model for $\phi = 0.5$ and $\theta \in [0.1, 0.9]$. Comparison with the continuous poles +

This study confirms the results obtained by [9] for the case of the centered Preissmann scheme where $\phi = 0.5$.

Figure 4.14 depicts the equivalent continuous poles obtained with the Preissmann discretization scheme for various values of $\phi \in [0.1, 0.9]$, and $\theta = 0.5$. We observe that when $\phi < 0.5$, the poles are in the left half plane, therefore the system is stable. When $\phi = 0.5$, the poles obtained with the discrete-time model are on the imaginary axis, relatively close to the continuous poles for low frequencies. When $\phi > 0.5$, the poles move towards the right in the complex plane, contrary to the case depicted in Fig. 4.13 where the poles move to the left when θ increases.

As an example, in Fig. 4.15 we depict the equivalent continuous poles obtained for various values of $\phi \in [0.1, 0.9]$, and $\theta = 0.6$, which is a standard value for this parameter. In that case, it appears that the standard choice $\phi = 0.5$ is not necessarily the best one, since one could choose larger value of ϕ to represent the dynamics of the channel more accurately.

4.3.2.2 Uniform Flow Case

In the case where the friction and the slope are not zero, the poles can also be computed explicitly, since σ is the zero of a fourth order polynomial given by (4.18),

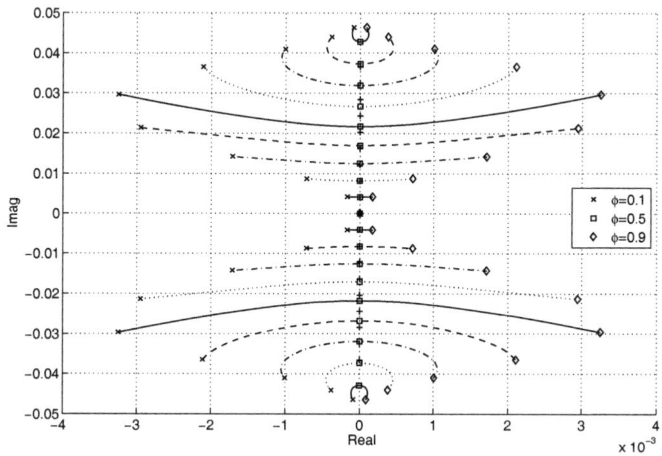

Fig. 4.14 Poles of the discretized model for $\theta = 0.5$ and $\phi \in [0.1, 0.9]$. Comparison with the continuous poles $+$

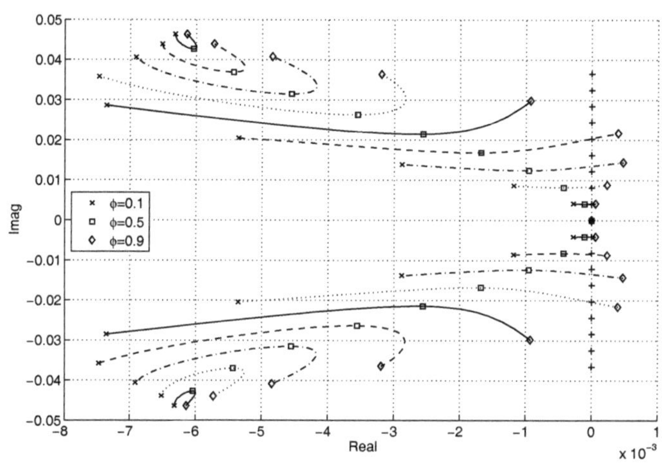

Fig. 4.15 Poles of the discretized model for $\theta = 0.6$ and $\phi \in [0.1, 0.9]$. Comparison with the continuous poles $+$

but the analytical form does not provide more insight, this is why we use a numerical approach.

This equation can be solved numerically, to obtain the poles in the uniform flow case, with nonzero slope and friction. As an example, we have computed the poles of the example canals 1 and 2 around uniform flow.

In both cases, the discretization step is chosen such that $\Delta x = L/10$, and $\Delta t = \tau_1/10$, which leads to a Courant number of 1. Figure 4.16 depicts the equivalent continuous poles of canal 1 obtained with the Preissmann discretization scheme for various values of $\theta \in [0.1, 0.9]$, and $\phi = 0.5$. We observe a similar pattern to the one already seen in the horizontal frictionless case.

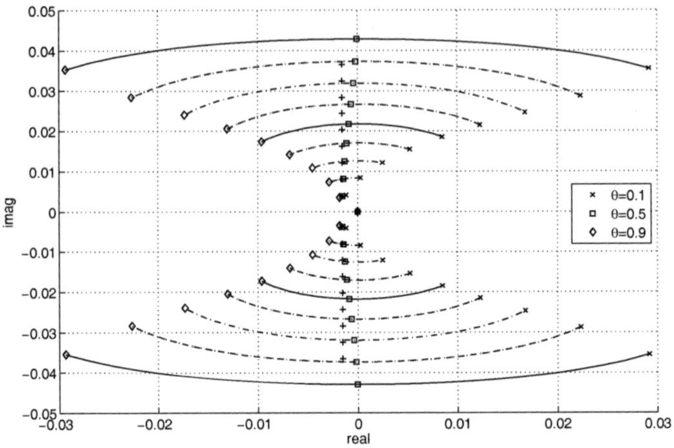

Fig. 4.16 Poles of the discretized model of canal 1 for $\phi = 0.5$ and $\theta \in [0.1, 0.9]$. Comparison with the continuous poles +

Figure 4.17 depicts the module of the error for each pole, as a function of the parameter θ, when $\phi = 0.5$. We see that the value $\theta = 0.53$ minimizes the error on all the modes. This study therefore provides a way to choose the parameters θ and ϕ in order to correctly reproduce the dynamics of the open channel.

Figure 4.18 depicts the equivalent continuous poles of canal 2 obtained with the Preissmann discretization scheme for various values of $\theta \in [0.1, 0.9]$, and $\phi = 0.5$. In that case, there are two negative real poles. One is correctly modeled by the discretization scheme. The other one is more sensitive to changes in the parameter θ.

We now use the above analysis to study the effect of the time and space discretization on the poles of the system. We study the effect of the Courant number by varying the time step while keeping the same space step, which leads to the same number of poles.

4.3.2.3 Effect of the Courant Number C_r

We fix the space step equal to $\Delta x = L/10$ and vary the time step to obtain different Courant numbers $C_r = \alpha \Delta t / \Delta x$. The Courant number varies between 0.1 and 10.

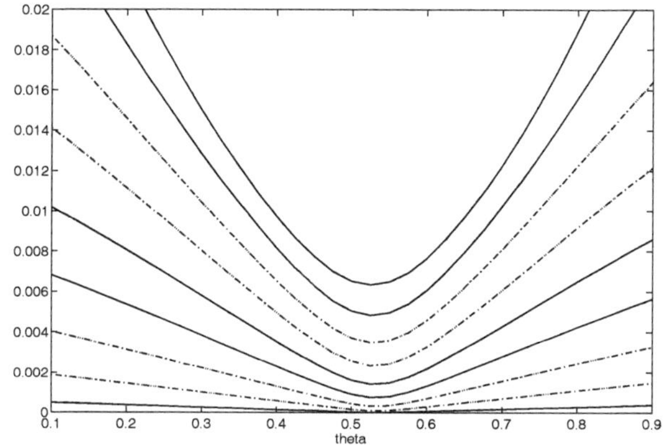

Fig. 4.17 Distance between the continuous poles of canal 1 around uniform flow and the poles of the discretized model for $\phi = 0.5$ and $\theta \in [0.1, 0.9]$

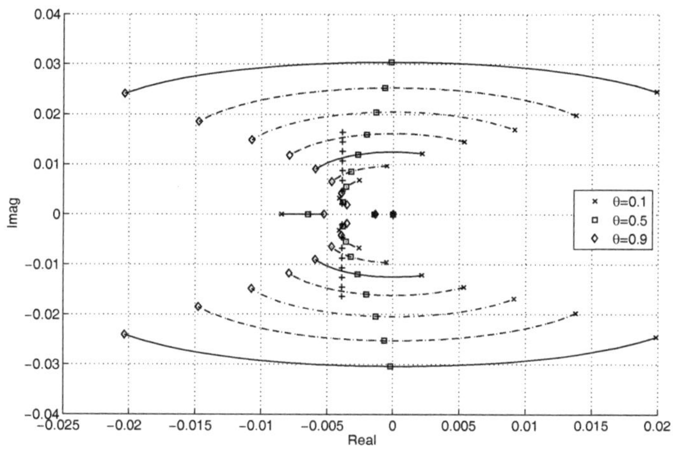

Fig. 4.18 Poles of the discretized model of canal 2 for $\phi = 0.5$ and $\theta \in [0.1, 0.9]$. Comparison with the continuous poles $+$

This corresponds to a time step varying from $\Delta t = \tau_1/100$ to $\Delta t = \tau_1$. Since the space step does not change, we keep the same number of poles (nine pairs of poles plus the integrator).

As we have seen above, it is well-known that a Courant number close to 1 ensures a correct representation of the system's dynamics in the horizontal frictionless case. We study here the uniform case with nonzero slope and friction, to see whether this result also holds in this case.

We fix $\theta = 0.5$ and $\phi = 0.5$ and study the effect of C_r on the poles of the system. Figure 4.19 depicts the error for each pole as a function of C_r for canals 1 and 2 around uniform flow.

We see that for canal 1, the minimum is obtained for $C_r = 1.2$, while for canal 2, the minimum is obtained for $C_r = 2$. Therefore, it appears that the friction modifies the classical results obtained for horizontal frictionless channels. A graphic representation such as the one in Fig. 4.19 enables us to choose Δt such that the modes are correctly reproduced.

Contrarily to what happens in classical continuous-time systems, reducing the time step does not increase the accuracy. In that case, with a fixed value of the space step, reducing the time step corresponds to reducing the Courant number, and this leads to a large error in the high frequency modes.

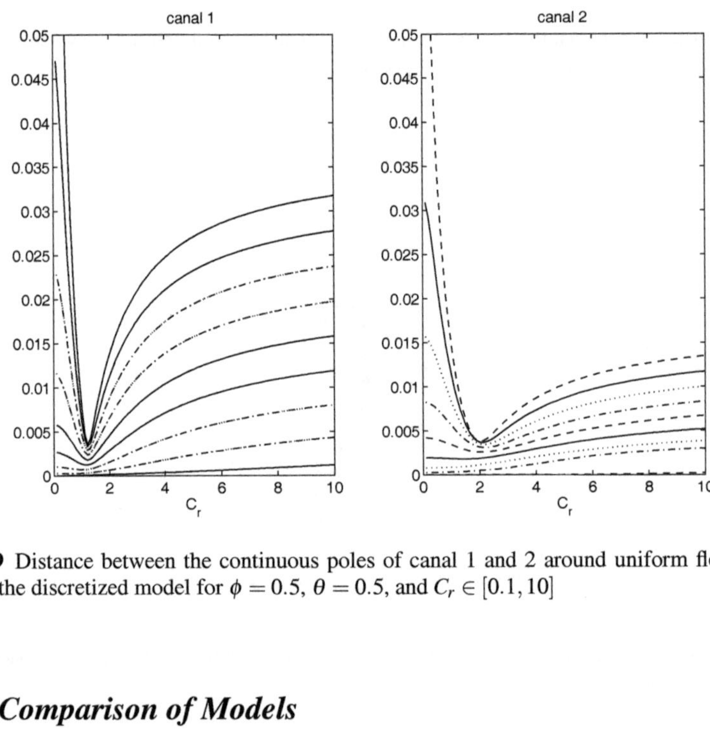

Fig. 4.19 Distance between the continuous poles of canal 1 and 2 around uniform flow and the poles of the discretized model for $\phi = 0.5$, $\theta = 0.5$, and $C_r \in [0.1, 10]$

4.3.3 Comparison of Models

The finite difference model obtained with the Preissmann scheme is compared to the transfer matrix model on both example canals. We compare the models obtained for two different pairs of time and space steps. The space and time steps are chosen in order to have a Courant number $C_r = (V_0 + C_0)\Delta t/\Delta x$ close to 1.

Figure 4.20 represents the Bode plot of Saint-Venant transfer function $p_{21}(s)$ of canal 1, comparing the continuous transfer function to those obtained with two sets of time and space samplings.

Figure 4.21 represents the Bode plot of Saint-Venant transfer function $p_{21}(s)$ of canal 2, also comparing the continuous transfer function to those obtained with two sets of time and space samplings.

For discrete-time models, the frequency response is only valid up to the Nyquist frequency $\omega_N = \pi/\Delta t$. The frequency response is therefore plotted up to the Nyquist frequency (denoted by a vertical bar in the figures). There is, however, a discrepancy observed for frequencies close to ω_N, which is due to the numerical discretization.

We observe that for small time and space steps, the discretized system is close to the continuous one. This is normal, since the Preissmann scheme is consistent: the discretized system converges towards the solution of Saint-Venant equations as the time and space steps approach zero.

It is clear from the figures that the smaller Δt, the better the fit between the Bode plot of the finite difference model and the Bode plot of the continuous system. However, it is difficult to obtain an accurate frequency response with the discrete-time model, since the order of the model increases as Δx decreases, which may lead to numerical problems.

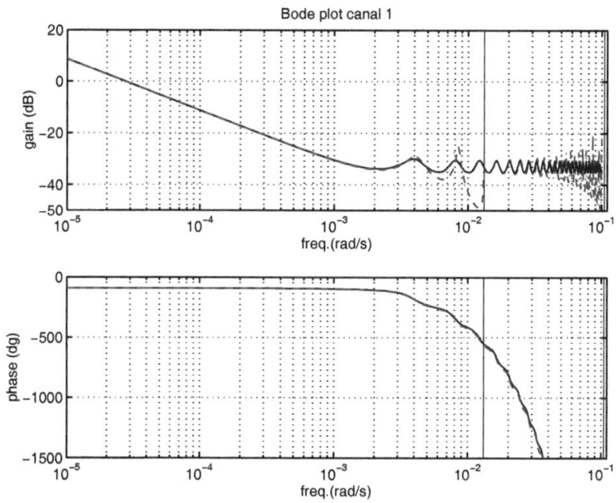

Fig. 4.20 Bode plots of $p_{21}(s)$ for canal 1 around $Q_0 = Q_{max}/2$, comparison between the continuous model (—) and two finite difference models with different pairs of time and space steps: $\Delta t = 240$ s, $\Delta x = 1000$ m (– –), and $\Delta t = 30$ s, $\Delta x = 126$ m (– · –)

This illustrates a potential use of the proposed method for providing a frequency validation of finite difference numerical schemes used either to simulate or to obtain a model for controller design from the Saint-Venant equations.

We have extended our frequency domain approach to the case of a discretized model obtained from the Preissmann discretization scheme. The method enables us to compute the poles of the discretized model and evaluate their distance to the poles

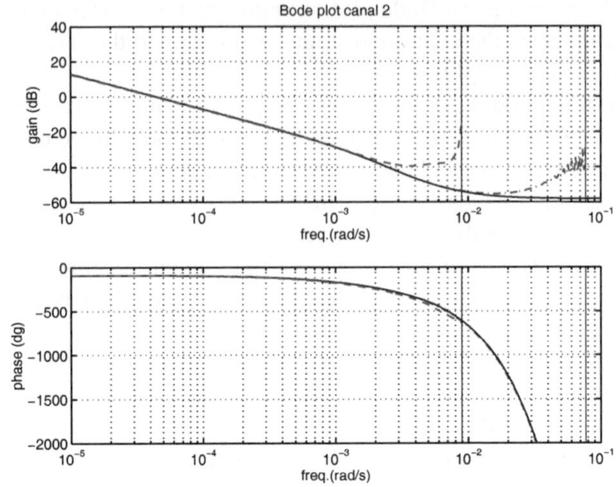

Fig. 4.21 Bode plots of $p_{21}(s)$ for canal 2 around $Q_0 = Q_{max}/2$, comparison between the continuous model (—) and two finite difference models with different pairs of time and space steps: $\Delta t = 350$ s, $\Delta x = 2000$ m (– –), and $\Delta t = 40.8$ s, $\Delta x = 234$ m (– · –)

of the continuous system. This also provides guidance in the choice of the time step Δt and the parameters θ and ϕ.

4.4 Summary

In this chapter, we developed various methods to obtain a rational approximation of open channel flow described by linearized Saint-Venant equations. The method based on modal decomposition enables us to compute a distributed approximate model, either with or without delays. The order of the model depends on the number of poles included in the approximation. The finite difference Preissmann scheme was studied in the frequency domain, and we obtained an explicit expression of the poles in the horizontal frictionless case. We finally proposed a way to evaluate the discrepancy between the discretized system and the continuous one. The results can be used to determine the best choice of discretization time and space steps and parameters θ and ϕ in order to correctly reproduce the system's dynamics.

References

[1] Abramowitz M, Stegun IA (1972) Handbook of mathematical functions, Applied Mathematics Series, vol 55, 10th edn. National Bureau of Standards

[2] Balogun O, Hubbard M, DeVries J (1988) Automatic control of canal flow using linear quadratic regulator theory. J Hydraul Eng 114(1):75–102
[3] Cunge J (1966) Etude d'un schéma de différences finies appliqué à l'intégration numérique d'un certain type d'équations hyperboliques d'écoulement. PhD thesis, Faculté des Sciences de Grenoble, (in French)
[4] Cunge J, Holly F, Verwey A (1980) Practical aspects of computational river hydraulics. Pitman Advanced Publishing Program
[5] Dulhoste JF (2001) Contribution à la commande non linéaire de systèmes d'irrigation. PhD thesis, INPG, Grenoble
[6] Dulhoste JF, Besançon G, Georges D (2001) Nonlinear control of water flow dynamics by input-output linearisation based on a collocation model. In: European Control Conference, Porto, Portugal
[7] Litrico X, Fromion V (2002) Infinite dimensional modelling of open-channel hydraulic systems for control purposes. In: Proc. 41st IEEE Conf. decision and control, Las Vegas, pp 1681–1686
[8] Lyn D, Goodwin P (1987) Stability of a general Preissmann scheme. J Hydraul Eng 113(1):16–28
[9] Malaterre PO (1994) Modélisation, analyse et commande optimale LQR d'un canal d'irrigation. PhD thesis, ENGREF–Cemagref
[10] Malaterre PO (1998) Pilote: linear quadratic optimal controller for irrigation canals. J Irrig Drain Eng 124(4):187–194
[11] Ooi S, Weyer E (2003) Control design for an irrigation channel from physical data. In: European control conference ECC'03, Cambridge (UK)
[12] Preissmann A (1961) Propagation des intumescences dans les canaux et rivières. In: 1er Congrès de l'Association Française de Calcul, Grenoble, France, pp 433–442
[13] Samuels P, Skeels C (1990) Stability limits for Preissmann's scheme. J Hydraul Eng 116(8):997–1012
[14] Venutelli M (2002) Stability and accuracy of weighted four-point implicit finite difference schemes for open channel flow. J Hydraul Eng 128(3):281–288
[15] Weyer E (2001) System identification of an open water channel. Contr Eng Pract 9:1289–1299

Chapter 5
A Simplified Model of Open Channel Flow

Abstract This chapter develops a simple model for open channel flow, called the IDZ model, for integrator delay zero. After the models developed in the previous chapters, we show that such a simple model can capture the main physical properties of the open channel dynamics. The main interest of this model is that it can be computed analytically from the physical parameters of the channel such as its geometry, slope and roughness coefficient, and the flow characteristics such as the average flow and the downstream water level. It is a simple yet accurate low frequency model of open channel flow, which will be used in Chap. 7 to design robust tuning rules for PI controllers.

5.1 Introduction

The rational models developed in Chap. 4 have been obtained by simplifying the equations using various mathematical tools (modal decomposition, convex optimization, numerical schemes). However, our physical intuition leads us to think that an open channel can be accurately modeled using an integrator, a delay, and a high frequency component. In this chapter, we develop a way to obtain the parameters of such a simple model as explicit functions of the physical parameters. This allows us to:

- handle the dynamics of the system with few parameters,
- understand the impact of physical parameters on the dynamics, and
- facilitate the development of a systematic design method.

This problem has already been considered in the literature. Simple models have been obtained for the Saint-Venant equations linearized around a uniform regime (see [16, 4, 5, 7]). In that case, analytical expressions are available, which allow one to make simple approximations, leading to approximate rational models (possibly including a time-delay).

Unfortunately, the uniform regime is seldom observed in practical situations, since it requires constant discharges and water depths along the pool. Moreover, hydraulic structures are generally used in irrigation canals to raise the water level at some control points, which influences the flow upstream (i.e., there is a backwater curve). This is why it is necessary to develop simple models for realistic cases, i.e., for canal pools under backwater. A possible way to take into account the nonuniform regime is to derive linear models from numerical schemes used to solve the Saint-Venant equations [2, 12]. However, it seems difficult and questionable to use this approach to derive a simple model. Indeed, this numerical approach leads to complex models with a large number of parameters. More theoretically, from the robust control point of view, it is better to approximate the frequency response of a system than its time-response, because it allows one to quantify the model error in an appropriate way [18].

Schuurmans et al. [15] proposed an integrator delay (ID) model that applies to a canal under backwater flow conditions, which with two parameters captures the low frequency behavior of the system, i.e., a delayed integrator for the transfer between upstream discharge and downstream level. This model is a notable step towards the solution, but it is built on a series of assumptions that are not always fulfilled [9]. Therefore, using this method to compute the ID model parameters can lead to a bad approximation of the system's dynamics.

Our objective is to get a model that is valid for any regime, which gives accurate approximate values for the delay and the integrator gain, which are essential for control design. To do this, the method developed in Chap. 3 to obtain the frequency domain representation of Saint-Venant transfer matrix [10] has been simplified, leading to a simple approximate model, the IDZ model [9]. This model is able to represent the canal behavior in low and high frequencies: the integrator delay accounts for low frequencies, while the zero represents the direct influence of the discharge on the water level in high frequencies. The proposed model is obtained from mathematical approximations of the exact transfer matrix. It aims at reproducing the behavior of the system in any flow condition (including backwater flow configurations).

The IDZ model can therefore be summarized as follows:

- It provides accurate approximate values of the delay and the integrator gain for a classical ID model.
- It extends the ID model by adding a zero, leading to a better fit in high frequencies. This also improves the accuracy of the time domain simulations, since this zero accounts for the direct influence of the discharge on the water levels.
- All parameters can be computed analytically, which enables a simple implementation.

The proposed model is multivariable, i.e., it can represent the upstream influence of a downstream discharge deviation, and accurately reproduces the behavior of the system in backwater flow configurations. The model parameters (integrator gains, delays and zeros) are explicitly given as functions of the steady-state discharge Q_0 and the downstream water level Y_L, and change continuously with those parameters.

Before detailing this approximate model, we first discuss the notion of the delay of an open channel.

5.2 On the Delay of an Open Channel

It is well-known that open channel systems include propagation delays between two distant points: the discharge delivered at one upstream point will take some time to propagate downstream. However, the question concerning the definition of this time-delay is still unclear in the literature. Based on the detailed frequency domain analysis of the linearized Saint-Venant equations performed in Chap. 3, we can distinguish three different notions:

- The mathematical time-delay of a canal pool, which is obtained from the characteristics and corresponds to the wave behavior.
- The group delay, which is an equivalent delay, specifying the variable delay according to the signal frequency.
- The response time, which expresses the time for a discharge perturbation to reach the end of a pool. This is the hydraulic delay considered in canal management, and it can be shown that this delay depends on the downstream boundary condition (see [17, 14]).

Those three delays have different values and different meanings, as will be shown below.

5.2.1 Mathematical Delay

As we have seen in Sect. 3.3.2.1 of Chap. 3, the linearized Saint-Venant transfer functions can be factorized as a product of an outer part and an inner part. The outer part is invertible, and in this case the inner part consists of a pure delay. We have shown that this delay corresponds to the one of the characteristics lines.

In the case of the transfer function $p_{21}(s)$, relating the downstream water elevation to the upstream discharge, this delay τ_1 can be interpreted as the minimum time for a change in the upstream discharge to have an effect on the downstream water level. In the uniform flow case, this delay is equal to $\tau_1 = \frac{L}{\alpha} = \frac{L}{V_0+C_0}$.

For transfer function $p_{12}(s)$, relating the upstream water elevation to the downstream discharge, this delay τ_2 can be interpreted as the minimum time for a change in the downstream discharge to have an effect on the upstream water level. In the uniform flow case, this delay is equal to $\tau_2 = \frac{L}{\beta} = \frac{L}{C_0-V_0}$.

Transfer functions $p_{11}(s)$ and $p_{22}(s)$ have no time-delay, which means that the local effect of the discharge on the water level is instantaneous.

Let us now introduce the group delay of a transfer function.

5.2.2 Group Delay

The group delay of a transfer function $P(s)$, denoted $\tau_g(P)(\omega)$ is associated to the phase of $P(j\omega)$ and corresponds to its first derivative with respect to the frequency ω. The group delay is therefore defined by [6]:

$$\tau_g(P)(\omega) = -\frac{\partial \arg(P(j\omega))}{\partial \omega}, \tag{5.1}$$

where $\arg(P(j\omega))$ denotes the phase of the complex number $P(j\omega)$.

It is easy to show that the group delay of an integrator delay system is equal to the delay. Indeed, the phase of an integrator delay system is given by:

$$\arg\left(\frac{e^{-j\tau\omega}}{j\omega}\right) = -\frac{\pi}{2} - \tau\omega,$$

and the group delay is in this case equal to the delay τ.

However, when the system has nonzero poles or zeros, the group delay varies with the frequency. The group delay $\tau_g(P)(\omega)$ represents the apparent delay for perturbations occurring around the frequency ω.

Let us compute the group delay of the linearized Saint-Venant transfer matrix in the uniform flow case. To do this, we need to compute the phase of each transfer function $p_{kl}(j\omega)$, and compute its derivative with respect to ω. We end up with complex expressions depending on ω, which will not be displayed here.

Figure 5.1 depicts the variation of the group delay $\tau_g(p_{21})(\omega)$ along the frequency.

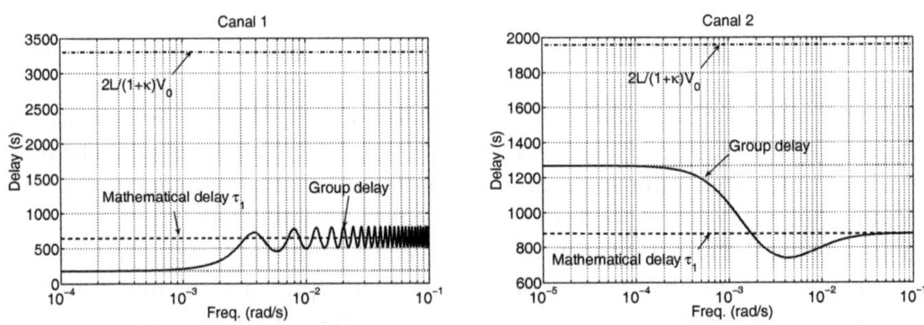

Fig. 5.1 Variation of the group delay of $p_{21}(s)$ with frequency, canals 1 and 2

We notice a different pattern for canal 1 and canal 2: in high frequencies, the group delay of canal 1 oscillates around the value of the mathematical delay $\tau_1 = L/\alpha$. This is due to the oscillating modes already studied in Chap. 3. For canal 2, the group delay tends towards the value of the mathematical delay. This is coherent with our physical intuition: the characteristics correspond to the description of high frequency perturbations.

For low frequencies, we have again a different pattern: the group delay is lower than the mathematical delay for canal 1, while it is higher for canal 2. As we will show later, this peculiar behavior of canal 1 is due to the direct term between the discharge and the water level. In the case of canal 2, the negative real poles add some phase, which here is integrated in the group delay, leading to a higher delay than the mathematical delay obtained in high frequency corresponding to the characteristics.

Let us compute the value of the group delay at $\omega = 0$, since we are interested in a low frequency approximation of open channel flow. After some calculations, one obtains the value of the group delay of the linearized Saint-Venant transfer functions $p_{12}(s)$ and $p_{21}(s)$ for $\omega = 0$:

$$\tau_g(p_{12})(0) = L\left(\frac{\alpha-\beta}{\alpha\beta}+\frac{\delta}{\gamma}\right) - \frac{[(\alpha-\beta)\gamma+2\alpha\beta\delta]}{\gamma^2}\left(1-\psi\frac{e^{-\psi}}{1-e^{-\psi}}\right), \quad (5.2a)$$

$$\tau_g(p_{21})(0) = \frac{\delta L}{\gamma} - \frac{[(\alpha-\beta)\gamma+2\alpha\beta\delta]}{\gamma^2}\left(1-\psi\frac{e^{-\psi}}{1-e^{-\psi}}\right), \quad (5.2b)$$

with $\psi = \frac{\gamma L}{\alpha\beta}$.

If we express this result as a function of the physical variables, we obtain:

$$\frac{\delta L}{\gamma} = \frac{2L}{(1+\kappa)V_0}, \quad (5.3a)$$

$$\frac{(\alpha-\beta)\gamma+2\alpha\beta\delta}{\gamma^2} = \frac{2A_0[2+(\kappa-1)F_0^2]}{T_0V_0S_b(1+\kappa)^2}. \quad (5.3b)$$

We note that the approximation given in [15] for the delay of a pool at uniform flow includes only the first term in (5.2b), corresponding to the exponential term $e^{(\lambda_1(s)-\lambda_2(s))L}$. The second term corresponds to the phase added by the factor $(\lambda_1(s)-\lambda_2(s)$ and by the poles $1/(1-e^{(\lambda_1(s)-\lambda_2(s))L})$. Therefore, the actual apparent delay will in fact be smaller than the one corresponding only to the first term. This explains why the approximate delay given in [15] is generally larger than the apparent delay corresponding to the low frequency group delay.

We now introduce another notion, which is used for practical applications to estimate the time necessary for a change in upstream discharge to reach the downstream end of a canal.

5.2.3 Hydraulic Response Time

The hydraulic response time is usually referred to as the time for a change in upstream discharge to reach the downstream end of the pool. This is the time-delay that is important for water distribution, when water delivery needs to be done at a given time at a certain point. As we will see below, this response time depends on the downstream boundary condition.

Canal pools are usually controlled with hydraulic structures, which impose a static relationship between discharge and water level. The effect of such a structure on the flow will be examined in detail in Chap. 6. Here, we only show the effect of such a structure on the response time.

In the literature, some approximate formulas are used to estimate the response time when the discharge changes from an initial value Q_i to a final value Q_f. The classical one is given by the ratio of the variation of the total volume $\Delta V = V_f - V_i$ in the stretch over the discharge variation $\Delta Q = Q_f - Q_i$:

$$T_{R0} = \frac{\Delta V}{\Delta Q}.$$

Another formula was proposed by [1]:

$$T_{RA} = 2T_{R0} - T_w,$$

where T_w is the wave travel time, i.e., the mathematical delay. The accuracy of these formulas is difficult to assess, since the response time is not clearly defined.

We use here the frequency domain approach to define the response time of a channel. As will be shown in Chap. 6, a hydraulic structure induces a local feedback between the water level and the discharge. Then, the flow routing in the open channel with the hydraulic structure can be computed in the Laplace domain as follows:

$$\hat{q}(L,s) = G_k(s)\hat{q}(0,s), \tag{5.4}$$

with $G_k(s) = \frac{kp_{21}(s)}{1-kp_{22}(s)}$ and k the value of the feedback, depending on the geometry of the hydraulic structure.

This transfer function $G_k(s)$ is infinite dimensional, and one cannot easily compute its impulse or step response. However, this system can be approximated by a first-order with delay.

Following the same line as that developed in [13, 14], we use the moment matching method to obtain a first-order with delay approximation of this transfer function:

$$G_k(s) \approx \frac{e^{-Ts}}{1+Ks}. \tag{5.5}$$

The nth logarithmic moment of a transfer function $h(s)$ expressed in the Laplace domain is defined by:

$$M_n(h) = (-1)^n \frac{\mathrm{d}^n}{\mathrm{d}s^n}\{\log[h(s)]\}|_{s=0}.$$

Equating the two first moments of transfer functions (5.4) and (5.5) leads to:

$$T + K = M_1,$$
$$K^2 = M_2,$$

where M_1 and M_2 are the first-order and second-order logarithmic moments of the transfer function (5.4).

Solving for K and T enables us to obtain an analytical expression for these coefficients as function of the parameters of the pool (see [13] for details).

The first-order with delay model provides a way to accurately predict the hydraulic response time of a canal pool. In fact, the response time is the sum of two terms: the delay and the rising time (see Fig. 5.2).

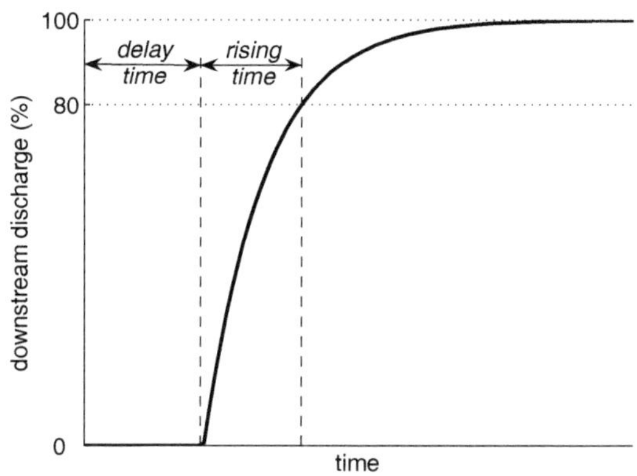

Fig. 5.2 The response time at 80% is the sum of the mathematical delay and the rising time

Using the first-order with delay approximation model, we have an analytical expression for both terms. For a step response, the rising time at 80% of a first-order linear model is easily computed as:

$$t_{80} = -K\log(1-0.80),$$

where K is the time constant of the first-order model. Therefore, the hydraulic response time at 80% can be computed as:

$$T_R = T + K\log 5.$$

This can be generalized to another level of response, or to the case of a ramp input instead of a step input, if one is interested in evaluating the effect of the rate of change in upstream discharge.

Therefore, the hydraulic response time depends on the downstream boundary condition and can be accurately evaluated using a moment matching method to identify a first-order with delay model for the flow transfer.

We now introduce the IDZ model, which captures the low frequency behavior of the open channel, assuming boundary conditions in terms of discharges.

5.3 Integrator Delay Zero Model

We first consider the uniform case, then extend the model to the nonuniform case.

5.3.1 Uniform Flow Case

Our objective is to obtain a model whose parameters can be computed analytically as function of the physical parameters of the canal. We begin by a way to explicitly compute the uniform depth Y_n as a function of the geometry of the channel and the discharge Q_0.

5.3.1.1 Computation of the Uniform Depth Y_n

To obtain an analytical model, it is necessary to have an analytical expression for the uniform depth Y_n. The uniform depth is computed from the solution of the algebraic equation

$$S_{f0} = S_b, \tag{5.6}$$

with S_{f0} the friction slope given by the Manning formula (2.3) and S_b the slope. In the general case, S_{f0} is not a simple function of Y_0, therefore there is no analytical solution for this equation. The hydraulic exponent method [3] enables us to obtain a good analytical approximation of the uniform depth in prismatic channels. In this method, the conveyance $K_0 = Q_0/\sqrt{S_{f0}}$ is assumed to be a power function of the depth of flow Y_0:

$$K_0^2 = CY_0^N, \tag{5.7}$$

where C is a coefficient and N is the hydraulic exponent for uniform flow computation.

A good approximation of N is obtained by computing the uniform depth for two different discharges Q_1 and Q_2. N is then given by:

$$N = 2\frac{\log(Q_1/Q_2)}{\log(Y_n(Q_1)/Y_n(Q_2))}. \tag{5.8}$$

Given the uniform flow depth $Y_n(Q_1)$ corresponding to a discharge Q_1 (e.g., the maximum discharge for the considered pool), the value of the uniform flow depth $Y_n(Q_0)$ for another discharge Q_0 can be approximated by:

$$Y_n(Q_0) \approx \left(\frac{Q_0}{Q_1}\right)^{2/N} Y_n(Q_1). \tag{5.9}$$

Example 5.1 (Computation of the uniform depth for the example canals). The downstream limit condition Y_L is chosen equal to the uniform depth Y_n for Q_{max}. The uniform depth approximations are obtained using (5.8) and (5.9) where $Q_1 = Q_{\max}$ and $Q_2 = Q_{\max}/8$. This leads to $N = 3.51$ for canal 1 and $N = 3.56$ for canal 2. The obtained approximations of the uniform depths are very accurate for both canals (see Fig. 5.3).

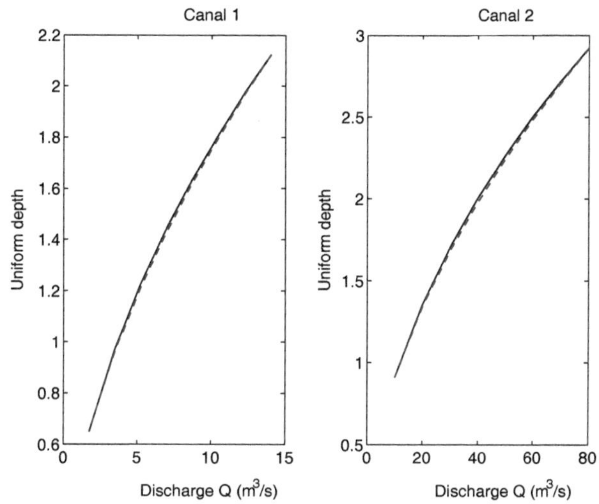

Fig. 5.3 Variation of the uniform depth with the discharge, canals 1 and 2. Comparison between the exact (—) and approximate (– –) values.

5.3.1.2 IDZ Model Parameters

Frequency Domain Expressions

The IDZ model is an approximation of the input–output transfer matrix $P(s)$, defined as follows:

$$p_{11}(s) = \frac{1}{A_u s} + \tilde{b}_u, \tag{5.10a}$$

$$p_{12}(s) = -\left(\frac{1}{A_u s} + b_u\right) e^{-\tau_u s}, \tag{5.10b}$$

$$p_{21}(s) = \left(\frac{1}{A_d s} + b_d\right) e^{-\tau_d s}, \tag{5.10c}$$

$$p_{22}(s) = -\left(\frac{1}{A_d s} + \tilde{b}_d\right), \tag{5.10d}$$

with A_u and A_d the equivalent backwater area for the upstream and downstream water levels, respectively, τ_d and τ_u the upstream and downstream propagation time-delays, and b_u, $\tilde{b}_u$, b_d and $\tilde{b}_d$ are positive constants obtained as the mean value of $|p_{ij}(s)|$ when $s \to \infty$.

For low frequencies, the behavior of the transfer matrix is dominated by the integrator and the delays.

For high frequencies, the delay and the gravity waves are predominant in the transfer matrix elements. For simplicity, the gravity waves are approximated by a constant gain in high frequencies. Such a static approximation implies that the oscillating modes are not modeled as such. Indeed, these modes are controlled by the hydraulic structure and, as we will point out at the end of Chap. 6, only the low frequency part of the canal pool has to be considered. The static term enables us to better represent the flow behavior at the hydraulic structures by describing the direct impact of the gate opening on the water levels.

In the uniform case, the IDZ model parameters are obtained as follows. First the integrator gain is obtained by computing the values $A_u = 1/[sp_{11}(s)]|_{s=0}$ and $A_d = 1/[sp_{22}(s)]|_{s=0}$, which directly gives:

$$A_u = \frac{\alpha\beta T_0}{\gamma}\left(e^{\frac{\gamma L}{\alpha\beta}} - 1\right), \tag{5.11a}$$

$$A_d = \frac{\alpha\beta T_0}{\gamma}\left(1 - e^{-\frac{\gamma L}{\alpha\beta}}\right). \tag{5.11b}$$

Then, the static terms b_u, $\tilde{b}_u$, b_d and $\tilde{b}_d$ are obtained as the mean value of $|p_{ij}(s)|$ when $s \to \infty$. To this end, we use the high frequency approximation of the eigenvalues given by

$$\lambda_1(s) \approx -r_1 - \frac{s}{\alpha},$$

$$\lambda_2(s) \approx r_2 + \frac{s}{\beta}.$$

This leads to:

$$\tilde{b}_u = \frac{1}{T_0\alpha}\sqrt{\frac{1+\frac{\alpha^2}{\beta^2}e^{-2(r_1+r_2)L}}{1+e^{-2(r_1+r_2)L}}}, \tag{5.12a}$$

$$b_u = \frac{(\alpha+\beta)}{T_0\alpha\beta}\frac{e^{-r_2 L}}{\sqrt{1+e^{-2(r_1+r_2)L}}}, \tag{5.12b}$$

$$b_d = \frac{(\alpha+\beta)}{T_0\alpha\beta}\frac{e^{-r_1 L}}{\sqrt{1+e^{-2(r_1+r_2)L}}}, \tag{5.12c}$$

$$\tilde{b}_d = \frac{1}{T_0\beta}\sqrt{\frac{1+\frac{\beta^2}{\alpha^2}e^{-2(r_1+r_2)L}}{1+e^{-2(r_1+r_2)L}}}, \tag{5.12d}$$

with $r_1 = \frac{\alpha\delta-\gamma}{\alpha(\alpha+\beta)}$ and $r_2 = \frac{\beta\delta+\gamma}{\beta(\alpha+\beta)}$.

In the IDZ model approximation, the coefficients b_u, $\tilde{b}_u$, b_d and $\tilde{b}_d$ include the static terms denoted by $b_{11}(x)$ and $b_{12}(x)$ in Chap. 3 and the static term linked to the oscillating modes.

Finally, the delays τ_u and τ_d are computed using the moment matching method, and are closely related to the group delays introduced earlier. The first-order logarithmic moment of the transfer functions $sp_{12}(s)$ and $sp_{21}(s)$ are given by:

$$M_1(sp_{12}(s)) = L\left(\frac{\alpha-\beta}{\alpha\beta}+\frac{\delta}{\gamma}\right) - \frac{[(\alpha-\beta)\gamma+2\alpha\beta\delta]}{\gamma^2}\left(1-\psi\frac{e^{-\psi}}{1-e^{-\psi}}\right),$$
$$M_1(sp_{21}(s)) = \frac{\delta L}{\gamma} - \frac{[(\alpha-\beta)\gamma+2\alpha\beta\delta]}{\gamma^2}\left(1-\psi\frac{e^{-\psi}}{1-e^{-\psi}}\right),$$

with $\psi = \frac{\gamma L}{\alpha\beta}$.

Let us note that these expressions are identical to the group delays (5.2).

The first-order logarithmic moment of the corresponding IDZ transfer functions are given by:

$$M_1(p_{12}) = \tau_u - A_u b_u,$$
$$M_1(p_{21}) = \tau_d - A_d b_d.$$

Equating the two moments leads to the expressions for the approximate delays

$$\tau_u = \tau_g(p_{12})(0) + A_u b_u, \quad (5.13a)$$
$$\tau_d = \tau_g(p_{21})(0) + A_d b_d, \quad (5.13b)$$

where $\tau_g(p_{12})(0)$ and $\tau_g(p_{21})(0)$ are given by (5.2).

Remark 5.1. In this approximation, the time-delays τ_u and τ_d include the mathematical delays and also the phase added by the zero. This delay approximation is different from the one originally used in [9, 11], where the mathematical delay was used in the IDZ approximation. The choice of an approximate delay depends on the objective of the approximate model: one may choose a different delay for control purposes or for simulation purposes.

Remark 5.2. The way to compute the IDZ model parameters is a trade-off between complexity and precision. We could also have used the moment matching method to compute the static gain and the delay. To do this, one needs to compute the second-order moment of transfer functions $sp_{12}(s)$ and $sp_{21}(s)$, and solve for the corresponding equations, following the line of [14]. This may certainly lead to a more accurate model, but to the expense of complexity. In the following, we keep the original formulations of the IDZ model, except for the delay term, which has been modified to incorporate the additional phase of the oscillating modes.

Bode Plots

In Figs. 5.4 and 5.5, we compare the Bode plots of the approximate model to the one of the linearized Saint-Venant equations for canals 1 and 2.

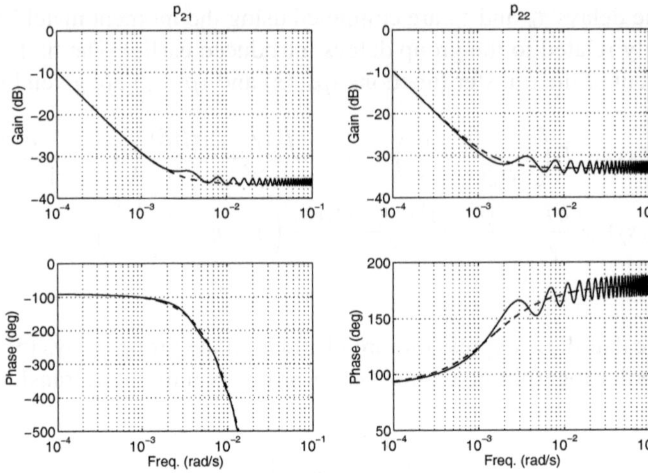

Fig. 5.4 Bode plot of transfer functions $p_{21}(s)$ and $p_{22}(s)$ (—), and the IDZ approximation (– –), canal 1 around uniform flow

The IDZ model approximation efficiently captures the low and high frequencies behavior of the open channel. We observe that in the case of canal 2, the transfer function $p_{21}(s)$ has a smaller delay than that used in the approximation. This is due to the fact that the group delay incorporates the phase shift due to the first negative real pole (see the poles of example canal 2 in Fig. 3.8). This ensures a better fit of the approximate model for low frequencies, and it is more efficient for PI controller design.

Time Domain Expressions

This frequency domain model can be written as an ordinary differential equation with input delay, given by:

$$\begin{cases} A_d \frac{dh(t)}{dt} = q(0,t-\tau_d) - q(L,t) \\ y(L,t) = h(t) + b_d q(0,t-\tau_d) - \tilde{b}_d q(L,t) \end{cases}, \tag{5.14}$$

where $h(t)$ is an intermediate variable (initialized by $h(0)=0$), $y(X,t)$ is the output and $q(0,t)$ and $q(X,t)$ are the inputs. The delay τ_d only occurs on the upstream discharge $q(0,t)$, while the downstream discharge has a direct impact on the downstream water level.

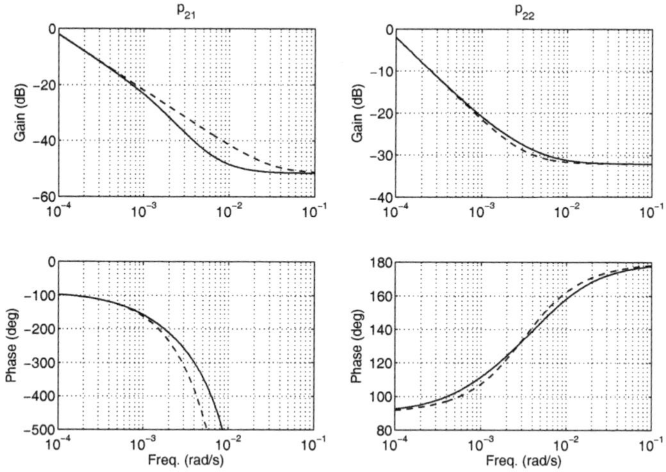

Fig. 5.5 Bode plot of transfer functions $p_{21}(s)$ and $p_{22}(s)$ (—), and the IDZ approximation (– –), canal 2 around uniform flow

Step Responses

Figure 5.6 depicts the step response of the IDZ model approximation, and compares it to the step response of the input–output transfer matrix of the linearized Saint-Venant equations, computed with the modal decomposition method of Chap. 4 with 21 poles.

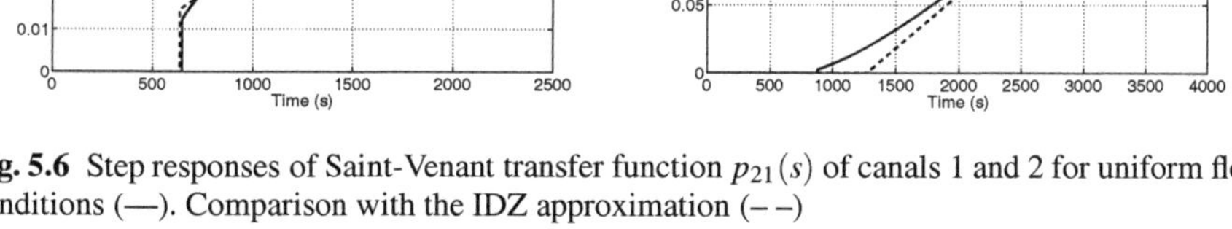

Fig. 5.6 Step responses of Saint-Venant transfer function $p_{21}(s)$ of canals 1 and 2 for uniform flow conditions (—). Comparison with the IDZ approximation (– –)

5.3.2 Nonuniform Flow Case

We now consider the nonuniform case, where we use an approximation of the backwater curve, in order to be able to compute the IDZ model parameters analytically.

5.3.2.1 Backwater Curve Approximation

The backwater curve defines the steady-state water level for a given discharge Q_0 and a downstream boundary water depth $Y_0(L)$. It is obtained as the solution of the differential equation (2.16b), which is rewritten here for convenience:

$$\frac{\mathrm{d}Y_0(x)}{\mathrm{d}x} = \frac{S_b - S_{f0}(x)}{1 - F_0(x)^2}, \tag{5.15}$$

where F_0 is the Froude number $F_0 = \frac{V_0}{C_0}$ with $C_0 = \sqrt{g\frac{A_0}{T_0}}$, $V_0 = \frac{Q_0}{A_0}$, and T_0 is the top width.

Using the backwater curve definition given by (5.15), let S_L denote the deviation from bed slope of the line tangent to the backwater curve at the downstream end of the pool:

$$S_L = \frac{S_b - S_{f0}(L)}{1 - F_0^2(L)}.$$

When $x \to -\infty$, we have $\frac{\mathrm{d}Y_0(x)}{\mathrm{d}x} \to 0$, and $Y_0(x) \to Y_n$, the uniform depth, which is the solution of (5.6). We therefore approximate the upstream part of the backwater curve by a straight line parallel to the bottom slope at the uniform water depth Y_n, and the downstream part by a straight line of slope S_L.

The intersection between the two lines occurs at abscissa x_1:

$$x_1 = \begin{cases} \max\{L - \frac{Y_L - Y_n}{S_L}, 0\} & \text{if } S_L \neq 0 \\ L & \text{if } S_L = 0 \end{cases}.$$

Finally, the approximate backwater curve satisfies the equation:

$$\tilde{Y}(x) = \begin{cases} Y_1 & \text{for } x \in [0, x_1] \\ Y_1 + (x - x_1)S_b & \text{for } x \in [x_1, L] \end{cases},$$

with

$$Y_1 = \begin{cases} Y_n & \text{if } x_1 \neq 0 \\ Y_L - LS_L & \text{if } x_1 = 0 \end{cases}. \tag{5.16}$$

Here, $Y_1 = Y_n$ when part of the pool is in uniform flow, and $Y_1 = Y_L - LS_L$ when the whole pool is affected by backwater, i.e., the water level is approximated by a straight line of slope S_L.

Example 5.2 (Backwater approximation for example canals 1 and 2). The approximate backwater curves corresponding to a discharge of $Q_{\max}/2$ and a downstream boundary condition equal to $Y_n(Q_{\max})$ are depicted in Fig. 5.7 for both canals.

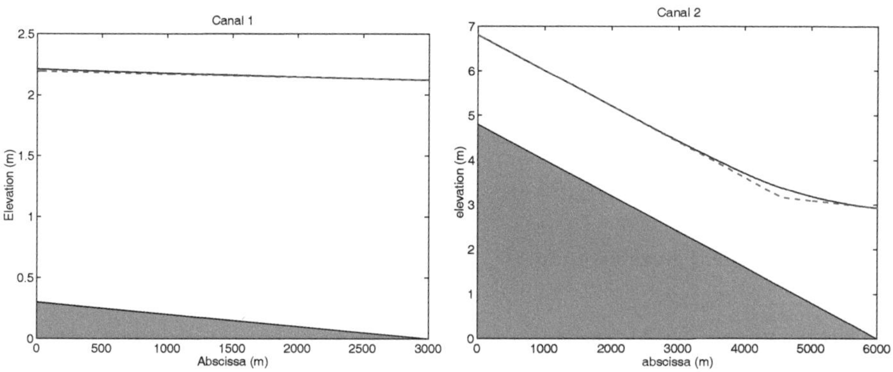

Fig. 5.7 Backwater curve (—) and its approximation (– –) for $Q_{\max}/2$, canals 1 and 2

In this case, canal 1 is completely under backwater, while the backwater part only affects the downstream end of canal 2.

5.3.2.2 IDZ Model Parameters

The global IDZ model is obtained following the same line as in the uniform flow case. We make use of the interconnection rules given by (3.86):

$$p_{11}(s) = p_{11}^{(1)}(s) + \frac{p_{12}^{(1)}(s)p_{21}^{(1)}(s)}{p_{11}^{(2)}(s) - p_{22}^{(1)}(s)}, \tag{5.17a}$$

$$p_{12}(s) = -\frac{p_{12}^{(1)}(s)p_{12}^{(2)}(s)}{p_{11}^{(2)}(s) - p_{22}^{(1)}(s)}, \tag{5.17b}$$

$$p_{21}(s) = \frac{p_{21}^{(1)}(s)p_{21}^{(2)}(s)}{p_{11}^{(2)}(s) - p_{22}^{(1)}(s)}, \tag{5.17c}$$

$$p_{22}(s) = p_{22}^{(2)}(s) - \frac{p_{12}^{(2)}(s)p_{21}^{(2)}(s)}{p_{11}^{(2)}(s) - p_{22}^{(1)}(s)}, \tag{5.17d}$$

where $p_{ij}^{(1)}(s)$ denotes the linearized Saint-Venant transfer functions of the upstream part of the pool between 0 and x_1, and $p_{ij}^{(2)}(s)$ denote the transfer functions of the downstream part of the pool between x_1 and L.

Using a similar method as in the uniform flow case, we obtain the IDZ model for the open channel in nonuniform flow as follows [9]:

$$p_{11}(s) = \frac{1}{A_u s} + \tilde{b}_u, \tag{5.18a}$$

$$p_{12}(s) = -\Big(\frac{1}{A_u s} + b_u\Big)e^{-\tau_u s}, \tag{5.18b}$$

$$p_{21}(s) = \Big(\frac{1}{A_d s} + b_d\Big)e^{-\tau_d s}, \tag{5.18c}$$

$$p_{22}(s) = -\Big(\frac{1}{A_d s} + \tilde{b}_d\Big), \tag{5.18d}$$

with the equivalent areas given by

$$A_u = A_u^{(1)}\left(1 + \frac{A_u^{(2)}}{A_d^{(1)}}\right), \tag{5.19a}$$

$$A_d = A_d^{(2)}\left(1 + \frac{A_d^{(1)}}{A_u^{(2)}}\right), \tag{5.19b}$$

where, for the upstream uniform flow part

$$A_u^{(1)} = \frac{\alpha\beta T_0}{\gamma}\left(e^{\frac{\gamma x}{\alpha\beta}} - 1\right),$$

$$A_d^{(1)} = \frac{\alpha\beta T_0}{\gamma}\left(1 - e^{-\frac{\gamma x}{\alpha\beta}}\right),$$

and the variables are evaluated at 0 and $x = x_1$. For the downstream backwater part, $A_u^{(2)}$ and $A_d^{(2)}$ are given by the same equations but with the variables evaluated at $x_2 = (x_1 + L)/2$ and $x = L - x_1$.

The high frequency gains are given by:

$$\tilde{b}_u = \tilde{b}_u^{(1)} + \frac{b_u^{(1)} b_d^{(1)}}{\tilde{b}_u^{(2)} + \tilde{b}_d^{(1)}}, \tag{5.20a}$$

$$b_u = \frac{b_u^{(1)} b_u^{(2)}}{\tilde{b}_u^{(2)} + \tilde{b}_d^{(1)}}, \tag{5.20b}$$

$$b_d = \frac{b_d^{(1)} b_d^{(2)}}{\tilde{b}_u^{(2)} + \tilde{b}_d^{(1)}}, \tag{5.20c}$$

$$\tilde{b}_d = \tilde{b}_d^{(2)} + \frac{b_u^{(2)} b_d^{(2)}}{\tilde{b}_u^{(2)} + \tilde{b}_d^{(1)}}, \tag{5.20d}$$

where $\tilde{b}_u^{(1)}$, $b_u^{(1)}$, $b_d^{(1)}$, and $\tilde{b}_d^{(1)}$ are given by (5.12), with the variables evaluated at 0 and L is replaced by x_1. The coefficients $\tilde{b}_u^{(2)}$, $b_u^{(2)}$, $b_d^{(2)}$, and $\tilde{b}_d^{(2)}$ are given by the same equations , with the variables evaluated at x_2 and L is replaced by $L-x_1$.

Finally, the equivalent delays are given by:

$$\tau_u = \tau_u^{(1)} + \tau_u^{(2)} - \frac{A_d^{(1)} A_u^{(2)}}{A_d^{(1)} + A_u^{(2)}} (\tilde{b}_d^{(1)} + \tilde{b}_u^{(2)}), \tag{5.21a}$$

$$\tau_d = \tau_d^{(1)} + \tau_d^{(2)} - \frac{A_d^{(1)} A_u^{(2)}}{A_d^{(1)} + A_u^{(2)}} (\tilde{b}_d^{(1)} + \tilde{b}_u^{(2)}), \tag{5.21b}$$

where the additional term accounts for the phase due to the interconnection.

5.3.3 *Summary*

The overall model is an IDZ model, whose parameters can be computed analytically as functions of the channel physical parameters. This simple model is useful for hypothetical channels that are not yet built. It enables us to derive automatic controller design methods based on these computations.

5.4 Model Evaluation

5.4.1 *Comparison of Frequency Responses*

The Bode plots of canals 1 and 2 are obtained in nonuniform flow conditions for $Q_0 = Q_{\max}/2$ using the complete numerical method presented in Chap. 3 (see Fig. 5.8).

The approximate model reproduces very well the global frequency behavior of the system, especially in the case of canal 1. Canal 2 has a pole that is not taken into account by the approximate model.

In order to show the accuracy of the method, it is important to see how the parameters vary with the hydraulic conditions, i.e., the discharge Q_0.

5.4.2 *Comparison of Parameters*

Let us consider a constant downstream limit condition Y_L corresponding to the uniform flow for $Q_{\max}$ and vary the discharge Q_0. The canal is therefore in uniform flow for $Q_0 = Q_{\max}$, and the backwater curve increases as the discharge diminishes.

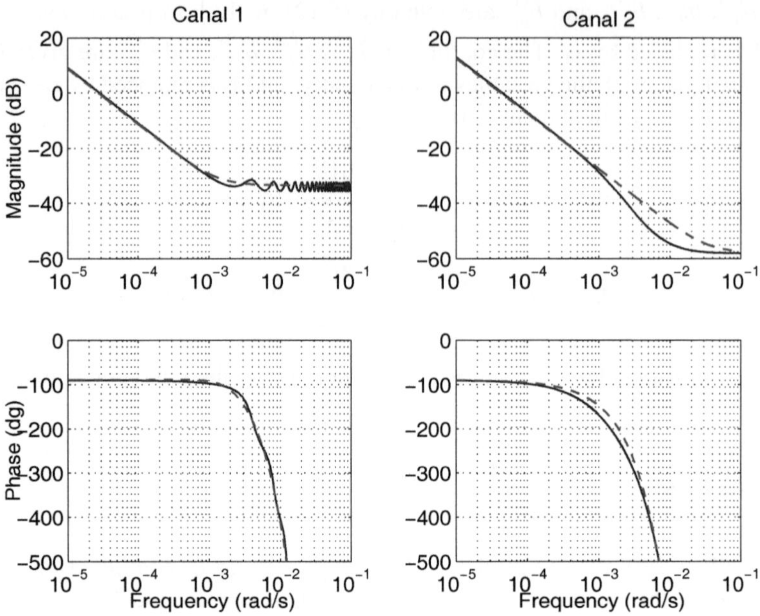

Fig. 5.8 Bode plot of transfer $p_{21}(s)$ (—) and its IDZ approximation (– –), canals 1 and 2 for $Q_0 = Q_{\max}/2$.

The approximate parameters are compared with their exact values when they are available.

Figures 5.9 and 5.10 respectively depict the exact and approximate values of A_d, τ_d, and $\limsup_{\omega\to\infty}|p_{21}(j\omega)|$ and $\liminf_{\omega\to\infty}|p_{21}(j\omega)|$ for both canals.

It is clear from the figures that the approximation is very accurate for the delay τ_d and for coefficient A_d. The gain at high frequencies b_d is rather well-estimated for canal 2 and is between min and max values for canal 1. In fact, the max value of the gain at high frequencies b_d increases a lot as the discharge tends towards zero, since the damping of resonant modes diminishes. This is not taken into account by the approximate model, which is sufficient for control purposes, since the resonant modes will not be controlled actively (see Chap. 6 for details).

5.4.3 Comparison of Time Domain Simulations

The approximate model is compared to a finite dimensional model obtained by rational approximation of the exact model (see Chap. 4). Simulations are done for canals 1 and 2, for $Q_0 = Q_{\max}/2$ (see Fig. 5.11). The models are connected with a downstream gate (equivalent to a local feedback) with a gain $k_v = 6.84$.

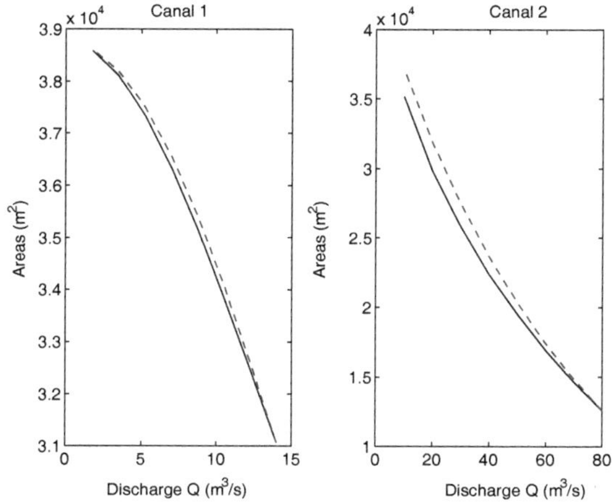

Fig. 5.9 Variation of coefficient A_d with the discharge, for a downstream limit condition Y_L corresponding to the uniform flow for $Q_{\max}$, canals 1 and 2. Comparison between the exact (—) and approximate (– –) values

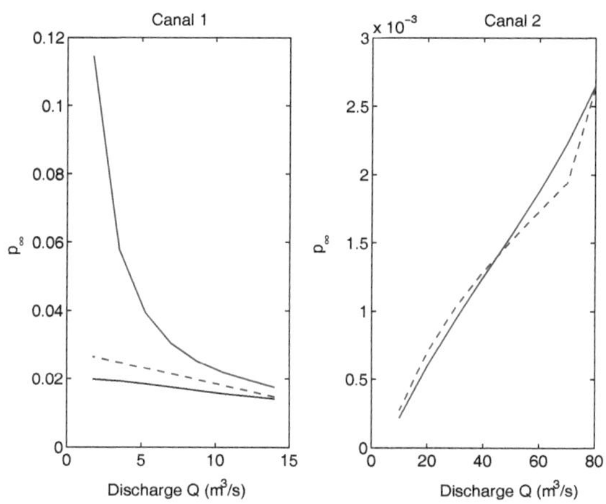

Fig. 5.10 Variation of gain b_d with the discharge, for a downstream limit condition Y_L corresponding to the uniform flow for $Q_{\max}$, canals 1 and 2. Comparison between min and max values (—) and the approximation (– –).

The simulation is very accurate for both canals: in fact, as can be seen from the Bode plot, the whole frequency response is very well fitted by the approximate model, except the resonant modes for canal 1 and a supplementary pole for canal 2. These discrepancies occur at rather high frequencies and at a small gain, which is why they are not clearly visible in the step response.

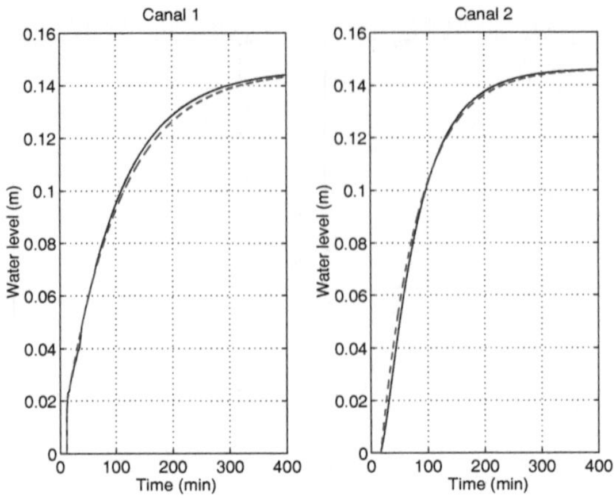

Fig. 5.11 Step responses of the rational approximation of the complete model (—) and the IDZ model (– –), canals 1 and 2.

5.5 Summary

The chapter proposed a new analytical approximate model for an open channel pool subject to backwater. This IDZ model is simple, and very accurate. It is obtained from a frequency domain approximation of the linearized Saint-Venant transfer matrix. It models the complete transfer matrix (i.e., the downstream influence is also taken into account). The inputs are the upstream and downstream discharges, and its outputs are the upstream and downstream water levels.

The model is inspired from the one originally proposed by the authors in [9, 11], with a modification concerning the approximation of the time-delay. In the present model, the group delay approximation is used instead of the mathematical delay.

The model has been validated in frequency and time domains using the complete transfer matrix obtained in Chap. 3 (see also [8, 10]) on two different example canals.

The approximate model can be used for controller design, such as distant downstream PI controllers, or even for advanced controller design (e.g., multivariable controllers).

References

[1] Ankum P (1995) Flow control in irrigation and drainage. Communications of the Water Management Department, Delft University of Technology, Faculty of Civil Engineering, 294 p

[2] Balogun O (1985) Design of real-time feedback control for canal systems using linear quadratic regulator theory. PhD thesis, Department of Mechanical Engineering, University of California at Davis, 230 pp.
[3] Chow V (1988) Open-channel hydraulics. McGraw-Hill, New York, 680 p
[4] Corriga G, Patta F, Sanna S, Usai G (1979) A mathematical model for open-channel networks. Appl Math Mod 3:51–54
[5] Ermolin Y (1992) Study of open-channel dynamics as controlled process. J Hydraul Eng 118(1):59–71
[6] Freudenberg J, Looze D (1987) A sensitivity tradeoff for plants with time delay. IEEE Trans Autom Contr 32(2):99–104
[7] Hancu S, Dan P (1992) Wave-motion stability in canals with automatic controllers. J Hydraul Eng 118(12):1621–1638
[8] Litrico X, Fromion V (2002) Infinite dimensional modelling of open-channel hydraulic systems for control purposes. In: Proc. 41st IEEE Conf. decision and control, Las Vegas, pp 1681–1686
[9] Litrico X, Fromion V (2004) Analytical approximation of open-channel flow for controller design. Appl Math Mod 28(7):677–695
[10] Litrico X, Fromion V (2004) Frequency modeling of open channel flow. J Hydraul Eng 130(8):806–815
[11] Litrico X, Fromion V (2004) Simplified modelling of irrigation canals for controller design. J Irrig Drain Eng 130(5):373–383
[12] Malaterre PO (1998) Pilote: linear quadratic optimal controller for irrigation canals. J Irrig Drain Eng 124(4):187–194
[13] Munier S, Litrico X, Belaud G (2007) Linear approximation of open-channel flow routing with backwater effect. In: 32nd IAHR Congress, Venice
[14] Munier S, Litrico X, Belaud G, Malaterre PO (2008) Distributed approximation of open-channel flow routing accounting for backwater effects. Adv Water Resourc 31:1590–1602
[15] Schuurmans J, Clemmens A, Dijkstra S, Hof A, Brouwer R (1999) Modeling of irrigation and drainage canals for controller design. J Irrig Drain Eng 125(6):338–344
[16] Shand M (1971) Automatic downstream control systems for irrigation canals. PhD thesis, University of California, Berkeley
[17] Strelkoff T, Deltour JL, Burt C, Clemmens A, Baume JP (1998) Influence of canal geometry and dynamics on controllability. J Irrig Drain Eng 124(1):16–22
[18] Zhou K, Doyle J (1998) Essentials of robust control. Prentice Hall, Upper Saddle River, NJ

Part II
Control of a Canal Pool

In this part, we focus on the control of a single canal pool.

In Chap. 6, we study the behavior of an open channel controlled with hydraulic structures, such as gates, weirs, or hydromechanical automatic gates.

Chapter 7 then turns towards the classical feedback control policies of a canal pool: distant downstream and local upstream control. We focus on the design of PI controllers, because of their simplicity and their widespread use in the industry. We develop robust tuning rules for distant downstream, local upstream PI controllers. We also develop more advanced controllers obtained using H_∞ optimization.

Chapter 8 introduces the mixed local upstream/distant downstream control policy, with PI and H_∞ controllers. This mixed control policy enables us to mix the two control policies: local upstream control and distant downstream control in a way that ensures good performance and at the same time efficient water management.

In Chap. 9, we derive open-loop feedforward controllers for a canal pool. This is important when we know in advance the perturbations affecting the channel, such as water withdrawals for irrigation, or discharge perturbations generated at one boundary.

Chapter 6
Control of a Canal Pool with Hydraulic Structures

Abstract The equations of open channel flow do not apply at sections where the flow varies very rapidly. This is the case at cross-sections equipped with hydraulic structures, such as gates or weirs. These structures are generally used to control the water level, or to deliver a discharge. In this chapter, we mainly deal with static hydraulic structures and study the behavior of the interaction between the flow and the structure, which is viewed as a local boundary controller imposing a feedback between the flow and the water level.

6.1 Introduction

Open channels are interconnected with hydraulic structures, such as gates or weirs, which are generally used to control the water level, or to deliver a discharge. In fact, we will consider here hydraulic structures as controllers imposing a local feedback between the discharge and the water level at one boundary of the pool. This feedback can be static if the structure is not operated, or dynamic, if it is operated, either with an electric motor, or with a mechanical system. The latter refers to hydromechanical gates such as AVIS/AVIO, AMIL, Mixte gates designed by NEYRPIC in the 1950s or Begemann/Vlugter gates designed by Dutch engineers in the 1940s (see [5, 9, 6, 12, 3]). In this chapter, we mainly deal with static hydraulic structures.

A static hydraulic structure can be efficiently modeled by a static nonlinear relation between the discharge and the water level. This relation can be obtained based on simple assumptions on the local flow properties using the Bernoulli (or Torricelli) theorem.

To study the linearized system, the hydraulic structure is modeled by a proportional gain between the discharge and the water level.

As we have seen in the first part of the book, a specific feature of open channel flow is that, in some cases, it possesses a high frequency behavior, leading to high frequency oscillations. These high frequencies can usually not be controlled actively by classical actuators, i.e., hydraulic structures operated with electric motors. However, as we will show below, the static part of the actuator, i.e., the local feedback induced by the gate or the weir, can itself control the oscillating modes.

More specifically, we show that in the uniform flow case, a dynamic boundary controller can suppress the oscillating modes of a canal pool, and that a static controller can efficiently dampen the modes. In the nonuniform flow case, it is no longer possible to cancel the oscillating modes using only a boundary controller. We show that some residual modes may exist due to reflections inside the domain, where the boundary controller has no effect.

6.2 Modeling of Hydraulic Structures

There are many different ways to model hydraulic structures, which depend on the geometry of the structure and of the flow characteristics. The reader interested in the details of hydraulic structures modeling is referred to [11], which is a good reference on this subject. We are interested in structures since they are used to control open channels. Most structures are modeled by a static nonlinear equation linking the discharge and the water level.

6.2.1 Discharge Equations of Hydraulic Structures

6.2.1.1 Free Flow

A hydraulic structure is said to be in free flow condition if the flow is critical or supercritical at the structure. In that case, the downstream level has no influence on the flow, and the discharge only depends on the upstream water level.

Weir

A weir is an overflow structure, which is usually described by a static nonlinear relation between flow and water elevation above the crest of the weir. This nonlinear relation is based on Bernoulli's theorem.

The free flow weir discharge equation is given by:

$$Q = C_d \sqrt{2g}(Y_1 - W_s)^{3/2}, \tag{6.1}$$

with W_s the weir sill elevation and Y_1 the upstream water level, and where C_d is a discharge coefficient, generally close to 0.4.

Undershot Gate

An undershot gate is an orifice (see Fig. 6.1), described by the following equation:

$$Q = C_d L_g W \sqrt{2gY_1}, \tag{6.2}$$

with W the gate opening, Y_1 the upstream water level, and where C_d is a discharge coefficient, generally close to 0.6.

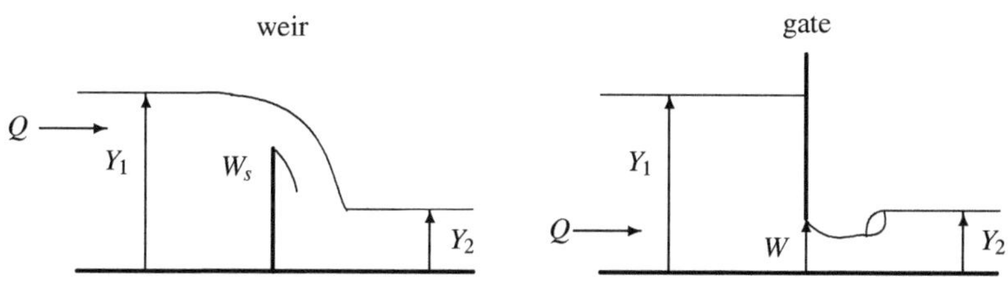

Fig. 6.1 Free flow structure: weir and gate

General Case

More generally, a free-flow structure is modeled by a static equation:

$$Q = f(Y_1, W),$$

where f is a nonlinear function of its arguments.

6.2.1.2 Submerged Flow

A hydraulic structure is said to be in submerged flow condition if the flow remains subcritical at the structure. In that case, the flow can be influenced by the downstream level (see Fig. 6.2).

Weir

A submerged weir can be described by the following equation:

$$Q = C_d \sqrt{2g}(Y_1 - Y_2)^{3/2}, \tag{6.3}$$

with Y_1 the upstream water level, and Y_2 the downstream water level.

Undershot Gate

A submerged undershot gate is usually described as follows:

$$Q = C_d L_g W \sqrt{2g(Y_1 - Y_2)}, \tag{6.4}$$

with W the gate opening, Y_1 the upstream water level, and Y_2 the downstream water level.

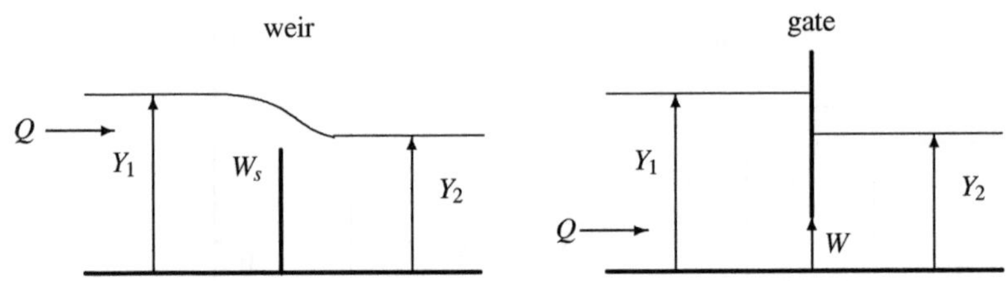

Fig. 6.2 Submerged flow structure: weir and gate

General Case

More generally, a submerged structure is modeled by a static equation:

$$Q = f(Y_1, Y_2, W),$$

where f is a nonlinear function of its arguments.

6.2.2 Equilibrium Regimes

The equilibrium regimes of a canal with a hydraulic structure at its downstream end are solutions of the backwater curve equations:

$$\frac{dQ_0(x)}{dx} = 0,$$
$$\frac{dY_0(x)}{dx} = \frac{S_b - S_{f0}(x)}{1 - F_0(x)^2},$$

with a downstream boundary verifying the structure equation:

$$Q_0(L) = f(Y_0(L), Y_2, W),$$

with Y_2 the structure downstream water elevation.

6.2.3 Linearized Equations

In the following, we consider small variations of water levels and discharge around a steady flow. This justifies considering linearized equations for the structures.

A hydraulic structure is therefore modeled as a static proportional feedback relation between the upstream and downstream water levels and the discharge. In the case of a gate, we have to take into account the possible variations in the gate opening.

Finally, the linearized structure is represented by the equation:

$$q(t) = k_u y_1(t) - k_d y_2(t) + k_w w(t),$$

with $q(t)$ the deviation in discharge from the steady flow value Q, y_1 the deviation in upstream water level from Y_1, and y_2 the deviation in downstream water level from Y_2. The coefficients k_u, k_d, and k_w are obtained as the derivatives of the function f introduced earlier with respect to Y_1, Y_2, and W, respectively:

$$\begin{aligned} k_u &= \frac{\partial f}{\partial Y_1}(Y_1, Y_2, W) \\ k_d &= \frac{\partial f}{\partial Y_2}(Y_1, Y_2, W) \\ k_w &= \frac{\partial f}{\partial W}(Y_1, Y_2, W) \end{aligned}$$

In the free-flow case, the coefficient k_d is zero. In the submerged flow case, it is strictly positive. In both cases, coefficients k_u and k_w are strictly positive.

6.3 Stability of a Canal Pool with Static Boundary Control

We now consider a canal pool interconnected with hydraulic structures at its upstream and downstream ends. As we have seen above, this can be studied as a static boundary controller for the hyperbolic system. We assume in the following that the upstream gate is submerged and that the upstream gate level is constant. The downstream gate is assumed to be free-flow. The gate openings are assumed to stay constant. Therefore, the open channel is connected with two proportional controllers at each boundary: $q(0,t) = k_0 y(0,t)$ at the upstream end and $q(L,t) = k_L y(L,t)$ at the downstream end, with $k_0 = -k_d$ for the upstream gate and with $k_L = k_u$ for the downstream gate.

The open channel is therefore connected with a static diagonal boundary controller defined by:

$$K = \begin{pmatrix} k_0 & 0 \\ 0 & k_L \end{pmatrix}, \tag{6.5}$$

where k_0, k_L are constant scalars.

Proof. For $|s| \gg \frac{2[(\alpha-\beta)\gamma+2\alpha\beta\delta]}{(\alpha+\beta)^2}$, the eigenvalues can be approximated by:

$$\lambda_1(s) = -r_1 - \frac{s}{\alpha} + O(1/s), \tag{6.13a}$$

$$\lambda_2(s) = r_2 + \frac{s}{\beta} + O(1/s). \tag{6.13b}$$

Then, using a continuity argument, one may show that if inequality (6.12) is not verified, there exists R such that the closed-loop poles with modulus larger than R are unstable. Therefore condition (6.12) is a necessary condition for stability. □

Now, using this property, we can restrict the domain where condition (6.9) needs to be tested. This is stated in the following corollary.

Corollary 6.1. *If condition (6.12) is verified, then there exists $R_0 > 0$ such that condition (i) of Theorem 6.1 need only be tested on a finite range $|s| < R_0$.*

Proof. Let us first note that:

$$\det(I - KP(s)) = \frac{f_1(s) - f_2(s)}{1 - e^{(\lambda_1(s)-\lambda_2(s))L}},$$

with

$$f_1(s) = \left(1 + k_0\frac{\lambda_2(s)}{s}\right)\left(1 + k_L\frac{\lambda_1(s)}{s}\right)e^{(\lambda_1(s)-\lambda_2(s))L},$$

$$f_2(s) = \left(1 + k_0\frac{\lambda_1(s)}{s}\right)\left(1 + k_L\frac{\lambda_2(s)}{s}\right).$$

Using the asymptotic approximations (6.13a–6.13b), we know that for any $\varepsilon > 0$ there exists R_0 such that for any s such that $|s| > R_0$ and $\Re(s) > 0$, we have:

$$\left|\left|\frac{f_1(s)}{f_2(s)}\right| - \left|\frac{(\beta+k_0)(\alpha-k_L)}{(\alpha-k_0)(\beta+k_L)}\right| e^{-(r_1+r_2)L-\tau\Re(s)}\right| \le \varepsilon,$$

with $\tau = L\left(\frac{1}{\alpha} + \frac{1}{\beta}\right)$.

If inequality (6.12) is verified, there exists $\varepsilon > 0$ such that:

$$\left|\frac{(\beta+k_0)(\alpha-k_L)}{(\alpha-k_0)(\beta+k_L)}\right| e^{-(r_1+r_2)L} \le 1 - 2\varepsilon,$$

and then for $|s| > R_0$ we have:

$$\left|1 - \frac{f_1(s)}{f_2(s)}\right| \ge \left|1 - \left|\frac{f_1(s)}{f_2(s)}\right|\right| \ge \varepsilon.$$

We then conclude that there exists R_0 such that $|\det(I - KP(s))| > 0$ when condition (6.12) is fulfilled. □

Therefore, one may use the classical Nyquist graphical criteria to test condition (i) of Theorem 6.1 on a finite range of frequencies.

To summarize, we have obtained a necessary and sufficient condition of stability that can be tested using classical methods such as the Nyquist plot for finite dimensional systems, and two algebraic conditions that can easily be tested numerically.

6.3.3 Nonuniform Flow Case

In the nonuniform flow case, one can extend the results by providing a sufficient condition for stability, using the high frequency asymptotic result on the poles of (3.91) together with the Nyquist criteria of stability. This point is not detailed here for space reasons.

6.3.4 Local Upstream Boundary Control

Let us now study the case of a pool interconnected with a hydraulic structure located at its downstream end. This corresponds to the so-called "local upstream control" policy, where the hydraulic structure is used to control its upstream water level. As an example, a duckbill weir can be used to impose a high gain feedback on the water level at the downstream end of a canal pool.

In the following, we study the effect of such control on the poles of the system.

6.3.4.1 Closed-loop Transfer Matrix

Connecting the open-loop distributed transfer matrix of (3.51) with a downstream boundary controller $q(L,s) = k_u(s)y(L,s)$ leads to the closed-loop distributed transfer matrix:

$$\begin{pmatrix} y(x,s) \\ q(x,s) \end{pmatrix} = \begin{pmatrix} G^{(1)}_{k_u}(x,s) \\ G^{(2)}_{k_u}(x,s) \end{pmatrix} q(0,s), \tag{6.14}$$

with $G^{(1)}_{k_u}(x,s)$ and $G^{(2)}_{k_u}(x,s)$ given by:

$$\begin{aligned} G^{(1)}_{k_u}(x,s) &= \frac{\lambda_2 e^{\lambda_2 x+\lambda_1 L} - \lambda_1 e^{\lambda_1 x+\lambda_2 L} + k_u \frac{\lambda_1\lambda_2}{T_0 s}(e^{\lambda_2 x+\lambda_1 L} - e^{\lambda_1 x+\lambda_2 L})}{T_0 s(e^{\lambda_2 L} - e^{\lambda_1 L}) + k_u(\lambda_2 e^{\lambda_2 L} - \lambda_1 e^{\lambda_1 L})}, \\ G^{(2)}_{k_u}(x,s) &= \frac{T_0 s(e^{\lambda_1 x+\lambda_2 L} - e^{\lambda_2 x+\lambda_1 L}) + k_u(\lambda_2 e^{\lambda_1 x+\lambda_2 L} - \lambda_1 e^{\lambda_2 x+\lambda_1 L})}{T_0 s(e^{\lambda_2 L} - e^{\lambda_1 L}) + k_u(\lambda_2 e^{\lambda_2 L} - \lambda_1 e^{\lambda_1 L})}. \end{aligned}$$

The oscillating poles of the closed-loop system (6.14) are solutions of the following equation:

$$e^{(\lambda_2(s)-\lambda_1(s))L} = \frac{T_0 s + k_u(s)\lambda_1(s)}{T_0 s + k_u(s)\lambda_2(s)}. \tag{6.15}$$

This equation cannot be solved analytically in the general case, but we can use it to determine the optimal controller for the oscillating modes.

6.3.4.2 Root Locus

According to (6.14), the closed-loop system poles for a static boundary proportional controller of gain $k_u \in \mathbb{R}_+$ are given by the solutions of the following equation:

$$\psi(s) := e^{(\lambda_2(s)-\lambda_1(s))L} - \frac{T_0 s + k_u\lambda_1(s)}{T_0 s + k_u\lambda_2(s)} = 0. \tag{6.16}$$

This equation has no closed-form solution in general. Numerical resolution for different values of k_u leads to the root locus depicted in Fig. 6.4. For $k_u = +\infty$, the poles coincide with the open-loop zeros of the Saint-Venant transfer matrix. We observe that the closed-loop poles negative real parts reach a minimum for the optimal static controller value and that the modes' damping increases with the frequency (i.e., higher frequency modes are more damped than low frequency modes). The following proposition provides a closed-form result explaining this behavior for high frequency poles:

Proposition 6.3. *When $|s| \gg 1$, the solutions of (6.16) tend asymptotically towards*

$$\tilde{p}_k = -\frac{(r_1+r_2)L}{\tau_1+\tau_2} - \frac{1}{\tau_1+\tau_2}\log\left(\frac{T_0 L + k_u\tau_2}{T_0 L - k_u\tau_1}\right) \pm \frac{2jk\pi}{\tau_1+\tau_2}, \tag{6.17}$$

and the approximation error is at the first-order given by:

$$p_k \approx \tilde{p}_k - \frac{\psi(\tilde{p}_k)}{\psi'(\tilde{p}_k)}. \tag{6.18}$$

Proof. The first equality is a direct consequence of the open mapping theorem (see [13]).

The approximation error can be evaluated by writing the Taylor expansion of ψ around $p_k = \tilde{p}_k + p'_k + o(p'_k)$:

$$\psi(p_k) = \psi(\tilde{p}_k) + \psi'(\tilde{p}_k)p'_k + o(p'_k) \tag{6.19}$$

Since $\psi(p_k) = 0$, and $\psi'(\tilde{p}_k) \neq 0$, (6.19) leads to the first-order approximation (6.18)
□

Equation (6.17) recovers the open-loop poles approximation given by (3.55) when $k_u = 0$. When k_u increases, the poles real part diminishes towards $-\infty$ for

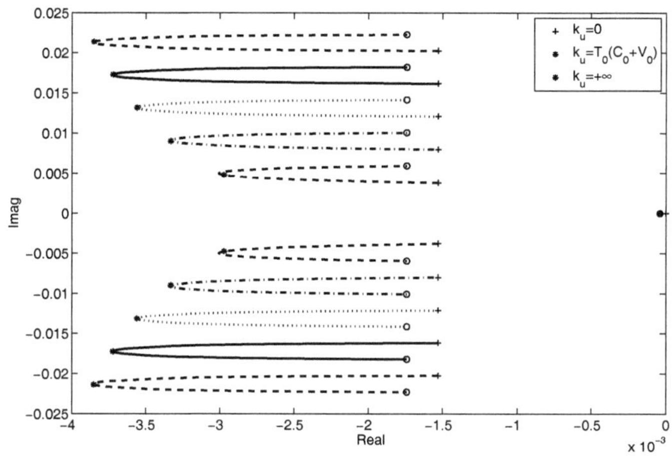

Fig. 6.4 Roots locus for example canal 1: poles (+) obtained for $k_u = 0$, zeros (∘) obtained for $k_u = +\infty$, and closed-loop poles location (∗) for $k_u = T_0(C_0 + V_0)$

$k_u < T_0(C_0+V_0)$. Then, it increases when $k_u > T_0(C_0+V_0)$, to finally tend towards $-\frac{(r_1+r_2)L}{\tau_1+\tau_2} - \frac{1}{\tau_1+\tau_2}\log\left(\frac{\tau_2}{\tau_1}\right)$ when $k_u \to \infty$, which corresponds to the high frequency approximation of the open-loop zeros of the Saint-Venant transfer function. Zeros have a real part smaller than that of the open-loop poles (because $\tau_1 < \tau_2$), and their imaginary part is given by $\pm(2k+1)\pi/(\tau_1+\tau_2)$, because the complex logarithm verifies $\log(-1) = \pm j\pi$.

6.4 Boundary Control of Oscillating Modes

As we have already discussed in Sect. 3.3.1.3, oscillating modes can be observed in small canal pools. These modes are due to the interaction of upstream and downstream propagating waves that occur in subcritical flow, when the wave celerity is larger than the water velocity. Their amplitude can be very large for some hydraulic conditions. These oscillating modes can lead to overtopping, which is highly undesirable for irrigation canals.

The objective of this section is to investigate linearized Saint-Venant equations modes and their control. We show that it is possible to suppress these oscillating modes *over the entire canal pool* by a well-designed boundary dynamic controller using only the water level measurement at the downstream end of the pool. However, this controller is infinite dimensional, and also not strictly proper. This property makes it difficult to implement on a real canal, since the actuators usually have a low bandwidth. We show that this difficulty can be bypassed in a real canal, which is usually controlled using hydraulic cross-structures such as gates or weirs. Such a hydraulic structure has an interesting feature: it structurally induces a local feedback between the discharge and the water level, whatever the frequency. This leads us to investigate the effect of a proportional boundary controller on the oscillating modes.

6.4.1 Horizontal Frictionless Case

Let us first consider the case of a rectangular horizontal frictionless canal, for which the characteristics are given by (see Sect. 2.4.2):

$$\chi_1(x,t) = q(x,t) + \beta T_0 y(x,t),$$
$$\chi_2(x,t) = q(x,t) - \alpha T_0 y(x,t).$$

The variables χ_1 and χ_2 are called the Riemann invariants of (2.28) and are easily shown to be are constant along the characteristics curves defined by:

$$\frac{dx_1}{dt} = \alpha$$
$$\frac{dx_2}{dt} = -\beta.$$

A way to eliminate the oscillating modes is to suppress the reflection of downstream propagating waves on the downstream boundary, i.e., to ensure that a perturbation reaching the boundary does not generate an upstream propagating perturbation. This can be done by specifying a boundary controller such that the Riemann invariant at the boundary $\chi_2(L,t)$ remains equal to zero for any $t > 0$:

$$\chi_2(L,t) = 0. \tag{6.20}$$

Then, the characteristic variable $\chi_2(x,t)$ remains constant for all $x \in [0,L]$ and all $t \geq \tau_1 + \tau_2$. In this case, the channel behaves as if it were semi-infinite, since all waves arriving at the downstream boundary "cross" it without reflection. Therefore, no oscillating modes can occur in the canal pool. The corresponding downstream boundary controller is given by

$$q(L,t) = \alpha T_0 y(L,t). \tag{6.21}$$

Therefore, a proportional boundary controller of gain $k_u = \alpha T_0 = T_0(C_0 + V_0)$ linking the discharge to the water elevation eliminates the oscillating modes in the special case of a rectangular horizontal frictionless canal pool. In this case, the system does not possess any pole.

6.4.2 Uniform Flow Case

In the uniform flow case, it is also possible to cancel oscillating modes over all the canal pool by using a dynamic boundary controller.

Theorem 6.2. *With a downstream boundary control $q(L,s) = k_u^*(s)y(L,s)$ defined by*

$$k_u^*(s) = -\frac{T_0 s}{\lambda_1(s)}, \tag{6.22}$$

the canal pool represented by the closed-loop distributed transfer matrix (6.14) has no oscillating modes.

Proof. Connecting the open-loop distributed transfer matrix (3.51) with a downstream boundary controller $q(L,s) = k_u(s)y(L,s)$ leads to the closed-loop distributed transfer matrix (6.14). The oscillating poles of the closed-loop system (6.14) are solutions of (6.15). With the controller $k_u^*(s)$, we get:

$$e^{(\lambda_2(s)-\lambda_1(s))L} = 0,$$

which has no finite solution and thus the system has no oscillating modes. □

Remark 6.1. This result is similar to the classical concept of "impedance matching" for electrical networks [2]. Indeed, with the controller $k_u^*(s)$ given by (6.22), the distributed transfer functions are given by $G_{k_u^*}^{(1)}(x,s) = -\frac{\lambda_1(s)}{T_0 s}e^{\lambda_1(s)x}$ and $G_{k_u^*}^{(2)}(x,s) = e^{\lambda_1(s)x}$, and thus only the downstream propagating waves remain.

The optimal dynamic controller (6.22) can be interpreted as a nonreflexive downstream boundary condition. With this controller, the canal behaves as if it were semi-infinite, i.e., the waves propagating downstream do not reflect on the downstream boundary and the oscillating modes then disappear.

In the horizontal frictionless case, we recover the result obtained with the characteristic form. Indeed, in this case, the first eigenvalue is equal to $\lambda_1(s) = -s/\alpha$, and the optimal controller $k_u^*(s)$ given by (6.22) becomes a static controller:

$$k_u^*(s) = T_0(C_0 + V_0), \tag{6.23}$$

which is exactly the one obtained in (6.21) with a Riemann invariant approach.

Moreover, this gain corresponds to the high frequency asymptotic value of the optimal controller in the uniform flow case given by

$$\lim_{\omega \to \infty} |k_u^*(j\omega)| = T_0(C_0 + V_0). \tag{6.24}$$

The Bode diagram of $k_u^*(s)$ is depicted in Fig. 6.5 for the example canal 1. This stable, infinite-dimensional controller strongly looks like a lead-lag filter. It is not strictly proper, since it has a constant gain in high frequencies.

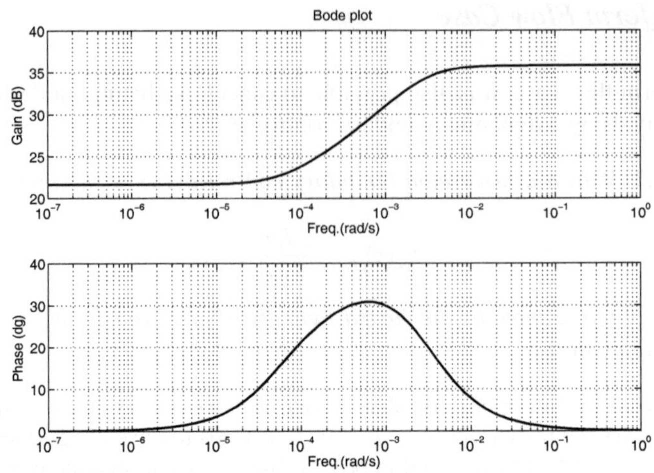

Fig. 6.5 Bode plot of $k_u^*(s)$ for canal 1

Analyzing the Bode diagram of Fig. 6.5 at a frequency corresponding to the first oscillating mode ($\omega_r = 4.1 \times 10^{-3}$ rad/s) shows that the amplitude of the optimal controller has almost reached its asymptotic value given by (6.24). Using a root-locus technique, we have seen that this value leads to the best damping of oscillating modes.

Example 6.1. Let us examine the pool behavior in the three extreme situations by plotting the spatial Bode diagram of the controlled canal pool. Figures 6.6, 6.7, and 6.8 represent the Bode diagram of the controlled canal along the longitudinal abscissa x, with different gains: $k_u = 0$ in Fig. 6.6, $k_u = +\infty$ in Fig. 6.7, and $k_u^* = T_0(V_0 + C_0)$ in Fig. 6.8. It is clear that the constant controller k_u^* dramatically dampens the oscillation modes over all the canal pool. In fact, a constant controller leads to a performance very close to that obtained with a dynamic controller $k_u(s)$. This result has been tested experimentally on a small-scale open channel, and experimental results are presented in Chap. 11.

6.4.3 Nonuniform Flow Case

We now extend the results for the control of oscillating modes to the case of nonuniform flow. As a first analysis, we consider the case of the interconnection of two horizontal frictionless pools. This will provide a first approximation of a nonuniform flow channel and enable us to analyze the physical behavior of the channel.

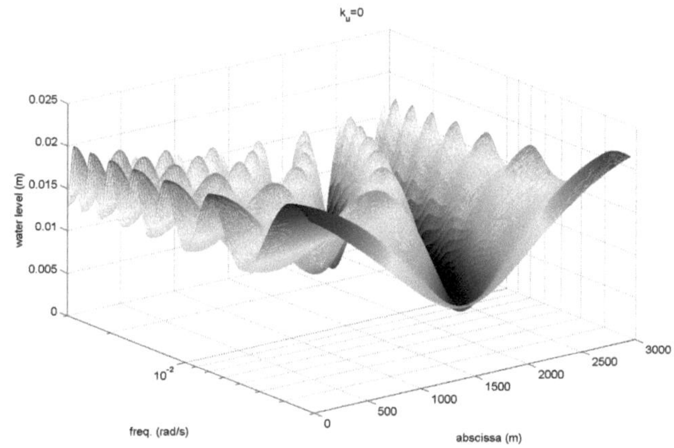

Fig. 6.6 Spatial Bode diagram of canal 1 in open-loop ($k_u = 0$)

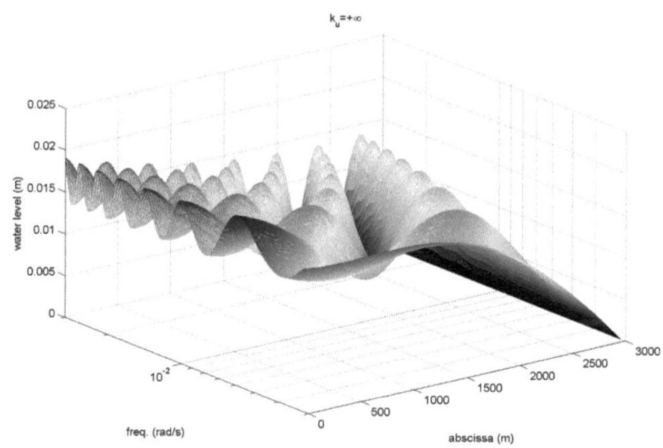

Fig. 6.7 Spatial Bode diagram of canal 1 with a boundary controller $k_u = +\infty$

6.4.3.1 Interconnection of Two Horizontal Frictionless Pools

Let us consider two interconnected horizontal frictionless pools. Let us denote with a superscript (1) the upstream part and with a superscript (2) the downstream one.

Each part of the pool is represented by its characteristic variable: $\chi^{(1)}(x,t)$ for the first part, i.e., when $x \in [0,x_1)$, and $\chi^{(2)}(x,t)$ for the second part, i.e., when $x \in (x_1,L]$. The system in the characteristic variables is diagonal in each part of the pool. At the interconnection, we have $q^{(1)}(x_1) = q^{(2)}(x_1)$ and $y^{(1)}(x_1) = y^{(2)}(x_1)$, which, in terms of the characteristic variables, leads to:

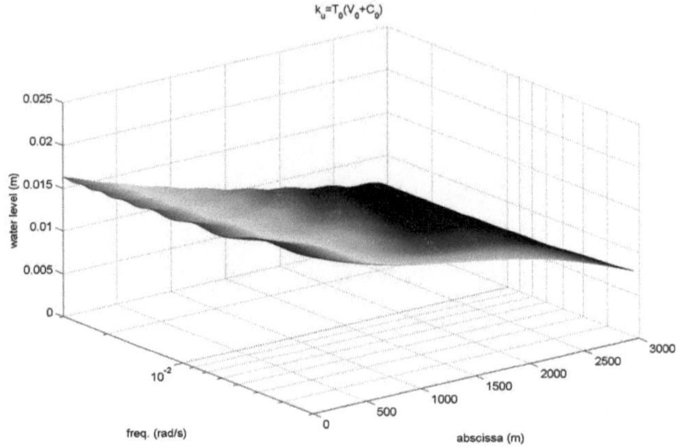

Fig. 6.8 Spatial Bode diagram of canal 1 with a boundary controller $k_u^* = T_0(C_0 + V_0)$

$$\begin{pmatrix} \beta^{(1)}T_0^{(1)} & 1 \\ -\alpha^{(1)}T_0^{(1)} & 1 \end{pmatrix}^{-1} \chi^{(1)}(x_1,t) = \begin{pmatrix} \beta^{(2)}T_0^{(2)} & 1 \\ -\alpha^{(2)}T_0^{(2)} & 1 \end{pmatrix}^{-1} \chi^{(2)}(x_1,t). \qquad (6.25)$$

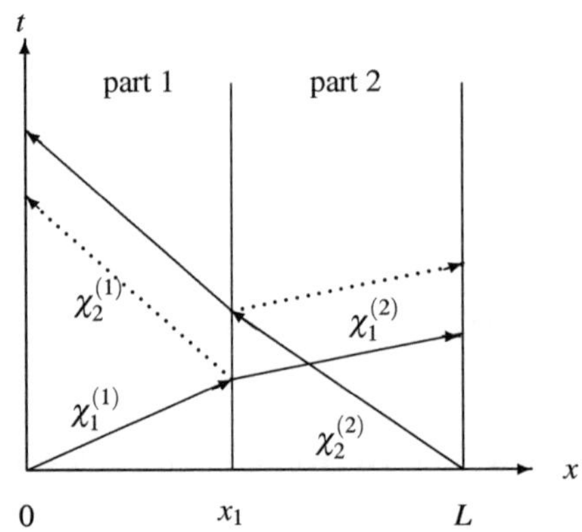

Fig. 6.9 Characteristic curves for the interconnection of two horizontal frictionless rectangular channels. The reflected characteristics are shown with *dotted lines*

Figure 6.9 depicts the characteristics curves at the interconnection of the pools. There is a part of the characteristics that is reflected at the internal boundary located in x_1. The reflection coefficients can be evaluated using (6.25). Indeed, using this equation, and after rearranging, we have:

$$\chi_1^{(2)}(x_1,t) = k_{t1}\chi_1^{(1)}(x_1,t) + k_{r1}\chi_2^{(1)}(x_1,t),$$
$$\chi_2^{(1)}(x_1,t) = k_{t2}\chi_2^{(2)}(x_1,t) + k_{r2}\chi_1^{(2)}(x_1,t),$$

with $k_{t1} = \frac{T_0^{(1)}\alpha^{(1)}+T_0^{(2)}\beta^{(2)}}{T_0^{(1)}(\alpha^{(1)}+\beta^{(1)})}$, $k_{r1} = \frac{T_0^{(1)}\beta^{(1)}-T_0^{(2)}\beta^{(2)}}{T_0^{(1)}(\alpha^{(1)}+\beta^{(1)})}$, $k_{t2} = \frac{T_0^{(1)}\alpha^{(1)}+T_0^{(2)}\beta^{(2)}}{T_0^{(2)}(\alpha^{(2)}+\beta^{(2)})}$, and $k_{r2} = \frac{T_0^{(2)}\alpha^{(2)}-T_0^{(1)}\alpha^{(1)}}{T_0^{(2)}(\alpha^{(2)}+\beta^{(2)})}$. The coefficients k_{ti} and k_{ri} are, respectively, the transmission and reflection coefficients for characteristic i. We observe that since $T_0^{(1)}\alpha^{(1)} \neq T_0^{(2)}\alpha^{(2)}$ and $T_0^{(1)}\beta^{(1)} \neq T_0^{(2)}\beta^{(2)}$, the reflection coefficients are different from zero, and therefore oscillating modes can be generated due to this reflection inside the domain.

Finally, due to the superposition principle, we can write the outgoing characteristics $\chi_1^{(2)}$ and $\chi_2^{(1)}$ as a function of the incoming ones $\chi_1^{(1)}$ and $\chi_2^{(2)}$ as follows:

$$\chi_1^{(2)}(x_1,t) = \frac{k_{t1}}{1-k_{r1}k_{r2}}\chi_1^{(1)}(x_1,t) + \frac{k_{r1}k_{t2}}{1-k_{r1}k_{r2}}\chi_2^{(2)}(x_1,t),$$
$$\chi_2^{(1)}(x_1,t) = \frac{k_{r2}k_{t1}}{1-k_{r1}k_{r2}}\chi_1^{(1)}(x_1,t) + \frac{k_{t2}}{1-k_{r1}k_{r2}}\chi_2^{(2)}(x_1,t).$$

We can check that the determinant of this matrix relation is equal to $k_{t1}k_{t2}$, i.e., the product of the transmission coefficients. Therefore, the relation between the incoming and the outgoing characteristics at the junction can be viewed as a kind of rotation with attenuation.

Let us now study the case where the two pools are considered in uniform flow.

6.4.3.2 Interconnection of Two Uniform Pools

We use the high frequency approximations obtained in the uniform flow case and the interconnection equations (3.86) to compute the transfer matrices for the interconnection. The denominator is given by:

$$D(s) = D^{(1)}(s)N_{11}^{(2)}(s) - D^{(2)}(s)N_{22}^{(1)}(s),$$

with $N_{kl}^{(i)}(s)$ and $D^{(i)}(s)$ the numerator and denominator of transfer functions $p_{kl}(s)$ for pool i.

Now, we use the high frequency approximation of the eigenvalues $\lambda_1(s)$ and $\lambda_2(s)$, leading to:

$$D^{(i)}(s) \approx T_0^{(i)}(1 - e^{-r^{(i)}x^{(i)}-\tau^{(i)}s}),$$
$$N_{11}^{(2)}(s) \approx \alpha^{(2)} + \beta^{(2)}e^{-r^{(2)}x^{(2)}-\tau^{(2)}s},$$
$$N_{22}^{(1)}(s) \approx -(\beta^{(1)} + \alpha^{(1)}e^{-r^{(1)}x^{(1)}-\tau^{(1)}s}),$$

where $r^{(i)} = r_1^{(i)} + r_2^{(i)}$ and $\tau^{(i)} = \tau_1^{(i)} + \tau_2^{(i)}$.

Using the expressions of the transfer function in the uniform flow case, the denominator of the interconnection is given by:

$$D(s) = a_0 - a_1 e^{-\tau^{(1)}s} - a_2 e^{-\tau^{(2)}s} - a_3 e^{-(\tau^{(1)}+\tau^{(2)})s},$$

with $a_0 = \alpha^{(2)}T_0^{(1)} + \beta^{(1)}T_0^{(2)}$, $a_1 = (\alpha^{(2)}T_0^{(1)} - \alpha^{(1)}T_0^{(2)})e^{-r^{(1)}x^{(1)}}$, $a_2 = (\beta^{(1)}T_0^{(2)} - \beta^{(2)}T_0^{(1)})e^{-r^{(2)}x^{(2)}}$, and $a_3 = (\alpha^{(1)}T_0^{(2)} + \beta^{(2)}T_0^{(1)})e^{-r^{(1)}x^{(1)}-r^{(2)}x^{(2)}}$.

We see that the oscillating modes of the interconnection are linked to:

- the oscillating modes of the first part, linked to the term $a_1 e^{-\tau^{(1)}s}$,
- the oscillating modes of the second part, linked to the term $a_2 e^{-\tau^{(2)}s}$, and
- the oscillating modes of the whole canal, linked to the term $a_3 e^{-(\tau^{(1)}+\tau^{(2)})s}$.

Therefore, the rate of additional oscillating modes induced by the separation depends on the value of a_1 and a_2. If these values are negligible with respect to a_0 and a_3, we obtain the following high frequency approximation of oscillating modes:

$$\begin{aligned}
\hat{p}_k &\approx -\frac{1}{\tau^{(1)}+\tau^{(2)}}\log\left(\frac{a_0}{a_3}\right) \pm \frac{2jk\pi}{\tau^{(1)}+\tau^{(2)}} \\
&\approx -\frac{r^{(1)}x^{(1)}+r^{(2)}x^{(2)}}{\tau^{(1)}+\tau^{(2)}} - \frac{1}{\tau^{(1)}+\tau^{(2)}}\log\left(\frac{\alpha^{(2)}T_0^{(1)}+\beta^{(1)}T_0^{(2)}}{\alpha^{(1)}T_0^{(2)}+\beta^{(2)}T_0^{(1)}}\right) \\
&\quad \pm \frac{2jk\pi}{\tau^{(1)}+\tau^{(2)}}.
\end{aligned}$$

These poles are close to the high frequency approximation of the poles of the whole pool.

Let us now study the poles behavior when the downstream part is much larger than the upstream part. This corresponds to a canal pool with a small uniform part upstream and a large backwater part downstream.

In this case, when $T_0^{(2)} \gg T_0^{(1)}$, the denominator $D(s)$ tends to:

$$D(s) \approx -D^{(2)}(s)N_{22}^{(1)}(s).$$

The poles of the system are then linked to the zeros of $p_{22}^{(1)}(s)$ and to the open-loop poles of the downstream part. This is coherent with the physical explanation: the downstream part, with $T_0^{(2)} \gg T_0^{(1)}$ imposes a high gain feedback that constrains the water level. A root locus would show that the closed-loop poles tend toward the zeros of $N_{22}^{(1)}(s)$, as we have seen earlier in the chapter. For the downstream part, there is no constraint, therefore one recovers the open-loop poles of the downstream part.

6.4.3.3 General Case of Nonuniform Flow

As we have seen above, in the nonuniform case, reflections occur inside the domain. Moreover, a boundary controller is only able to act on oscillating modes that are created by reflections on the boundary. Therefore, contrarily to the uniform case, it is not possible to eliminate all the oscillating modes in the nonuniform case. One can only attenuate the modes linked to the reflection on the boundaries.

The amount of oscillating modes that are created inside the domain is usually negligible as compared to those due to reflections on the boundaries. Therefore, a good strategy in nonuniform flow is to use a boundary controller that adapts to the local flow behavior. This local impedance matching ensures that the waves that reach that point are not reflected by the boundary condition. According to this remark, the optimal gain for a static proportional boundary controller is equal to:

$$k_u^* = T_0(L)(C_0(L) + V_0(L)),$$

i.e., it is equal to the proportional gain obtained with the equivalent uniform flow corresponding to the local hydraulic conditions.

6.5 Implementation Issues

Every controller is nowadays implemented digitally. That means that the output of the system is sampled at a given period T_s; this sampling signal is provided to a discrete-time controller, which is followed by a zero-order hold (ZOH) (see Fig. 6.10) in order to provide an input at each time to the continuous system.

In the sequel, we recall some aspects related to the control of continuous system by digital controllers. We mainly discuss the impact of the discretization on a continuous system and focus on problems related to the so-called aliasing effect. We conclude by recalling the necessity of a high frequency attenuation on the continuous system, through its strict properness, or by the addition of a suitable anti-aliasing filter. We then conclude by pointing out the specific status of the open channel control.

6.5.1 Discretization of Continuous Systems: The Aliasing Issue

In order to understand the possible problems induced by the discretization of a continuous-time system, we first recall the so-called aliasing effect. To illustrate this effect, let us consider the sampling of the sinusoidal signal $x(t) = \sin(2\pi t)$ at the period $T_s = 0.9524$, which leads to the sampling sequence $x_s(k) = x(kT_s)$ for all $k \in \mathbb{N}$.

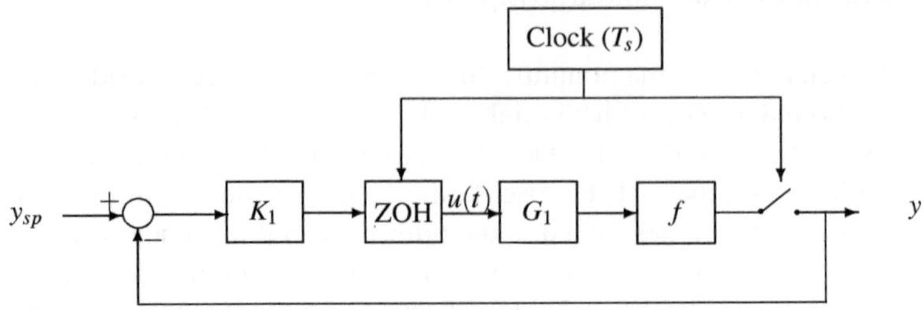

Fig. 6.10 Sampled-time control

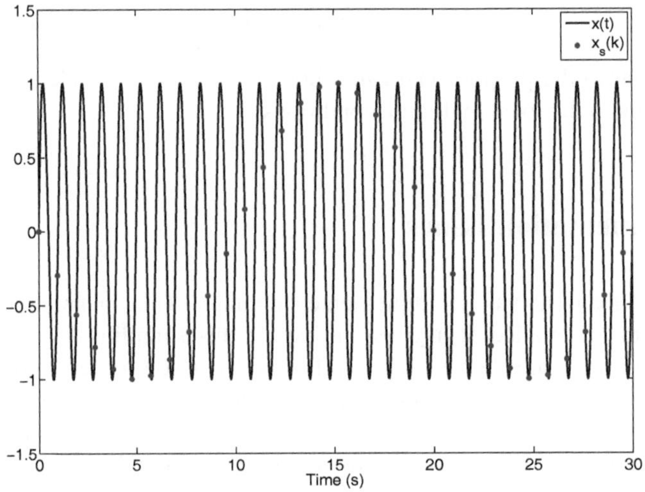

Fig. 6.11 Continuous signal $x(t) = \sin(2\pi t)$ and its sampling at period $T_s = 0.9524$

Figure 6.11 illustrates the problem of sampling a signal with a frequency that is larger than the so-called Nyquist–Shannon frequency given by $\omega_{\text{Nyquist}} \stackrel{\Delta}{=} \pi/T_s$ (in rad/s). Indeed, Fig. 6.11 allows us to consider that the sampling sequence associated to $x(t)$ may be also interpreted as the sampling sequence associated to another sinusoidal signal with a larger time period of $T = 20s$ (0.1π rad/s). This observation shows that the knowledge of the sampling sequence is not sufficient to discriminate whether the initial continuous signal corresponds to a sinusoidal signal at frequency 2π rad/s or 0.1π rad/s. In order to have an equivalence between the continuous signal and its associated sampling sequence, it is necessary to assume that the spectrum of the continuous signal satisfies the Nyquist–Shannon theorem condition that its spectrum is null outside the set of frequencies $[-\omega_{\text{Nyquist}}, \omega_{\text{Nyquist}}]$. In practice, in order for the previous condition to be satisfied, it is necessary to use a suitable low pass filter, called an anti-aliasing filter, which attenuates the spectrum of the signal for all frequencies larger than the Nyquist–Shannon frequency.

Let us now focus on the possible problems attached to the discretization of a continuous system. We have seen in the preliminary discussion that the aliasing effect is an important issue in this context. Another important issue is linked to the presence of a zero-order hold at the input of the system. As an illustration, let us consider the following continuous system:

$$G_m(s) = \frac{s^2 + 2s + 100}{s^2 + 0.01s + 100}$$

and its associated discretization at $T_s = 0.5$ s

$$G_{md}(z) = \frac{z^2 - 0.7563z + 1.185}{z^2 - 0.5659z + 0.995}.$$

Figure 6.12 compares the output of the continuous system associated to $G_m(s)$ when the input is, respectively, a sinusoidal signal with $\omega = 2.56$ rad/s and the same sinusoidal signal after the application of the zero-order holder. The zero-order holder generates at the sampling period a strong perturbation of the input, which itself strongly perturbs the output of the system. In this specific example, the strong perturbation induced by the zero order hold is at the frequency of the undamped mode of $G_m(s)$ and this explains why its effects on the system outputs is so strong.

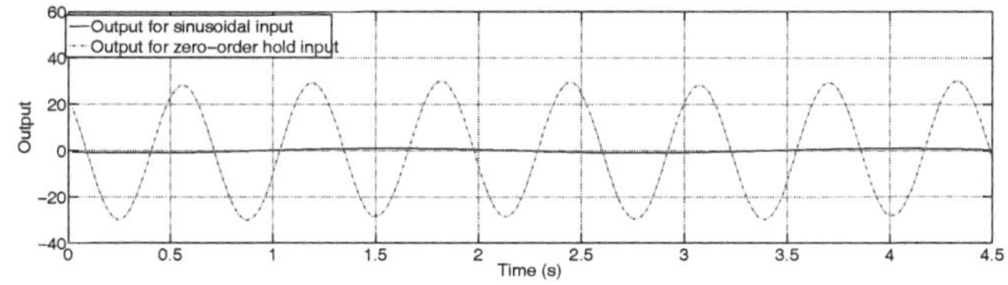

Fig. 6.12 Effect on the system output of the zero-order hold of a sinusoidal input

This example shows that the combination of the zero-order hold effect with the aliasing effect leads to the apparition of a "fictitious bending mode" (from a physical point of view) at the frequency $\omega_{md} = 2.56$ rad/s on the Bode plot of the discretization (see Fig. 6.13). Actually, ω_{md} and $\omega_m = 10$ rad/s are symmetric frequencies with respect to the Nyquist–Shannon frequency associated to $T = 0.5$ s. Indeed, we have $10 - \omega_{\text{Nyquist}} = \omega_{\text{Nyquist}} + 2.56$ where $\omega_{\text{Nyquist}} = \pi/0.5$ rad/s.

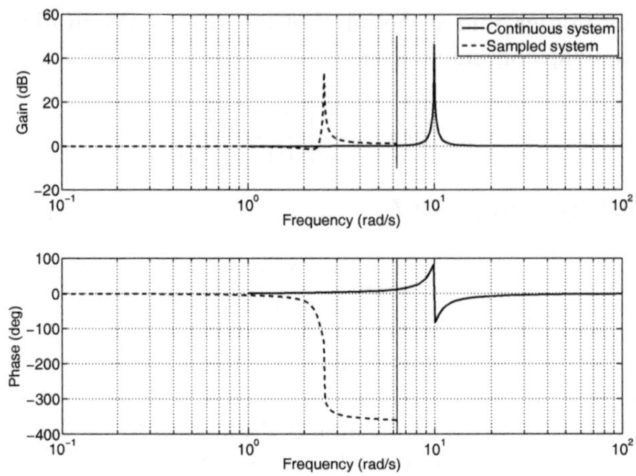

Fig. 6.13 Bode plots of $G_m(s)$ and its discretization, $G_{md}(z)$ for a sampling period $T_s = 0.5$ s

6.5.2 The Butterworth Filter as an Anti-aliasing Filter

In practice, an anti-aliasing filter $F(s)$ has to be added to the system. $F(s)$ is either a low-pass analogous filter that is designed to cut frequencies higher than the Nyquist frequency $\omega_N = \pi/T_s$, or a mechanical filter that averages the water level (e.g., a hydraulic filter). This filter is put before sampling (see Fig. 6.10). Another important role of the anti-aliasing filter is to ensure that noise will not influence the frequency response of the system.

We present hereafter the classical anti-aliasing filter, the so-called Butterworth filter. This filter has the advantage of a maximal flatness gain in the low frequencies and therefore it does not greatly modify the frequency response of the system in the low frequencies. The main feature of the n-order Butterworth filter is given by its module characteristic:

$$|F_B(j\omega)|^2 = \frac{1}{1+\left(\frac{\omega}{\omega_c}\right)^{2n}},$$

where n is its order and ω_c is its cut-off frequency ($|F_B(j\omega_c)| \approx -3$ dB). By definition, the poles of $F_B(s)$ are equally spaced points on the circle of radius ω_c and they are given by

$$p_k = \omega_c e^{\frac{j(2k+n-1)\pi}{2n}} \quad \text{for } k \in \{1,\ldots,n\}$$

which leads us to define the transfer function of a Butterworth filter as:

$$F_B(s) \triangleq \frac{\omega_c^n}{\Pi_{k=1}^n (s-p_k)}.$$

Let us now illustrate the role of Butterworth filter with respect to the discretization problem. For this purpose, let us consider that the system is given by

$$G(s) = \frac{0.01s+0.001}{s} + \frac{0.1s+0.01}{s^2+0.01s+100},$$

and its associated discretization (without any anti-aliasing filter) for a sampling period of $T_s = 0.5$ s:

$$G_d(z) = \frac{0.01z^3 - 0.02464z^2 + 0.03444z - 0.01908}{z^3 - 1.566z^2 + 1.561z - 0.995}.$$

Since we have an undamped mode at $\omega_m = 10$ rad/s, we then choose to use a Butterworth filter of order 8, with $w_c = 2\pi$ rad/s which offers an attenuation of -30 dB at the undamped mode frequency and which is given by:

$$F_B(s) = \frac{2419230}{(s^2+12.32s+39.44)(s^2+10.44s+39.44)(s^2+6.978s+39.44)(s^2+2.45s+39.44)}.$$

Figure 6.14 depicts the Bode plots associated to the continuous plant, i.e., $G(s)$, its discretization without anti-aliasing filter, i.e., $G_d(z)$, and the continuous plant filtered by the eighth order Butterworth filter $G(s)F_B(s)$ filter and its associated discretization. As expected, without an anti-aliasing filter, the undamped mode of the continuous system appears at a lowest frequency, while the Butterworth filter strongly attenuates its impact in the low frequencies.

We conclude this section by noting that the use of an anti-aliasing filter has an impact on the Bode plot of the initial system. Actually, there exists a trade-off between the attenuation level obtained in the high frequencies and the associated phase lag introduced by the anti-aliasing filter. This trade-off can be modulated by changing the order of the anti-aliasing filter or the cut-off frequency. It is also possible to use other kinds of anti-aliasing filters, with different features than those of the Butterworth filter. In Fig. 6.15, we illustrate the effect induced by a cut-off frequency modification where we have considered $\omega_c = 1$ rad/s, $\omega_c = \pi$ rad/s, and $\omega_c = 2\pi$ rad/s. We observe, as expected, that a better attenuation of high frequencies leads to a degradation of the phase lag.

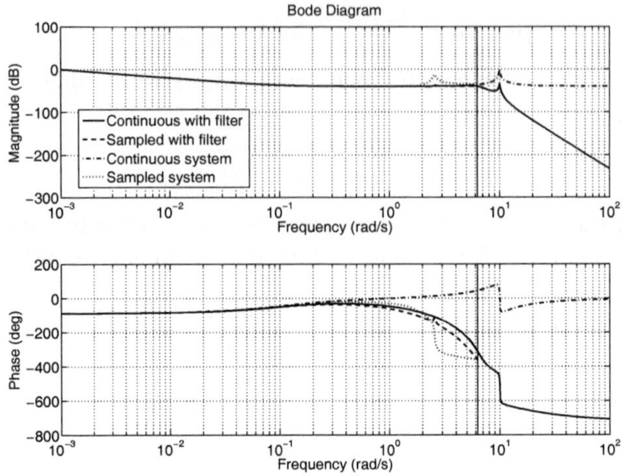

Fig. 6.14 Bode plots associated to the continuous plant, i.e., $G(s)$, its discretization without anti-aliasing filter, i.e., $G_d(z)$, and the continuous plant filtered by the eighth order Butterworth filter, i.e., $G(s)F_B(s)$ filter and its associated discretization

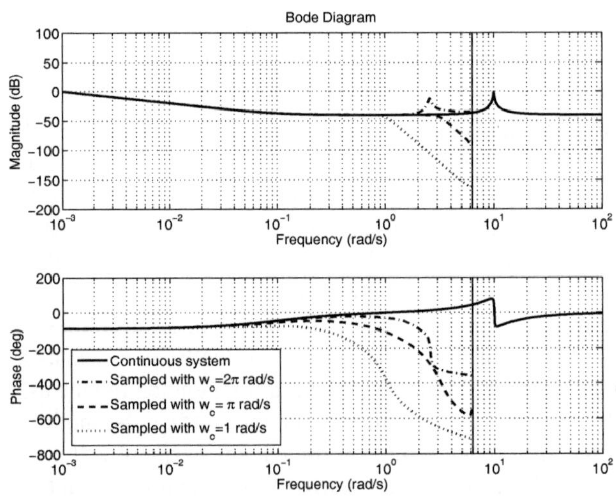

Fig. 6.15 Discretization of $G(s)$ associated to a Butterworth filter with various cut-off frequencies of 1 rad/s, π rad/s, and 2π rad/s

6.5.3 Sampling Time Selection

By definition, the frequency spectrum of a discretized system is limited (at least by the Nyquist-Shannon frequency), and thus the value of the sampling period has a major impact on the control of the system. Actually there exist two cases: (i) how to choose the sampling time when it is free and (ii) how to take account of T_s in the controller design when it is imposed. If there are no constraints on the sampling period, the classical way is to design a continuous-time controller and to sample it in order to recover the continuous-time performance. In that case, the sampling period can be selected according to the bandwidth of the controlled system, given by the crossover frequency ω_c. A classical rule is that the Nyquist frequency $\omega_N = \pi/Ts$ should be 6 to 20 times larger than the crossover frequency [1]. This rule leads to:

$$0.15 \leq T_s\omega_c \leq 0.5.$$

In the case where the sampling period is larger than that attached to the previous inequality, some care is necessary. In this context, a continuous design could be also made but this necessitates some manipulations. The main idea is then to discretize the continuous plant (with a suitable anti-aliasing filter) in order to obtain the discrete-time plant. On this basis, it is possible to return to the continuous domain by use of the so-called Tustin transformation. The Tustin transformation allows us to define an "equivalent" continuous plant, which can be used to design an "equivalent" continuous controller. The final discrete controller is then obtained after the application of the inverse Tustin transformation on the "equivalent" continuous controller. It is clearly beyond the scope of this book to explain why this approach is especially effective for the design of discrete robust controllers. Let us point out, for example, that the use of the Tustin transformation allows us to recover the input margins obtained on the continuous closed-loop system equivalent to the discrete one.

6.5.4 Impact of Bandwidth Limitations

As we have shown in the previous sections, the control of the oscillating mode of a canal pool is achieved through the use of a dynamical controller with a "high frequency control" or by a simple constant gain. Such a controller cannot be implemented digitally since as has already been mentioned, a digital controller has necessarily a limited bandwidth. The gate opening $w(t)$ is typically controlled by an electrical actuator with finite bandwidth, and thus even a continuous implementation is not possible. This explains why, in general, it is only possible to use the structural static feedback k_u that directly links the water level y to the discharge q to achieve a high frequency control policy.

The set of possible controllers can then be decomposed as the sum of two main terms:

$$q(t) = \underbrace{k_u y_1(t) - k_d y_2(t)}_{\text{high frequency}} + \underbrace{k_w w(t)}_{\text{low frequency}} ,$$

where the high frequency component is directly related to the static gain associated to the linearization of the weir or gate, and the low frequency component can be designed by a digital controller.

In most cases, the constraint imposed by the actuator dynamics (its bandwidth) or by the sampling period leads to design active controllers possessing a limited bandwidth that is incompatible to an active control of oscillating modes. If nothing else is done (assuming, for example, that the actuator of the system is not a weir but a pump), then the damping and the frequency of the oscillating modes remain unchanged since they are actually in "open-loop". In this context, any perturbation acting on their frequencies is able to strongly perturb the system output. It is easy to understand the strong interest of the use of a weir or a gate in the canal control problem. The conclusion of this simple remark is the following: even if from a control point of view, a gate or a weir is complicated to handle, since it is strongly nonlinear and adds a coupling term between canal pools, it is for practical reasons the simplest way to handle problems related to the oscillating modes. This, moreover, means that control designers have to focus their efforts on the low frequency part, which explains why in most cases, a simple model of a canal pool (delay + integrator, for example) is enough for design purposes.

6.6 Summary

We have shown that it is possible to dampen the resonant modes over the entire canal pool by using a dynamic boundary controller at the downstream end of the channel. This controller uses only downstream boundary water level measurement. In this control problem, the overall performance is not only linked to the behavior at the downstream boundary of the canal pool, since the problem is a distributed one. This is usually hidden in the classical input–output view of the problem. A remarkable fact is that a simple proportional boundary controller enables us to already dramatically dampen the oscillating modes *over the entire canal pool*. Experimental results on a small-scale channel are presented in Chap. 11.

Finally, in this chapter, we have highlighted the effect of hydraulic structures on the poles of the system. The open channel dynamics are strongly affected by the boundary conditions. Hydraulic structures modify the boundary conditions by introducing a local feedback between discharge and water level. Depending on the gain of this feedback, the hydraulic structure can dampen the oscillating modes. The static feedback performed by gates or weirs can provide control in frequency bandwidth where motorized actuators are no longer active, because of the bandwidth limitation of motorized actuators. These hydraulic structures are therefore essential tools in the control of an open channel.

References

[1] Aström K, Wittenmark B (1990) Computer controlled systems: theory and design. Prentice Hall, London, 544 p
[2] Beaty H (2001) Handbook of Electric Power Calculations. McGraw-Hill, New York
[3] Belaud G, Litrico X, De Graaff B, Baume JP (2008) Hydraulic modeling of an automatic upstream water-level control gate for sumerged flow. J Irrig Drain Eng 134(3):315–326
[4] Bounit H (2003) The stability of an irrigation canal system. Int J Appl Math Comp Sci 13(4):453–468
[5] Brouwer S (1987) Design and application of automatic check gate for tertiary turnouts. In: 13th ICID Congress, Rabat, Morroco, pp 671–683
[6] Burt C, Angold R, Lehmkuhl M, Styles S (2001) Flap gate design for automatic upstream canal water level control. J Irrig Drain Eng 127(2):84–91
[7] Curtain R, Zwart H (1995) An Introduction to Infinite-Dimensional Linear Systems Theory. Springer
[8] Desoer CA, Vidyasagar M (1975) Feedback systems: input ouput properties. Academic Press, New York
[9] de Graaff B (1998) Stability analysis of the Vlugter gate. Master's thesis, Delft University of Technology, Faculty of Civil Engineering and Geosciences
[10] de Halleux J, Prieur C, Coron JM, d'Andréa Novel B, Bastin G (2003) Boundary feedback control in networks of open-channels. Automatica 39:1365–1376
[11] Lencastre A (1996) Hydraulique générale. Eyrolles, SAFEGE, (in French)
[12] Litrico X, Belaud G, Baume JP, Ribot-Bruno J (2005) Hydraulic modeling of an automatic upstream water-level control gate. J Irrig Drain Eng 131(2):176–189
[13] Rudin W (1974) Real and complex analysis, 2nd edn. McGraw-Hill, New York
[14] Weiss G (1994) Regular linear systems with feedback. Math Control Signals Systems 7:23–57

Chapter 7
Classical Control Policies for a Canal Pool

Abstract In this chapter, we examine the classical control policies distant downstream control and local upstream control. We develop tuning methods for feedback control of a canal pool using PI controllers. We also cast the problem into the H_∞ optimization method, which naturally incorporates robustness constraints in the design. Finally, we compare the PI controller and the H_∞ controller in the distant downstream case.

7.1 Introduction

Static boundary control can stabilize open channel flow. However, this is not sufficient to efficiently deliver water according to a given schedule and reject perturbations. To do this, one needs to implement a dynamic feedback control. In this chapter, we deal with the control policies classically implemented in the management of an irrigation canal: distant downstream and local upstream control.

We begin by the most widespread and simplest feedback controller, namely the PI controller. Such a controller has only two parameters, the proportional gain, denoted k_p, and the integral time, denoted T_i. Even if it is a rather simple controller, it has the ability to react to errors (proportional action) and to reject static perturbations (integral action).

We propose tuning rules for distant downstream and local upstream PI controllers and we also introduce a more advanced control method, the H_∞ optimization.

The main interest of the H_∞ design method is that it enables us to express the design of robust and efficient controllers as the solution to an optimization problem. In the case of a simple system, the H_∞ controller is close to a PI, but it also enables us to handle more complicated cases. H_∞ can be used as a computer aided design method, which provides an alternative to PI controller design for more complex dynamical systems.

In practice, an irrigation canal is classically a series of pools, with each pool representing a portion of the canal in between two controlled hydraulic structures (gates or weirs, for example). We nevertheless assume in the following that the control action of a canal pool is the discharge at each cross-structure. For this purpose, we have to use a slave controller on each hydraulic structure that can be used to deliver a required discharge.

In view of the discussions about the bandwidth limitation mainly induced by the actuators and the digital implementation of controllers, we can deduce that the slave controller on each hydraulic structure is only able to eliminate the effects of the hydraulic structure in low frequencies (see Chap. 6). We then conclude that generally a canal pool controller is the sum of a passive controller, provided by the hydraulic structures, and an active (digital) controller acting only in low frequencies.

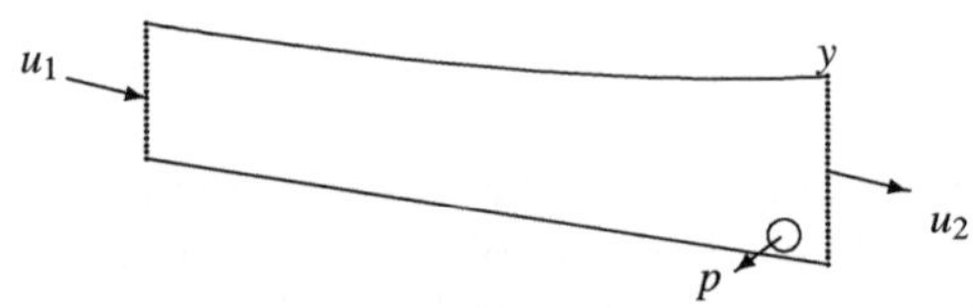

Fig. 7.1 Definition of control action variables for a canal pool

We denote by y the downstream water level, u_1 the upstream control (the upstream discharge $q(0,t)$), u_2 the downstream control (the downstream discharge $q(L,t)$), and p the downstream perturbation, corresponding to the unknown withdrawal (see Fig. 7.1).

In the following, $G_1(s)$ and $\tilde{G}_1(s)$, respectively, denote the transfer functions from u_1 to y and from u_2 to y. s is the Laplace variable. The transfer function from the perturbation p to y is equal to $\tilde{G}_1$, since p acts as an additive perturbation on the downstream discharge u_2 (the unknown perturbation is supposed to be located at the downstream end of the canal pool). The canal pool is therefore represented by:

$$y(s) = G_1(s)u_1(s) + \tilde{G}_1(s)(u_2(s) + p(s)). \tag{7.1}$$

In this chapter, we focus on a way to design the active controller acting on (low) frequencies compatible with the actuator bandwidth (or the constraint induced by a low sampling period). In this context, the main goal of the slave controller on each hydraulic structure is then to remove the effects induced by the hydraulic structures in low frequencies. This low frequency compensation of the hydraulic structures actually leads to recovering (in low frequencies) the model associated to the canal pool controlled by the discharge at each cross-structure.

This is why we assume in the following that the transfer functions $G_1(s)$ and $\tilde{G}_1(s)$ are given by an integrator delay model (see [11, 6]):

$$G_1(s) = \frac{e^{-\tau_d s}}{A_d s}, \tag{7.2a}$$

$$\tilde{G}_1(s) = -\frac{1}{A_d s}, \tag{7.2b}$$

with τ_d the propagation delay of the pool in seconds and A_d the backwater area in m^2. As we have seen in the first part of the book, this simple model is usually sufficient to capture the main dynamic properties of a canal pool in low frequencies.

7.2 Classical Policies from a Control Point of View

There are two classical decentralized control policies for a canal pool: distant downstream control and local upstream control (see Fig. 7.2). The mixed control policy allows us to obtain the advantages of both classical methods [5].

Let us recall the main points concerning performance and management of water resources for these three control policies.

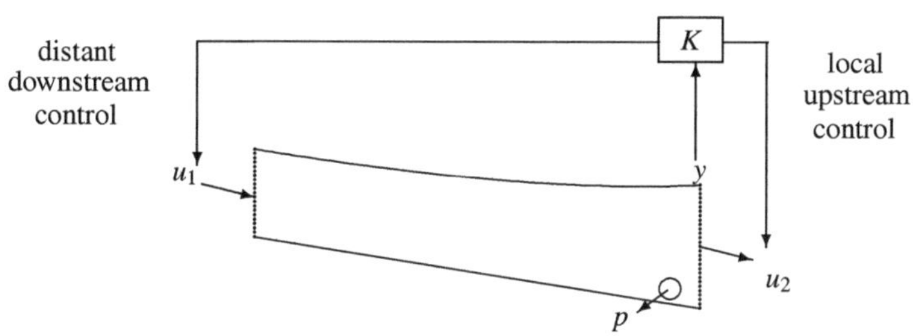

Fig. 7.2 Local upstream and distant downstream control of a canal pool

7.2.1 Distant Downstream Control

Distant downstream control of a canal pool consists in controlling the downstream water level y using the upstream control variable u_1. Let us denote by r the reference signal, e the tracking error, and K_1 the transfer function of the distant downstream controller.

The tracking error e can then be expressed by:

$$e = (1 + G_1 K_1)^{-1}(r - \tilde{G}_1 p). \tag{7.3}$$

The disturbance rejection is then directly characterized by the modulus of the transfer function $\tilde{G}_1(1+G_1K_1)^{-1}$. The control objective is to find a linear controller K_1 such that $\left|\tilde{G}_1(j\omega)(1+G_1(j\omega)K_1(j\omega))^{-1}\right| \approx 0$ over the largest frequency bandwidth.

Transfer function $G_1(s)$ has a time-delay, which imposes a limitation on the achievable bandwidth, therefore on the performance of the controlled system. Indeed, it is well-known that a time-delay limits the achievable bandwidth to about $1/\tau_d$. Unpredicted perturbations occurring at a frequency larger than $1/\tau_d$ will not be attenuated by the controller.

Therefore, distant downstream control leads to a high water efficiency, since it is "demand-driven", but has a low performance with respect to unpredicted perturbations, because of the time-delay. Indeed, when water withdrawal affects the pool, the controller will adapt the upstream discharge. This leads to a parsimonious water management, since only the necessary volume will be withdrawn from the resource, but to a low efficiency with respect to the water user, since water needs some time to travel downstream.

7.2.2 Local Upstream Control

Local upstream control of a canal pool consists in controlling the downstream water level y using the downstream control variable u_2. In the local upstream control case, the controller is denoted K_2, and the closed-loop is given by:

$$e = (1+\tilde{G}_1K_2)^{-1}(r+\tilde{G}_1p).$$

The disturbance rejection specification is related to the modulus of $\tilde{G}_1(1+\tilde{G}_1K_2)^{-1}$, which has to be small over the largest frequency bandwidth. But since there is no time-delay in $\tilde{G}_1(s)$, the achievable bandwidth is only limited by the actuators' limitations.

Therefore, local upstream control leads to a high performance with respect to unpredicted perturbations, but has a low water efficiency, since all perturbations are propagated downstream without adapting the upstream discharge. Indeed, faced with a decreasing demand, the only way to maintain the downstream water level is to let the superfluous discharge go downstream, leading to expensive water management.

7.2.3 Mixed Local Upstream/Distant Downstream Control

In the general case, one could use both control action variables u_1 and u_2 to control y_1. In that case, the controller $K(s)$ is given by:

$$K(s) = \begin{pmatrix} K_1(s) \\ K_2(s) \end{pmatrix}.$$

The open-loop transfer matrix is given by

$$G(s)K(s) = \left(G_1(s)K_1(s) + \tilde{G}_1(s)K_2(s) \right).$$

The structure of this mixed controller corresponds to the addition of two classical controllers:

$$G(s)K(s) = \underbrace{G_1(s)K_1(s)}_{\text{distant downstream}} + \underbrace{\tilde{G}_1(s)K_2(s)}_{\text{local upstream}}.$$

One may think in a naive approach that designing a local upstream controller and a distant downstream controller would be sufficient. This is not the case, and it is necessary to carefully design both controllers with an integrated approach. Indeed, we can use the multivariable structure of this controller by mixing both distant downstream and local upstream control in order to achieve a desired trade-off between water management and performance with respect to water users.

The functioning of the mixed controller can be represented as in Fig. 7.3, where the control action variables (upstream and downstream discharges) are depicted along time in response to a downstream withdrawal.

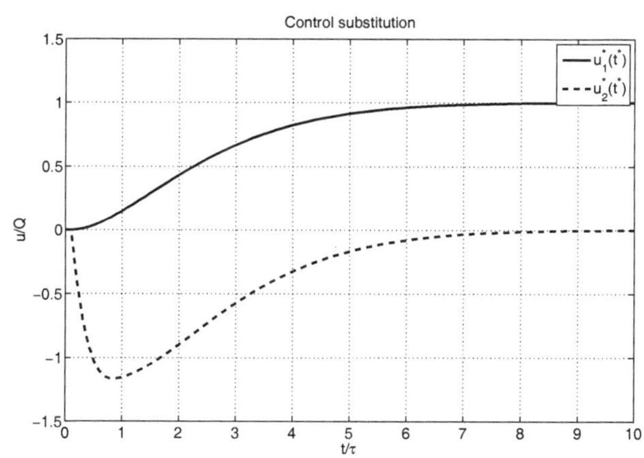

Fig. 7.3 Substitution of upstream and downstream discharges along time for a mixed controller

First, the downstream discharge is decreased to compensate for the drop in the water level due to the withdrawal. At the same time, the upstream discharge increases to compensate for this change in downstream discharge. Finally, when the upstream release reaches the downstream end, the downstream discharge goes back to its value. This can be viewed as an actuator substitution to recover the performance of the local upstream control and the water management of the distant downstream control.

Contrarily to the classical control structures, the mixed controller leads to a real multivariable problem, where robustness is more complicated to handle. The mixed controller design can be cast into the H_∞ framework in order to guarantee performance and robustness. Due to the structure of the system, one may also use simple PI controllers and evaluate robustness margins a posteriori.

The mixed controller design will be detailed in Chap. 8.

7.3 Tuning of Distant Downstream PI Controllers

7.3.1 Dimensionless Problem Formulation

7.3.1.1 Dimensionless ID Model

The system (7.1–7.2) can be normalized to dimensionless form by an adequate change of variables. Let Q_r be a reference discharge. In our case, the reference discharge is the maximum considered perturbation Q_p. Let us define $s^* = \tau_d s$ as the dimensionless Laplace variable, $y_r = \tau_d Q_r / A_d$ the reference water level deviation, $y^* = y/y_r$ the dimensionless water level deviation, $u_1^* = u_1/Q_r$ the dimensionless upstream discharge deviation, $u_2^* = u_2/Q_r$ the dimensionless downstream discharge, and $p^* = p/Q_r$ the dimensionless water withdrawal. Then, the system of equations (7.1) is given by:

$$y^*(s^*) = G_1^*(s^*)u_1^*(s^*) + \tilde{G}_1^*(s^*)(u_2^*(s^*) + p^*(s^*)), \tag{7.4}$$

with $G_1^*(s^*) = e^{-s^*}/s^*$ and $\tilde{G}_1^*(s^*) = -1/s^*$. This dimensionless system has a time-delay equal to 1 and an integrator gain also equal to 1.

7.3.1.2 Dimensionless PI Controller

Let the dimensionless controlled system be schematized as in Fig. 7.4.

The corresponding dimensionless PI controller will be denoted by:

$$K_1^*(s^*) = k_p^*\left(1 + \frac{1}{T_i^* s^*}\right),$$

with k_p^* the dimensionless proportional gain, and T_i^* the dimensionless integral time.

This corresponds to a continuous controller where the control u_1^* is obtained by the equation:

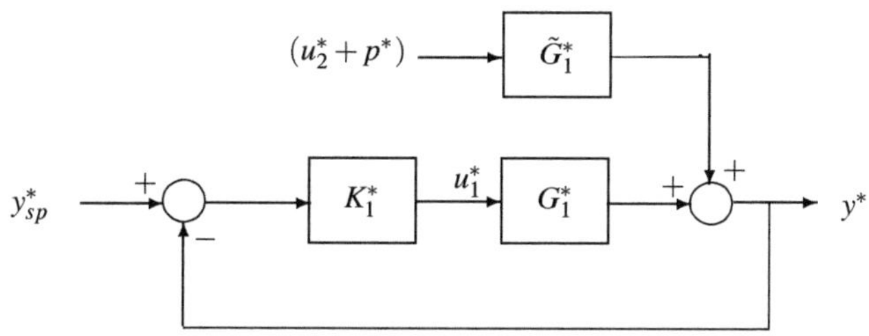

Fig. 7.4 Schematic representation of distant downstream control of a dimensionless canal pool with block diagrams

$$u_1^*(t^*) = k_p^*(y_{sp}^*(t^*) - y^*(t^*)) + \frac{k_p^*}{T_i^*}\int_0^{t^*}(y_{sp}^*(v) - y^*(v))\mathrm{d}v$$

where y_{sp}^* is the dimensionless set-point for the downstream water elevation.

Once a dimensionless PI with parameters (k_p^*, T_i^*) is designed on the dimensionless system, the dimensional PI controller parameters are obtained by:

$$k_p = k_p^* \frac{A_d}{\tau_d} \tag{7.5}$$

and

$$T_i = \tau_d T_i^*. \tag{7.6}$$

7.3.2 Nominal Distant Downstream PI Tuning

Most classical textbooks on automatic control emphasizes the importance of gain and phase margins for controller design and analysis (see [3]). These two key concepts are used by control engineers to analyze the performance and robustness of automatic controllers. The chapter proposes a distant downstream PI controller design method for irrigation canal pools based on these two quantities. First, an analytical biunivocal relation is derived between gain and phase margins and the PI controller parameters. Second, the time domain performance of the controlled canal pool is shown to be directly related to the gain and phase margins.

7.3.2.1 Definition of Robustness Margins

Let us first recall the definition of the considered robustness margins. The absolute gain margin δg is the maximum multiplicative increase in the gain of the system such that the closed-loop remains stable. The gain margin is denoted ΔG when expressed in dB, i.e., $\Delta G = 20\log_{10}(\delta g)$. The phase margin $\Delta\Phi$ expressed in degrees is the maximum additive phase to the system such that the closed-loop remains stable.

The chapter focuses on developing a method for a dimensionless canal pool using prespecified robustness margins, because the chosen margins are independent of the dimensional variables. Indeed, the open-loop K_1G_1 is dimensionless and it verifies the relation $K_1(s)G_1(s) = K_1^*(s^*)G_1^*(s^*)$. Therefore, if the dimensionless system has given gain and phase margins, so will the dimensional system.

7.3.2.2 Asymptotic Bode Diagram

For distant downstream control, the downstream water elevation is controlled using the upstream discharge. Then, the considered transfer function is $G_1^*(s^*)$, which relates the upstream discharge to the downstream water elevation. Let us denote by $K_1^*(s^*)$ the dimensionless distant downstream PI controller.

To study the frequency response of the controlled system, the open-loop transfer function $K_1^*(s^*)G_1^*(s^*)$ is evaluated on the imaginary axis $s^* = j\omega^*$, where ω^* is the dimensionless frequency and j the imaginary number such that $j^2 = -1$. This leads to:

$$|K_1^*(j\omega^*)G_1^*(j\omega^*)| = \frac{k_p^*}{\omega^*}\sqrt{1+\frac{1}{T_i^{*2}\omega^{*2}}}, \tag{7.7a}$$

$$\arg(K_1^*(j\omega^*)G_1^*(j\omega^*)) = -180 - \frac{180}{\pi}\omega^* + \frac{180}{\pi}\arctan(T_i^*\omega^*). \tag{7.7b}$$

The asymptotic Bode plot gives a good estimate of the frequency response of a given system [3]. The asymptotic Bode diagram of $K_1^*G_1^*$ is depicted in Fig. 7.5. The gain of the open-loop decreases with a slope of -40 dB per decade for frequencies lower than $1/T_i^*$ (denoted by (-2)), then with a slope of -20 dB per decade for frequencies higher than $1/T_i^*$ (denoted by (-1)). The phase of the open-loop starts at $-180°$ at low frequencies, increases due to the integral term of the controller, and then decreases due to the delay of the system.

Two frequencies are of great interest when studying the open-loop of a controlled system: the crossover frequency where the gain of the open-loop equals 1, denoted by ω_c^*, and the frequency where the phase of the open-loop equals $-180°$, denoted by ω_{180}^*. These two frequencies are critical, since the phase margin is evaluated for $\omega^* = \omega_c^*$ and the gain margin is evaluated for $\omega^* = \omega_{180}^*$.

The frequency ω_{180}^* can be approximated if we assume that $\arctan(T_i^*\omega_{180}^*) \approx \pi/2$. Such an approximation is valid if $1/T_i^* \ll \omega_{180}^*$, typically if $T_i^* \gg 1$. Then, using (7.7b), one gets:

$$\omega_{180}^* = \frac{\pi}{2}. \tag{7.8}$$

The frequency ω_c^* will depend on the chosen gain margin for the controlled system. In Fig. 7.5, increasing the gain margin will result in translating the line of

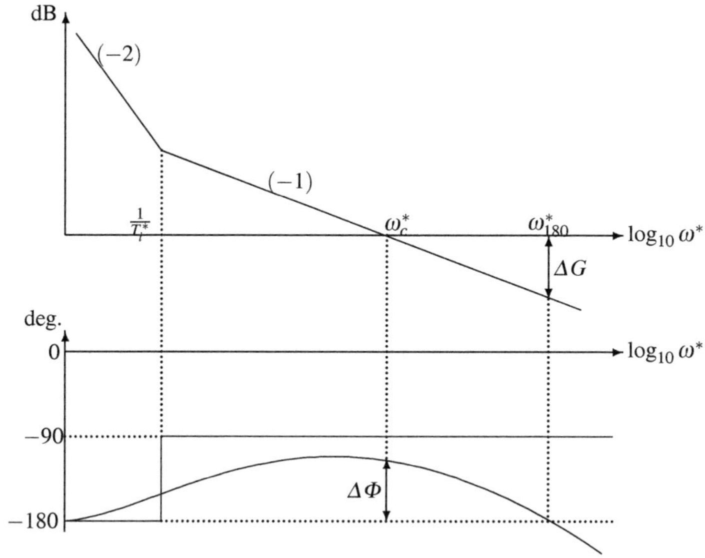

Fig. 7.5 Schematic Bode plot of $K_1^* G_1^*$ for distant downstream control

slope −20 dB/decade towards the bottom, and therefore will result in lowering the crossover frequency ω_c^*. This qualitative behavior can be precisely quantified.

7.3.2.3 PI Controller Tuning with Gain and Phase Margins

Let us specify a desired gain margin ΔG (in dB). As depicted in Fig. 7.5, between ω_c^* and ω_{180}^*, the Bode diagram is close to a line of slope −20 dB per decade. The relation between the crossover frequency ω_c^* and ω_{180}^* is then given by:

$$\omega_c^* = \omega_{180}^* 10^{-\frac{\Delta G}{20}}. \tag{7.9}$$

Therefore the crossover frequency is directly related to the gain margin of the controlled system.

Using the definition of the phase margin $\Delta\Phi$, the phase of the open-loop at the crossover frequency ω_c^* is given by:

$$\arg(K_1^*(j\omega_c^*)G_1^*(j\omega_c^*)) = -180 + \Delta\Phi.$$

Combined with (7.7b), this leads to:

$$\Delta\Phi = \frac{180}{\pi}\left(\arctan(T_i^*\omega_c^*) - \omega_c^*\right). \tag{7.10}$$

A delay in the open-loop limits the achievable phase margin. Indeed, for a simple integrator system without a delay, the maximum achievable phase margin is 90°. For an integrator with delay, the maximum phase margin is obtained when $\arctan(T_i^* \omega_c^*) \approx \pi/2$, which gives an upper bound of the achievable phase margin as a function of the desired gain margin:

$$\Delta\Phi_{\max} = 90 - \frac{180}{\pi}\omega_c^* = 90(1 - 10^{-\frac{\Delta G}{20}}). \tag{7.11}$$

Therefore, to tune the PI controller, one needs to specify a desired phase margin $\Delta\Phi$ smaller than the maximum achievable phase margin $\Delta\Phi_{\max}$. Then, T_i^* should be chosen as:

$$T_i^* = \frac{1}{\omega_c^*}\tan\left(\frac{\pi}{180}\Delta\Phi + \omega_c^*\right). \tag{7.12}$$

One can now compute the proportional gain of the controller. Since at the crossover frequency, one has $|G_1^*(j\omega_c^*)K_1^*(j\omega_c^*)| = 1$, (7.7a) gives:

$$k_p^* = \frac{T_i^* \omega_c^{*2}}{\sqrt{1 + T_i^{*2}\omega_c^{*2}}}. \tag{7.13}$$

Combining this with (7.12), one gets:

$$k_p^* = \omega_c^* \sin\left(\frac{\pi}{180}\Delta\Phi + \omega_c^*\right). \tag{7.14}$$

Then, with the approximate model (7.2a), one may compute the coefficients of a dimensionless distant downstream PI with desired gain and phase margins using (7.8–7.9) and (7.12–7.14).

Table 7.1 gives the values of dimensionless PI coefficients for different couples $(\Delta G, \Delta\Phi/\Delta\Phi_{\max})$.

Table 7.1 Dimensionless PI coefficients (k_p^*, T_i^*) for different values of ΔG and $\Delta\Phi/\Delta\Phi_{\max}$

	ΔG				
$\Delta\Phi/\Delta\Phi_{\max}$	6 dB	8 dB	10 dB	12 dB	14 dB
0.5	(0.728, 3.08)	(0.557, 3.13)	(0.427, 3.38)	(0.328, 3.80)	(0.254, 4.39)
0.6	(0.749, 3.92)	(0.581, 4.03)	(0.452, 4.39)	(0.352, 4.98)	(0.275, 5.80)
0.7	(0.766, 5.30)	(0.600, 5.49)	(0.471, 6.03)	(0.370, 6.88)	(0.291, 8.05)
0.8	(0.778, 8.04)	(0.614, 8.36)	(0.485, 9.23)	(0.384, 10.57)	(0.304, 12.42)
0.9	(0.785, 16.18)	(0.622, 16.86)	(0.494, 18.67)	(0.392, 21.45)	(0.311, 25.24)

The choice $\Delta G = 10$ dB and $\Delta\Phi = 0.7\Delta\Phi_{\max}$ is in general a good compromise between robustness and performance. In this case, the dimensionless PI coefficients are given by $k_p^* = 0.471$ and $T_i^* = 6.03$.

7.3.2.4 Delay Margin

The delay margin of a controlled system is the maximum additional delay in the loop such that the system remains stable. Indeed, a delay decreases the phase proportionally to the frequency. Therefore, since the phase margin gives the additional phase available before instability, the delay margin can be computed by the ratio of the phase margin in radians over the crossover frequency:

$$\Delta\tau^* = \frac{\pi}{180}\frac{\Delta\Phi}{\omega_c^*}.$$

Since there is a maximum available phase margin, the maximum dimensionless delay margin of the controlled system is given by:

$$\Delta\tau^*_{\max} = \frac{\pi}{180}\frac{\Delta\Phi_{\max}}{\omega_c^*}. \qquad (7.15)$$

Collecting (7.9), (7.11), and (7.15) leads to the following result: to one gain margin ΔG corresponds a dimensionless crossover frequency ω_c^*, a maximum phase margin $\Delta\Phi_{\max}$, and a maximum dimensionless delay margin $\Delta\tau^*_{\max}$ (see Table 7.2).

Table 7.2 Dimensionless crossover frequency ω_c^*, maximum phase margin $\Delta\Phi_{\max}$ and maximum dimensionless delay margin $\Delta\tau^*_{\max}$ for different gain margins ΔG

ΔG	4 dB	6 dB	8 dB	10 dB	12 dB	14 dB	16 dB
ω_c^*	0.991	0.787	0.625	0.497	0.395	0.313	0.249
$\Delta\Phi_{\max}$	33°	45°	54°	61°	67°	72°	76°
$\Delta\tau^*_{\max}$	0.58	1	1.51	2.16	2.98	4.01	5.30

The crossover frequency corresponds to the system bandwidth. This frequency is a measure of the time domain performance of the controlled canal pool, since perturbations occurring at a frequency higher than ω_c^*/τ_d rad/s will not be efficiently rejected by the controller. For example, for $\Delta G = 10$ dB, the dimensionless crossover frequency is about 0.5, which means that perturbations occurring at frequencies higher than $0.5/\tau_d$ rad/s cannot be efficiently attenuated by the controller.

The maximum phase and delay margins are a measure of the maximum achievable robustness of the controlled system for a given gain margin. For example, for $\Delta G = 10$ dB, the maximum phase margin is $\Delta\Phi_{\max} = 65°$. The maximum dimensionless delay margin is $\Delta\tau^*_{\max} = 2.16$, which means that for a phase margin ratio of 0.7, $\Delta\Phi = 0.7\Delta\Phi_{\max} = 43°$, the delay of the canal pool can be increased by 150% $(0.7 \times 2.16 = 1.5)$ without destabilizing the controlled system.

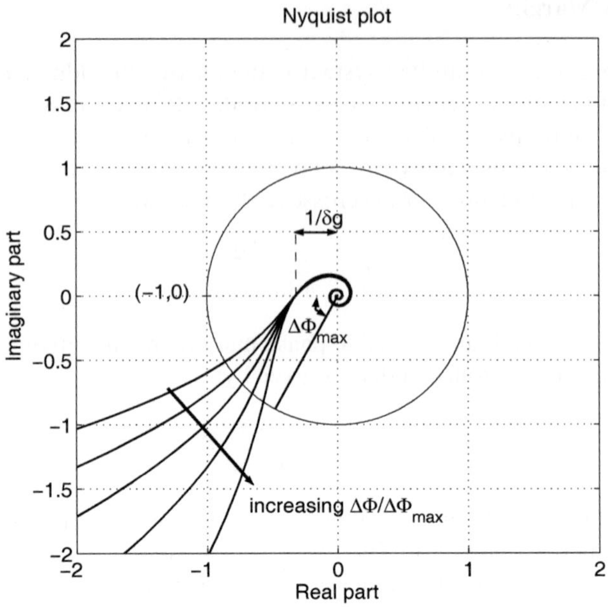

Fig. 7.6 Nyquist plot of the open-loop $K_1^*(j\omega^*)G_1^*(j\omega^*)$ for $\Delta G = 10$ dB and different values of phase margin ratio $\Delta\Phi/\Delta\Phi_{\max} \in [0.5, 0.9]$

7.3.2.5 Nyquist Plot

The gain and phase margins can also be depicted on a Nyquist plot, where the open-loop $K_1^*(j\omega^*)G_1^*(j\omega^*)$ is plotted in the complex plane for $\omega^* \in [0, +\infty)$. In this plot, the gain and phase margins can be directly evaluated, since they represent a measure of the distance between the open-loop and the critical point $(-1, 0)$. When the open-loop curve crosses this point, the system becomes unstable.

Figure 7.6 gives the Nyquist plot of the open-loop of the dimensionless system with different PI controllers having the same gain margin $\Delta G = 10$ dB, and different phase margins. One observes that the tuning method indeed leads to the desired gain margin: the open-loops cross the negative real axis around -0.320. This leads to a gain margin of $\delta g = 1/0.320 = 3.125$, corresponding to a gain margin in dB of $\Delta G = 20\log_{10}(3.125) = 9.9$ dB. The phase margin can be chosen between 0 and $\Delta\Phi_{\max}$. Figure 7.6 depicts the Nyquist plot of the open-loops $K_1^* G_1^*$ corresponding to controllers with a gain margin of 10 dB and phase margins ratios $\Delta\Phi/\Delta\Phi_{\max}$ between 0.5 and 0.9.

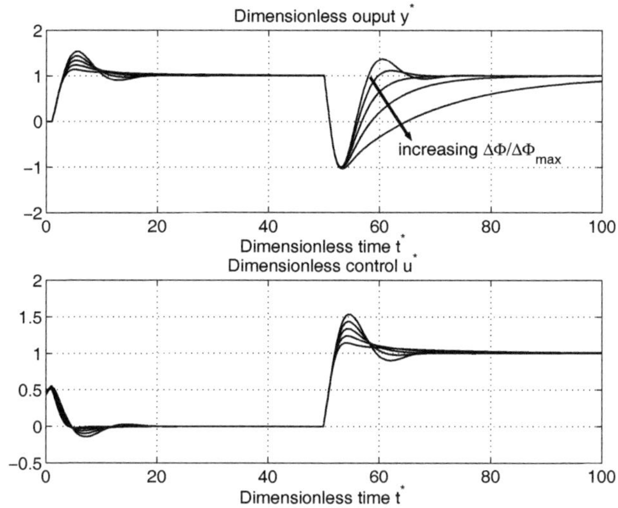

Fig. 7.7 Dimensionless linear simulation for $\Delta G = 10$ dB and different values of the phase margin ratio $\Delta\Phi/\Delta\Phi_{\max} \in [0.5, 0.9]$

7.3.3 Time Domain Performance

Let us now study the influence of the gain and phase margins on the time domain behavior of the controlled canal pool. This study is based on dimensionless simulation results, with a generic scenario including a step of reference downstream level (the reference water level changes from 0 to 1 at time $t^* = 0$), followed after 50 time units by a step of downstream perturbation ($p^* = 1$ at time $t^* = 50$). The simulation is stopped after 100 time units.

The system is simulated for different controllers, with different gain and phase margins. The gain margin is expressed in dB (from 6 to 14 dB, which corresponds to absolute gain margins from 2 to 5), and the phase margin is specified by the ratio $\Delta\Phi/\Delta\Phi_{\max}$.

7.3.3.1 Influence of the Gain Margin ΔG

The choice of a gain margin directly influences the minimum value of $y(t)$ in response to a unit step perturbation. This is illustrated in Fig. 7.7, where the time response of the dimensionless controlled system is depicted for different dimensionless PI controllers having the same gain margin $\Delta G = 10$ dB and different values of the phase margin. In this case, the maximum dimensionless deviation from equilibrium in response to a unit step perturbation is equal to 2.

The same behavior is observed when the system is simulated with various controllers having the same gain margin and different phase margins: the maximum

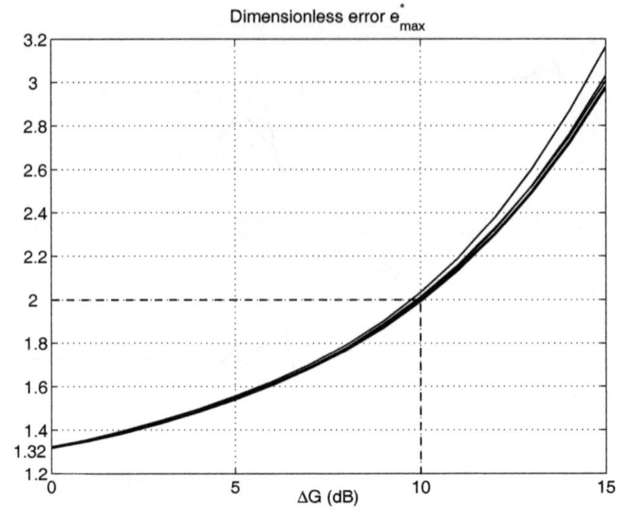

Fig. 7.8 Dimensionless maximum error as a function of ΔG and different values of the phase margin ratio $\Delta\Phi/\Delta\Phi_{\max} \in [0.5, 0.9]$

deviation of y^* in response to a unit step perturbation is almost the same whatever the chosen phase margin ratio. This is depicted in Fig. 7.8, where the maximum error on y^* is depicted as a function of ΔG for different values of $\Delta\Phi/\Delta\Phi_{\max}$.

A maximum dimensionless deviation from equilibrium $e^*_{\max}$ corresponds to a controller with a given gain margin. Table 7.3 gives the relation between the gain margin and the dimensionless maximum deviation from equilibrium corresponding to a unit downstream perturbation.

Table 7.3 Lower bound on $e^*_{\max}$ for different gain margins ΔG

ΔG	1 dB	2 dB	4 dB	6 dB	8 dB	10 dB	12 dB	14 dB	16 dB
$e^*_{\max}$	1.35	1.39	1.48	1.61	1.77	2	2.3	2.72	3.26

The gain margin therefore directly determines the maximum deviation of y^* in response to a unit step perturbation. For a canal pool controlled with a distant downstream PI controller with a given gain margin, an unknown step perturbation of Q_p m^3/s will lead to a dimensional error for the downstream water level of:

$$e_{\max} = \frac{e^*_{\max}\tau_d Q_p}{A_d}, \tag{7.16}$$

where $e^*_{\max}$ is given by Fig. 7.8. For $\Delta G = 10$ dB, one gets $e^*_{\max} = 2$.

Such a result enables us to fulfill the first point of the design requirements: there is a minimal dimensionless error of 1.32 (for a dimensionless withdrawal of 1), which cannot be reduced using any stable distant downstream PI controller. Therefore, if

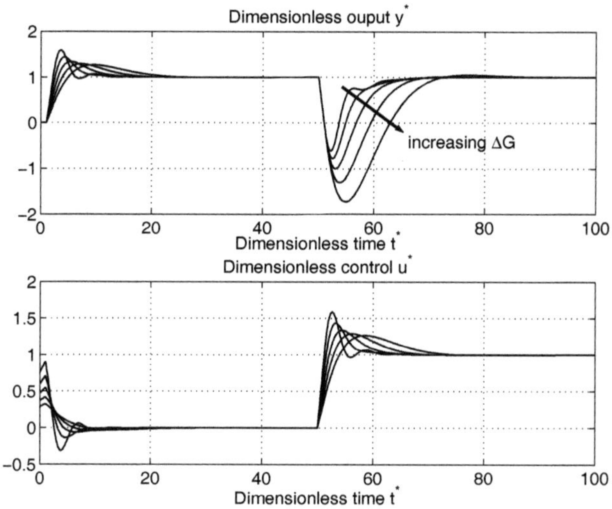

Fig. 7.9 Dimensionless linear simulation for $\Delta\Phi = 0.7\Delta\Phi_{max}$ and values of the gain margin ΔG from 6 to 14 dB with increment of 2 dB

the required maximum dimensionless error is greater than 1.32, the corresponding gain margin of the controller can be obtained from Fig. 7.8. If the required maximum dimensionless error is smaller than 1.32, the design specifications cannot be fulfilled with a PI controller.

7.3.3.2 Influence of the Phase Margin $\Delta\Phi$

Let us now focus on the influence of the phase margin on the time response. For a given gain margin, the response is all the more rapid than the phase margin ratio $\Delta\Phi/\Delta\Phi_{max}$ is small (see Fig. 7.7). Figure 7.9 depicts the time response of the dimensionless controlled system for a phase margin $\Delta\Phi = 0.7\Delta\Phi_{max}$ and different values of the gain margin, for the same scenario (step responses to reference level and perturbation). One observes that the response is all the more rapid than the gain margin is small, but the responses are qualitatively similar for controllers having the same phase margin ratio (no overshoot in the compensation of the unknown perturbation).

There are two interesting quantities to study: the time to reach the reference water level at $\pm 10\%$ in response to an unknown perturbation and the value of the overshoot in response to a step of reference water level. Usually for irrigation canal control, there are no changes in the reference water level, but it may happen in some cases, and the overshoot also gives a measure of the sensitivity to sensor noise: if the overshoot is too large, the system tends to be sensitive to noise affecting the water level sensors [3].

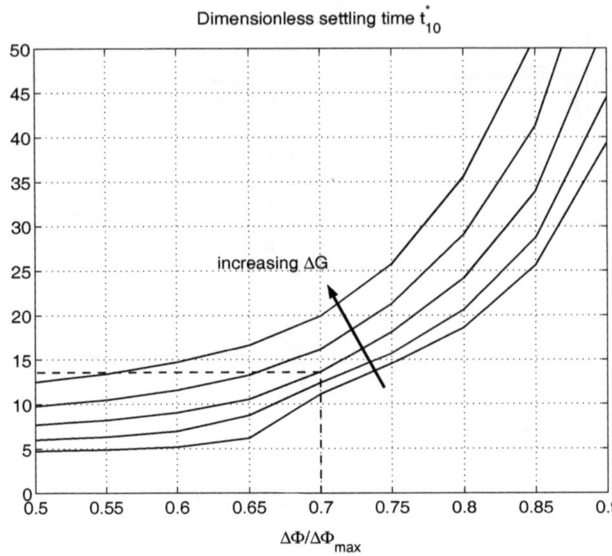

Fig. 7.10 Dimensionless time t^*_{10} as a function of $\Delta\Phi/\Delta\Phi_{\max}$ and values of the gain margin ΔG from 6 to 14 dB with increment of 2 dB

The dimensionless time to reach the reference water level after a step perturbation t^*_{10} is depicted in Fig. 7.10 as a function of the phase and gain margins. With $\Delta\Phi = 0.7\Delta\Phi_{\max}$ and $\Delta G = 10$ dB, one gets $t^*_{10} = 13$. This means that the settling time after a perturbation will be about 13 times the time-delay for such a controller.

Figures 7.8 and 7.10 enable us to predict the time domain performance of a given PI controller from the value of its gain and phase margins. They also provide a way to design controllers based on time domain design requirements.

As in the case of the maximum error, one may evaluate the minimum time to reach the reference level after a step perturbation (see Table 7.4). This time is obtained by defining a minimum phase margin ratio of 0.5, necessary for good performance.

Table 7.4 Lower bound on t^*_{10} for different gain margins ΔG

ΔG (dB)	1	2	4	6	8	10	12	14	16
t^*_{10}	3.25	3.40	3.90	4.75	6.0	7.70	9.80	12.5	15.9

Table 7.5 gives the dimensionless time t^*_{10} to reach $0.1 \times y_r$ after such a perturbation has occurred. Tables 7.3 and 7.5 are very useful for translating time domain requirements into robustness margins constraints.

Figure 7.11 gives the value of the overshoot for the response to a step of reference downstream water level as a function of the phase margin ratio and for different values of the gain margin. The larger the gain margin, the smaller the overshoot, and

Table 7.5 Dimensionless time t_{10}^* to reach $0.1 \times y_r$ after a step perturbation as function of ΔG and $\Delta\Phi/\Delta\Phi_{\max}$

	ΔG (dB)								
$\Delta\Phi/\Delta\Phi_{\max}$	1 dB	2 dB	4 dB	6 dB	8 dB	10 dB	12 dB	14 dB	16 dB
0.1	3.1	3.15	3.55	4.2	5.15	6.4	8.0	10.05	12.65
0.2	3.15	3.2	3.6	4.25	5.2	6.5	8.15	10.25	12.9
0.3	3.16	3.25	3.65	4.35	5.35	6.70	8.45	10.65	13.45
0.4	3.20	3.30	3.75	4.50	5.60	7.05	8.95	11.35	14.35
0.5	3.25	3.40	3.90	4.75	6.0	7.70	9.80	12.50	15.9
0.6	3.30	3.50	4.10	5.20	6.95	9.05	11.55	14.70	18.65
0.7	3.40	3.65	4.70	11.15	12.40	13.60	16.15	19.90	24.85
0.8	3.50	4.0	18.60	18.60	20.65	24.15	29.05	35.55	44.0
0.9	3.70	50.60	38.85	39.40	44.45	52.90	64.95	81.25	150.0

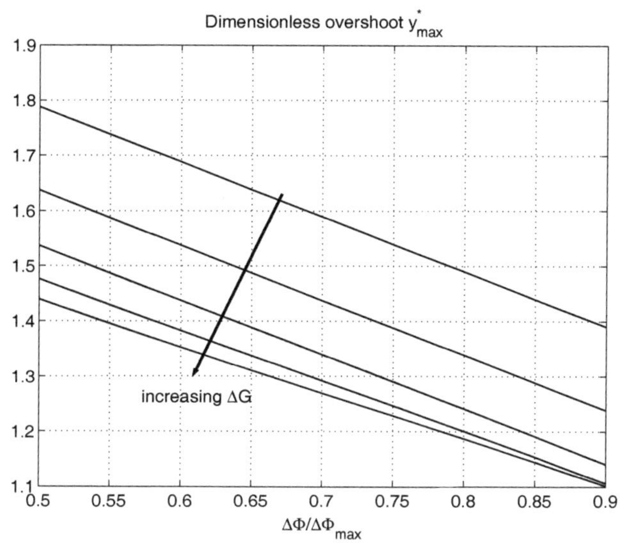

Fig. 7.11 Value of the overshoot for the step of reference downstream level as a function of $\Delta\Phi/\Delta\Phi_{\max}$ and values of the gain margin ΔG from 6 to 14 dB with increments of 2 dB

the same applies for the phase margin ratio. It should be noted that the overshoot can also be reduced by filtering the reference signal. Indeed, in the presented simulation results, a step input is used as the reference signal. In practical situations, the reference signal can be filtered in order to prevent large control actions.

7.3.4 Impact of Variation of Model Parameters

The above tuning rules apply for a canal pool approximated by an ID model with a given delay and backwater area. However, this model is an approximation of reality and its parameters vary when hydraulic parameters of the pool vary (e.g., the discharge, the friction coefficient, the downstream water elevation, the pool geometry, etc.). As already mentioned in the design specifications, the controller should be robust to take these variations into account. The time domain performance will be expressed in terms of minimum gain and phase margins.

The variations of discharge or other parameters modify the gain and the delay of the model. They will therefore modify the gain and phase margins of a PI controller controlling the system. These gain and phase margins variations can be quantified.

To this end, let us study the way gain and phase margins of a given PI controller vary with the parameters of the system: the delay and the integrator gain.

Let $\check{G}_1(s) = e^{-\check{\tau}_d s}/(\check{A}_d s)$ denote the modified system model. The PI controller is denoted by $K_1(s)$.

Let us express the *modified* system $\check{G}_1(s)$ in terms of dimensionless variables, to obtain:

$$\check{G}_1(s)K_1(s) = G_1^*(\check{s}^*)\check{K}_1^*(\check{s}^*),$$

where $\check{s}^* = \check{\tau}_d s$ and $\check{K}_1^*(\check{s}^*)$ denotes the dimensionless modified controller, with

$$\check{k}_p^* = k_p^* \frac{\check{\tau}_d}{\check{A}_d} \frac{A_d}{\tau_d},$$

$$\check{T}_i^* = T_i^* \frac{\tau_d}{\check{\tau}_d}.$$

In this way, the model parameters variations are transformed into controller parameter variations, which enable us to keep the dimensionless equations to study the controlled system.

Equations (7.9) and (7.12) enable us to express the modified gain and phase margins as functions of the modified crossover frequency $\check{\omega}_c^*$:

$$\Delta \check{G}^* = 20 \log_{10}\left(\frac{\check{\omega}_{180}^*}{\check{\omega}_c^*}\right), \tag{7.17a}$$

$$\Delta \check{\Phi}^* = \frac{180}{\pi}(\arctan(\check{T}_i^* \check{\omega}_c^*) - \check{\omega}_c^*), \tag{7.17b}$$

and $\check{\omega}_{180}^* = \pi/2$.

According to (7.13), the modified dimensionless crossover frequency $\check{\omega}_c^*$ is the positive solution of the following equation:

$$\check{\omega}_c^{*4} - \check{k}_p^{*2}\check{\omega}_c^{*2} - \frac{\check{k}_p^{*2}}{\check{T}_i^{*2}} = 0,$$

which leads to:

$$\check{\omega}_c^* = \frac{\check{k}_p^*}{\sqrt{2}}\sqrt{1 + \sqrt{1 + \frac{4}{\check{T}_i^{*2}\check{k}_p^{*2}}}}. \tag{7.18}$$

Therefore the gain and phase margins variations of a fixed PI controller can be directly linked to variations of the system's parameters:

$$\Delta\check{G} - \Delta G = 20\log_{10}\left(\frac{\omega_c^*}{\check{\omega}_c^*}\right),$$

and

$$\Delta\check{\Phi} - \Delta\Phi = \frac{180}{\pi}\left(\arctan(\check{T}_i^*\check{\omega}_c^*) - \arctan(T_i^*\omega_c^*) + \omega_c^* - \check{\omega}_c^*\right).$$

It is interesting to note that the gain margin variations are directly linked to the ratio $\omega_c^*/\check{\omega}_c^*$.

In a first approximation, assuming $\check{T}_i^{*2}\check{k}_p^{*2} \gg 4$ in (7.18), the dimensionless crossover frequency is close to the dimensionless proportional gain of the controller. Then, the gain margin variations are linked to the ratio $k_p^*/\check{k}_p^*$, since:

$$\Delta\check{G} \approx \Delta G + 20\log_{10}\left(\frac{\tau_d}{A_d}\frac{\check{A}_d}{\check{\tau}_d}\right).$$

Therefore, one way to guarantee a minimal gain margin is to choose as a nominal model the one with the largest ratio τ_d/A_d, since in this case the ratio $(\tau_d\check{A}_d)/(A_d\check{\tau}_d)$ is larger than 1, and the modified gain margin is larger than the nominal one.

Indeed, let us apply these results to evaluate the robustness margins of the robust PI tuning method of [12]. The following tuning rules for a sloping reach are proposed in [12]:

$$k_p = \frac{\min\{A_d\}}{3\max\{\tau_d\}} \text{ and } T_i = 6\max\{\tau_d\}.$$

Let us compute the robustness margins of this controller.

Assuming that $\min\{A_d\}/\max\{\tau_d\} = \min\{A_d/\tau_d\}$, one may compute the minimum gain margin and an average phase margin using (7.17a–7.18). With $k_p^* = 0.33$ and $T_i^* = 6$, this leads to $\Delta G = 12.7$ dB and $\Delta\Phi = 44.5°$. These are reasonable margins that lead to a stable PI controller, as attested by the experiments of [12]. However, this tuning rule does not guarantee a minimum time domain performance, since the maximum error and the settling time may vary with the hydraulic parameters.

7.3.5 Robust Performance Tuning Method

The robust performance design problem consists in tuning a unique PI controller such that the following design specifications are fulfilled for a set of discharges and friction coefficients:

- for a withdrawal of Q_p m^3/s, the controlled water level should not deviate more than $e_{\max}$ m from the target water depth, and

- after such a withdrawal has occurred, the time for the water level to reach its reference at $\pm 10\%$ should be at most t_{10} seconds.

Let us denote by $i = 1,\dots,N$ the number of models considered, each model corresponding to different discharges, friction coefficients, etc. To each model corresponds a couple $(\tau_d(i), A_d(i))$.

The maximum acceptable dimensionless error is given for each model by:

$$e^*_{\max}(i) = \frac{A_d(i)}{\tau_d(i)} \frac{e_{\max}}{Q_p}.$$

In order to fulfill the first design specification, one needs to design a controller that gives at most the lowest dimensionless error:

$$e^*_{\max}(i_M) = \min_i \{e^*_{\max}(i)\} = \min \left\{ \frac{A_d}{\tau_d} \right\} \frac{e_{\max}}{Q_p}.$$

Using Table 7.3, this gives the gain margin ΔG_M such that the first design specification is fulfilled. If ΔG_M is too small, one needs to increase $e_{\max}$. Else, the corresponding crossover frequency ω_{cM} is given by (7.9):

$$\omega^*_{cM} = \frac{\pi}{2} 10^{\frac{-\Delta G_M}{20}}.$$

Since $k^*_p \approx \omega^*_{cM}$, this provides a first approximation of the proportional gain of the controller:

$$k_p \approx \omega^*_{cM} \min \left\{ \frac{A_d}{\tau_d} \right\}. \tag{7.19}$$

With this proportional gain, the dimensionless crossover frequencies can be approximated for each model by:

$$\omega^*_c(i) = k_p \frac{\tau_d(i)}{A_d(i)}.$$

Using (7.9), the corresponding effective gain margins are given by:

$$\Delta G(i) = 20 \log_{10} \left(\frac{\pi}{2\omega^*_c(i)} \right). \tag{7.20}$$

The second design specification is transformed into a dimensionless form for each model:

$$t^*_{10}(i) = \frac{t_{10}}{\tau_d(i)}. \tag{7.21}$$

Using (7.20) and (7.21) with Table 7.5 and (7.11), one may compute for each model the phase margin $\Delta\Phi(i)$ such that the second design specification is fulfilled. In order to fulfill this second design requirement, the controller should have a phase margin lower than or equal to $\Delta\Phi(i_P) = \min_i \Delta\Phi(i)$. Finally, the controller parameters are obtained using (7.12–7.13):

$$T_i = \frac{\tau_d(i_P)}{\omega_{cM}^*} \tan\left(\frac{\pi}{180}\Delta\Phi(i_P) + \omega_{cM}^*\right), \tag{7.22}$$

and

$$k_p = \omega_{cM}^* \min\left\{\frac{A_d}{\tau_d}\right\} \sin\left(\frac{\pi}{180}\Delta\Phi(i_P) + \omega_{cM}^*\right). \tag{7.23}$$

7.3.6 Summary

The proposed tuning method is based on the above results about the influence of robustness margins on the time domain performance of the controlled system. The different steps are as follows:

1. Compute the ID model parameters for the considered canal pool, for different discharges (e.g., $[Q_{\min}, Q_{\max}]$) and/or different friction coefficients.
2. Specify a reference perturbation scenario (i.e., a discharge Q_p), a maximum dimensional error $e_{\max}$, a maximal time t_{10} to reach the reference level at $\pm 10\%$ after this perturbation, and minimum gain and phase margins $\Delta G_{\min}$ and $\Delta\Phi_{\min}$.
3. Compute the lowest dimensionless maximum error $e_{\max}^*(i_M)$ and the corresponding gain margin ΔG_M.
    - If $\Delta G_M < \Delta G_{\min}$, the required performance is too high. The desired maximum error $e_{\max}$ needs to be increased.
    - Else compute a first approximation of the proportional gain using (7.19).
4. Compute the dimensionless times $t_{10}^*(i)$ and the corresponding phase margins $\Delta\Phi(i)$. Compute the lowest phase margin $\Delta\Phi(i_P)$.
    - If $\Delta\Phi(i_P) < \Delta\Phi_{\min}$, the required performance is too high. The time t_{10} needs to be increased.
    - Else compute the integral time T_i and the proportional gain k_p using (7.22–7.23).

Let us now consider the case of a local PI upstream controller.

7.4 Tuning of Local Upstream PI Controllers

For local upstream control, the considered dimensionless transfer is $\tilde{G}_1^*(s^*)$ relating downstream discharge to downstream water elevation. We denote by $K_2^*(s^*)$ the dimensionless local upstream PI controller.

Figure 7.12 depicts the asymptotic Bode diagram of $K_2^*\tilde{G}_1^*$. In this case, there is no theoretical limit to the achievable bandwidth, because the frequency at $-180°$ is pushed towards $+\infty$. One can therefore obtain a theoretical phase margin of 90°.

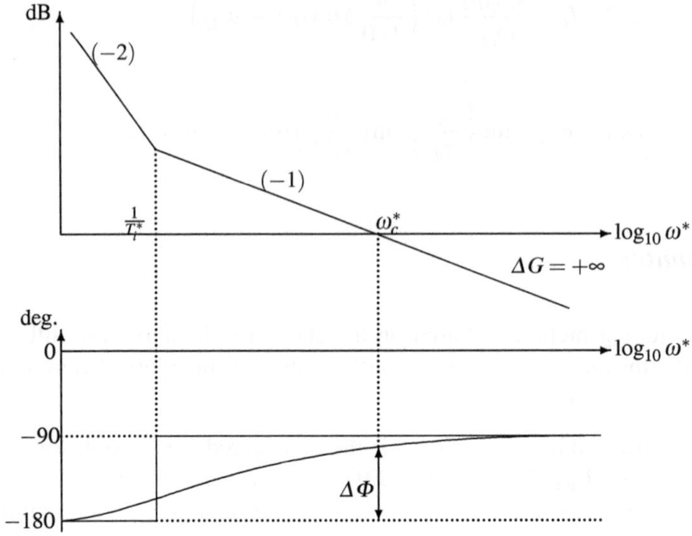

Fig. 7.12 Schematic Bode diagram of $K_2^*\tilde{G}_1^*$ for local upstream control

7.4.1 Theoretical Case

In the theoretical case, one may specify a phase margin $\Delta\Phi$, strictly smaller than 90°. The local upstream controller can then be computed by first choosing a proportional gain k_{p2}^*. Since $\tilde{G}_1^*$ is negative, the feedback should be done with a negative gain, then $k_{p2}^* < 0$. Equations (7.7a–7.7b) in this case become:

$$|K_2^*(j\omega^*)\tilde{G}_1^*(j\omega^*)| = \frac{|k_{p2}^*|}{\omega^*}\sqrt{1+\frac{1}{T_{i2}^{*2}\omega^{*2}}}, \qquad (7.24a)$$

$$\arg(K_2^*(j\omega^*)\tilde{G}_1^*(j\omega^*)) = -180 + \frac{180}{\pi}\arctan(T_{i2}^*\omega^*). \qquad (7.24b)$$

To have a desired phase margin $\Delta\Phi$, (7.24b) gives:

$$T_{i2}^*\omega_{c2}^* = \tan\left(\frac{\pi}{180}\Delta\Phi\right).$$

Then the crossover frequency ω_{c2}^* can be computed using (7.24a):

$$\omega_{c2}^* = \frac{|k_{p2}^*|}{\sin\left(\frac{\pi}{180}\Delta\Phi\right)}.$$

These equations provide the integral time for a PI controller with arbitrary proportional gain that ensures a desired phase margin $\Delta\Phi$ in the case of a perfect integrator. In theory, one may achieve any desired level of performance.

7.4.2 Practical Tuning

However, in practice, one should take actuator limitations and transmission delays into account, which limit the achievable bandwidth. Moreover, since the control system is implemented in discrete-time, there is a delay due to the sampling and first-order hold, which limits the achievable bandwidth. Let T_s denote the sampling time, then the additional delay is equal to $T_s/2$.

Then, in order to have a gain margin of ΔG and a phase margin of $\Delta\Phi$, we can choose the proportional gain and the integral time as follows:

$$k_{p2}^* = -\omega_{c2}^* \sin\left(\frac{\pi}{180}\Delta\Phi + \omega_{c2}^*\right), \tag{7.25}$$

$$T_{i2}^* = \frac{1}{\omega_{c2}^*} \tan\left(\frac{\pi}{180}\Delta\Phi + \omega_{c2}^* \frac{T_s^*}{2}\right), \tag{7.26}$$

with $\omega_{c2}^* = \frac{\pi}{T_s^*} 10^{-\frac{\Delta G}{20}}$.

These equations provide a tuning method in the case of a canal controlled with a discrete-time control system.

We now consider an advanced method to tune distant downstream and local upstream controllers, using the H_∞ optimization framework.

7.5 H_∞ Control of a Canal Pool

The H_∞ optimization method was introduced by Zames in the 1980s, in order to recover the properties of classical controllers such as PI as the solution of an optimization problem [15].

We detail here the design of a distant downstream H_∞ controller. The design of an H_∞ controller needs to be done in two steps: the first one is the expression of design requirements as constraints on the H_∞ norm of transfer functions of the feedback system. The second step then consists in formulating the problem as a minimization of the H_∞ norm of an augmented system including the design constraints. This leads to a so-called "standard problem", which can be solved using classical control software.

In our case, we will show that the controller design problem can be formulated as a "four-block" problem. We will then compare the PI and H_∞ controllers, to show that they are, in fact, very close. However, the H_∞ controller is the solution of a convex optimization problem, and can easily be modified to include supplementary constraints (typically to ensure robustness).

7.5.1 Expression of Design Requirements as H_∞ Constraints

In automatic control terms, the design requirements are as follows: the downstream water level should be kept close to its reference value, while limiting the control effort. The static error should be zero, even in the case of unpredicted water withdrawal, and we would also like to impose minimum gain and phase margins to ensure robustness to model uncertainties.

Let us now express these design requirements in terms of H_∞ constraints, as in [9].

7.5.1.1 Reference Tracking Specification

The water level tracking specification can be expressed as an H_∞ constraint on the closed-loop transfer function between the reference and the tracking error. This transfer function corresponds to the sensitivity function $S(s)$, which is such that:

$$e(s) = (1 + G_1(s)K_1(s))^{-1}r(s) \triangleq S(s)r(s).$$

Information about the frequency spectrum of $r(j\omega)$ can be used to specify the set of possible reference signals.

To have a static error lower than a given threshold g_s is equivalent to having:

$$|S(j0)| \leq g_s.$$

In the H_∞ framework, such a constraint can be imposed with a frequency weighting function usually denoted $W_1(s)$, and the H_∞ problem is as follows: find the smallest $\gamma > 0$ and the stabilizing controller K such that

$$||W_1 S||_\infty \leq \gamma.$$

Then, if $\gamma \leq 1$, we have

$$||W_1 S||_\infty \leq 1,$$

which ensures, by definition of the H_∞ norm that for all $\omega \in \overline{\mathbb{R}}$, we have:

$$|W_1(j\omega)S(j\omega)| \leq 1,$$

and therefore

$$|S(j\omega)| \leq |W_1(j\omega)|^{-1},$$

if $|W_1(j\omega)| \neq 0$.

This shows that the tracking error is directly proportional at a given frequency ω to the inverse of $|W_1(j\omega)|$. Therefore, the static error is lower than the inverse of the static gain of $W_1(s)$.

7.5.1.2 Perturbation Rejection Specification

The transfer function relating the perturbation input to the output error is given by:

$$e(s) = \tilde{G}_1(s)S(s)p(s).$$

The specification can therefore be ensured by using a frequency weighting function $W_p(s)$ in order to constrain the set of frequencies where the perturbations act on the system (typically low frequencies and also known frequencies linked to the behavior of water users).

This weighting function should be such that $|W_p(j\omega)| \gg 1$ for the considered set of frequencies (see [16]). Then, if a stabilizing controller K exists such that

$$||\tilde{G}_1 S W_p||_\infty \leq 1,$$

then for all $\omega \in \overline{\mathbb{R}}$, we have

$$|\tilde{G}_1(j\omega)S(j\omega)W_p(j\omega)| \leq 1,$$

and therefore

$$|\tilde{G}_1(j\omega)S(j\omega)| \leq |W_p(j\omega)^{-1}| \ll 1$$

for the considered set of frequencies.

Typically, we will ensure that constant perturbations are rejected by choosing $|W_p(j0)^{-1}| \ll 1$. We could also use this means to reject perturbations occurring at specific frequencies (typically water withdrawals occurring every 8 h, 12 h and 24 h).

7.5.1.3 Robustness Constraints

The gain or phase margin of a system can be specified via a constraint on the sensitivity function $S = (I + G_1K_1)^{-1}$ and the complementary sensitivity function $T = G_1K_1(I + G_1K_1)^{-1}$ (see [10]).

Indeed, using the small gain theorem [14], one may show that the gain and phase margins are bounded by:

$$\begin{aligned}
&\Delta G \in [\min\{1-\alpha_T,(1+\alpha_S)^{-1}\},\max\{1+\alpha_T,(1-\alpha_S)^{-1}\}],\\
&\Delta \Phi \in \left[-2\max\{\arcsin(\frac{\alpha_S}{2}),\arcsin(\frac{\alpha_T}{2})\},2\max\{\arcsin(\frac{\alpha_S}{2}),\arcsin(\frac{\alpha_T}{2})\}\right],
\end{aligned}$$

where $\alpha_S = ||S||_\infty^{-1}$ and $\alpha_T = ||T||_\infty^{-1}$.

Finally, we can link the input margins to the H_∞ norm of S and/or T. In our case, we use the lower bound of the weighting function W_1 to constrain S. Indeed, if we choose W_1 such that $\min_{\omega\in\bar{\mathbb{R}}}|W_1(j\omega)| \geq 1-\Delta G$ and the H_∞ minimization ends up with $\|W_1 S\|_\infty \leq 1$, then $|S(j\omega)| \leq |W_1(j\omega)|^{-1} \leq (1-\Delta G)^{-1}$, which by the above formula ensures that the closed-loop system has a gain margin larger than or equal to ΔG.

7.5.1.4 Control Effort Specification

The control effort needs to be constrained in order to reduce the effect of output noise on the control input. This can be specified by constraining the transfer function KS, which relates the reference r to the control input u (but also the transfer between a noise acting on the output and measurement and u) with a weighting function generally denoted $W_2(s)$. More generally, this weighting function can shape the controller in high frequencies. Indeed, since in high frequencies the sensitivity function verifies $|S(j\omega)| \approx 1$, the constraint $\|W_2 K_1 S\|_\infty \leq 1$ implies that

$$|K_1(j\omega)| \leq |W_2(j\omega)|^{-1}$$

for high frequencies.

7.5.2 A Solution as a Four-block Criteria

Now we use the following property of the H_∞ norm to collect all the constraints into a single criterion.

Proposition 7.1. *[17] Let $M(s) = [m_{ij}(s)] \in RH_\infty^{n\times m}$ such that $\|M\|_\infty \leq \gamma$. Then for all $i \in [1,n]$ and $j \in [1,m]$, one has*

$$\|m_{ij}\|_\infty \leq \gamma.$$

Let $H(s)$ denote the transfer matrix relating the reference $r(s)$ and the perturbation $p(s)$ to the output error $e(s)$ and the control input $u(s)$:

$$H(s) = \begin{pmatrix} (I+G_1K_1)^{-1} & \tilde{G}_1(I+G_1K_1)^{-1} \\ K_1(I+G_1K_1)^{-1} & \tilde{G}_1K_1(I+G_1K_1)^{-1} \end{pmatrix} \triangleq \begin{pmatrix} S & GS \\ KS & T \end{pmatrix}.$$

We now consider the augmented system including the weighting functions:

$$M(s) = W_o(s)H(s)W_i(s),$$

where W_i and W_o are, respectively, the input and output weighting functions, defined as follows:

$$W_i(s) = \begin{pmatrix} 1 & 0 \\ 0 & W_p(s) \end{pmatrix}, \; W_o(s) = \begin{pmatrix} W_1(s) & 0 \\ 0 & W_2(s) \end{pmatrix},$$

where $W_1(s)$, $W_2(s)$ and $W_p(s)$ are monovariable stable transfer functions that take into account the control specifications as explained above.

Then, the solution of the H_∞ problem leads to finding the smallest $\gamma > 0$ and the stabilizing controller $K(s)$ such that

$$\|M\|_\infty = \left\| \begin{matrix} W_1 S & W_1 GSW_p \\ W_2 KS & W_2 TW_p \end{matrix} \right\|_\infty \leq \gamma.$$

If γ is close to 1 then the above proposition guarantees that the H_∞ norm of each transfer function is close to 1. We have to note that the transfers functions considered in the criteria are not exactly the ones defined above. Consequently, this explains why we have to carefully consider how to obtain the desired requirements by using a four-block criterion. Generally, the reference tracking and the perturbation rejection specifications possess common features, which leads us to choose W_p as a constant weighting function and W_1 compatible with both requirements. There also appears an unspecified constraint on $T(s)$. It is generally possible to choose the weighing functions W_2 and W_p such that $\|W_2 TW_p\|_\infty \ll 1$.

7.5.2.1 Weighting Functions

In order to facilitate the frequency tuning, we use first-order frequency weighting functions parameterized as proposed by [2] (other forms are also available [13]):

$$W(s) = \frac{G_\infty \sqrt{|G_0^2 - 1|}s + G_0 \omega_c \sqrt{|G_\infty^2 - 1|}}{\sqrt{|G_0^2 - 1|}s + \omega_c \sqrt{|G_\infty^2 - 1|}}, \tag{7.27}$$

with

$$\begin{cases} (G_0 - 1)(G_\infty - 1) < 0 \\ \omega_c > 0 \end{cases},$$

which have the following properties [2]:

$$\begin{cases} |W(j0)| & = G_0 \\ \lim_{\omega \to \infty} |W(j\omega)| & = G_\infty \\ |W(j\omega_c)| & = 1 \end{cases}.$$

In the case of the weighting function W_1, which constrains the tracking error and the perturbation rejection, G_0^{-1} specifies the static tracking error, ω_c adjusts the frequency bandwidth of $\tilde{G}_1 S$, and G_∞^{-1} imposes a constraint on the maximal value of the sensitivity function, ensuring minimum robustness margins.

The weighting function W_2 is also chosen to be of the first-order following (7.27). Then, G_0 is such that W_2 be compatible with the low frequency constraint. Indeed, for low frequencies, one has $|S(j\omega)| \approx 0$ and therefore $|T(j\omega)| \approx 1$, since $S+T=1$. Then, to have a criteria close to 1, we need to have $|W_2(j0)W_p(j0)| < 1$, and therefore $G_0 < \frac{1}{|W_p(j0)|}$. For a constant W_p, this constraint is simply $G_0 < \frac{1}{W_p}$.

7.5.2.2 Application

We apply this methodology to the dimensionless delay integrator system. The weighting functions are obtained as in Table 7.6 and are displayed in Fig. 7.13 with $W_p = 0.2$.

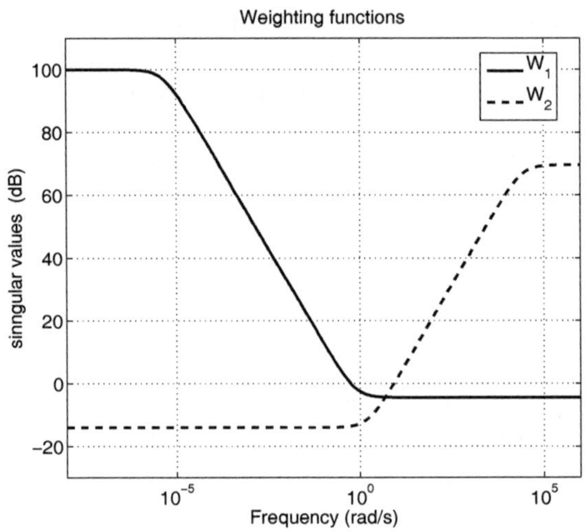

Fig. 7.13 Weighting functions W_1 and W_2

Table 7.6 Weighting function parameters

	W_1	W_2
G_0	1×10^5	0.2
G_∞	0.5	3000
ω_c	$\pi/2$	$15\times\pi/2$

In order to design an H_∞ controller, we have to use a rational approximation of the dimensionless delay integrator system. For this purpose, we use a first-order Padé approximation of the unitary delay, which leads us to define the rational plant used for the H_∞ synthesis as

$$\frac{e^{-s}}{s} \approx \frac{2-s}{s(s+2)}.$$

We choose a first-order approximation in order to obtain a controller with an order as low as possible. Indeed, in the H_∞ design, the order of the obtained controller is equal to the order of the model used for the synthesis plus the order of the weighting functions. In our case, the order of the H_∞ controller is 4, since we have 2 (for the plant) + 1 (for W_1) + 1 (for W_2). The H_∞ minimization leads to a criterion with an H_∞ norm equal to 1.14. The controller is given by

$$K_\infty(s) = 8.3\frac{(s+2)(s+0.14)(s+2.6\times 10^4)}{(s+0.0013)(s^2+6.6s+29.2)(s+1.7\times 10^3)}$$

and it then possesses a quasi-integrator pole $p_1 = -0.0013$, and a pair of conjugate poles with a pulsation $\omega_0 = 5.40$ rad/s and a damping factor equal to $\xi = 0.6$. The other pole at -1.7×10^3 only affects high frequencies and is rapidly compensated by the zero at -2.6×10^4. The H_∞ controller is then very close to a third-order controller.

Figure 7.14 depicts the maximal singular values of each transfer constrained by the H_∞ criteria, i.e., $S(s)$, $G(s)S(s)$, $K(s)S(s)$ and $T(s)$, and the inverse of the associated weighting function. This figure illustrates perfectly the consequence of Proposition 7.1 on each transfer function. Figure 7.14 also shows that the weighting function W_1 strongly constrains the transfer function $G(s)S(s)$ in low frequencies and, by contrast, has no effect on $S(s)$ in the same range of frequencies. Actually, since the system contains an integrator, it can be shown that a simple constant gain allows us to ensure perfect tracking of any constant reference (no static error). By contrast, the rejection of any constant perturbation acting on the system input (like the discharge perturbation) leads to a static error if the controller does not contain an integrator. This shows that the low frequency constraint induced by an input perturbation has a strong effect on the H_∞ design. Indeed, it imposes the necessity of an integrator in the controller. It is in fact a quasi-integrator, since $|W_1(j0)|$ is finite, which means that the static error will be less than $|W_p(j0)|^{-1}|W_1(j0)|^{-1}$. The constraint on $G(s)S(s)$ depicted in Fig. 7.14 reveals this simple fact. In the other hand, for the frequencies higher than the bandwidth of the system, $W_1(s)$ constrains the maximal value of $S(s)$ and imposes a minimal input margin to the closed-loop system.

7.5.3 Comparison with a PI Controller

We compare the H_∞ controller with the PI controller. We then consider a 10 dB gain margin PI controller (see the previous section), which has the following parameterization:

$$K_{PI}(s) = 0.47\frac{s+0.17}{s}.$$

We observe that the zero of the PI controller, i.e., $z = 0.17$ is close to the first zero of the H_∞ controller, i.e., $z_1 = 0.14$ and only the high frequency behavior of the two

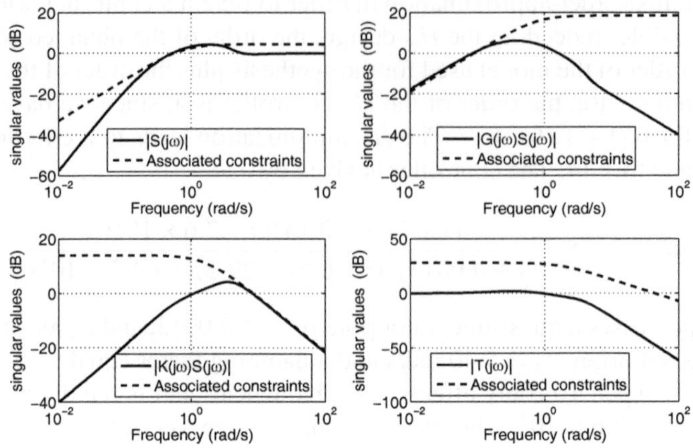

Fig. 7.14 The four blocks and the corresponding constraints

controllers are different (see Fig. 7.15). Indeed, in order to add high frequency roll-off in the H_∞ controller, it is necessary to compensate for the phase introduced by the high frequency filter, and this is done by a classical lead-lag filter effect. In conclusion, the open-loop transfers are similar (see Fig. 7.15), except the high frequency constraint imposed by $W_2(s)$. This constraint enables us to shape the controller by ensuring that $|K_1(j\omega)| \approx |\omega|^{-1}$ in high frequencies. This roll-off ensures that the effect of the measurement noise on the control input is filtered and that the control effort is limited.

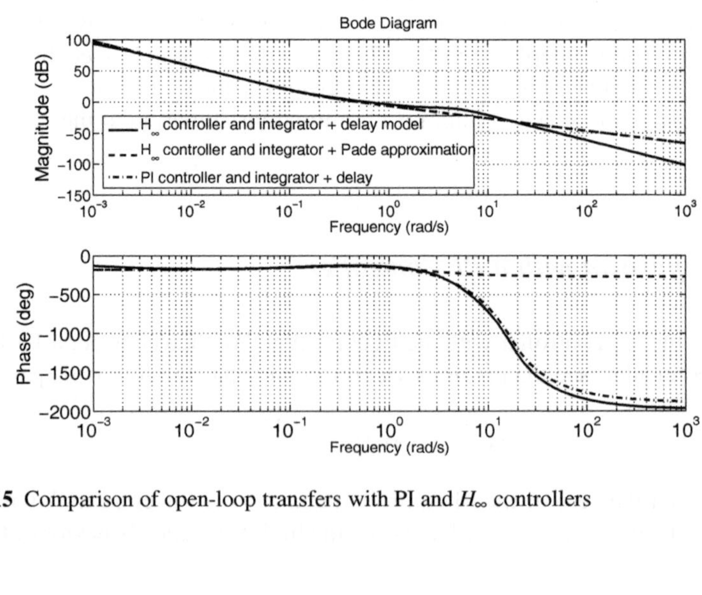

Fig. 7.15 Comparison of open-loop transfers with PI and H_∞ controllers

Finally, as shown in Fig. 7.16, the time domain behaviors of the closed-loop systems are very similar. Actually, since the input margin of the H_∞ controller is equal to 7.2 dB and it is then less than that of the PI controller (10 dB), following discussions of the first part of this chapter, we deduce that the H_∞ controller has a smaller tracking error than that of the PI controller with respect to a step perturbation.

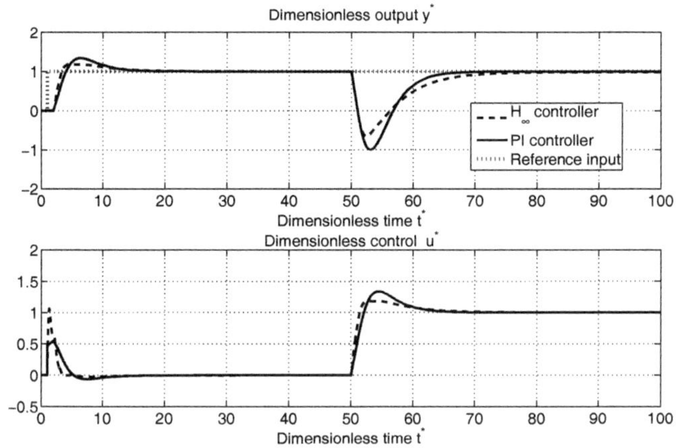

Fig. 7.16 Comparison of time domain response of the closed-loop system with a PI and an H_∞ controller. Reaction to a step reference input at $t^* = 0$ and perturbation at $t^* = 50$

7.5.4 Benefits of H_∞ for Handling Complex Requirements

We have shown that the H_∞ design method allows us in a simple case to recover classical controllers such as for example PI controllers. Clearly, for more complex requirements or also for a more complex plant, H_∞ offers a powerful alternative to the classical design. We now illustrate how the H_∞ method can handle an unmeasured periodic perturbation acting on the system.

We assume that the unmeasured perturbation is sinusoidal and has a known frequency $\omega_p = 0.01$ rad/s. From a classical point of view, the perfect rejection of such a periodic signal can be achieved if $|G(j\omega_p)S(j\omega_p)| = 0$. In order to achieve such a property, we then multiply the weighting function $W_1(s)$ defined previously by a resonant filter of the form:

$$W_f(s) = \frac{s^2 + 2\xi_n\omega_p s + \omega_p^2}{s^2 + 2\xi_p\omega_p s + \omega_p^2}$$

which, by taking $\xi_p = 10^{-3}$ and $\xi_n = 2$ leads to

$$W_f(s) = \frac{s^2 + 0.04s + 0.0001}{s^2 + 2 \times 10^{-5}s + 0.0001}$$

and a supplementary attenuation at ω_p equal to the inverse of $|W_f(j\omega_p)| = 2000$.

We synthesize the H_∞ controller and obtain the following value of the criteria: $\gamma = 1.15$. As expected, the constraint provided by the resonant filter $W_f(s)$ has a strong effect on $G(s)S(s)$ and a less important effect on $S(s)$ (see Fig. 7.17). The obtained controller possesses a resonance at ω_p, which ensures the required attenuation (see Fig. 7.18). Finally, the time domain simulation confirms the frequency domain analysis. Figure 7.19 compares the rejection of a sinusoidal perturbation at the frequency $\omega = 0.01$ rad/s with the PI controller and the H_∞ controller. As expected, the H_∞ controller achieves a strong attenuation of the perturbation, while it appears directly at the output of the system controlled with the PI controller. However, one should note that such a design is very specific, since it does not attenuate frequencies other than the one contained in $W_f(s)$.

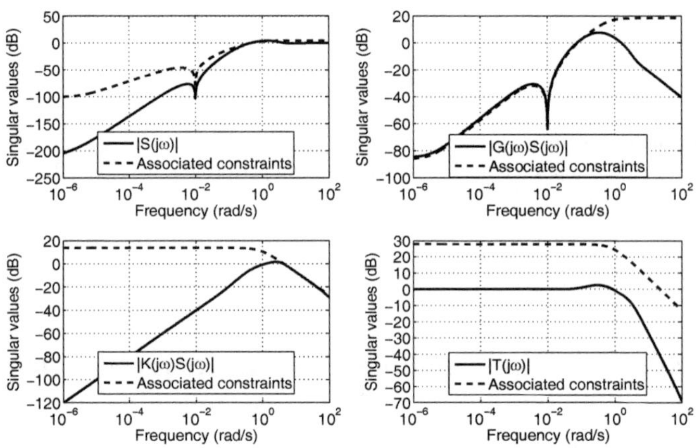

Fig. 7.17 The four blocks and the corresponding constraints

7.6 Summary

This chapter developed a PI controller tuning method to ensure robust performance: this means that the controller ensures a minimum performance level for a set of hydraulic conditions. The performance requirements include the maximum error after an unpredicted perturbation has occurred and the time to

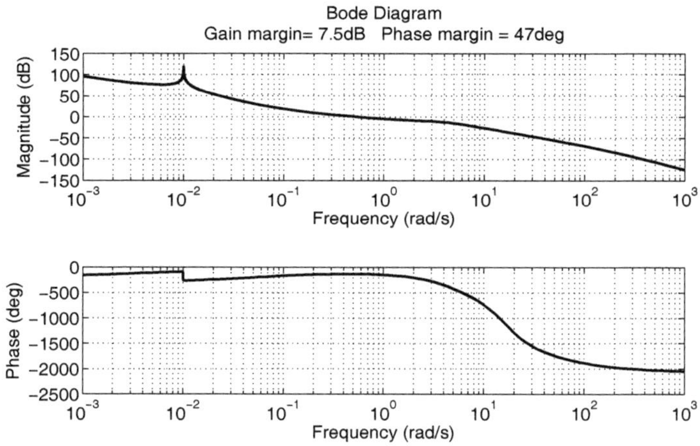

Fig. 7.18 The H_∞ controller associated to $W_f(s)$

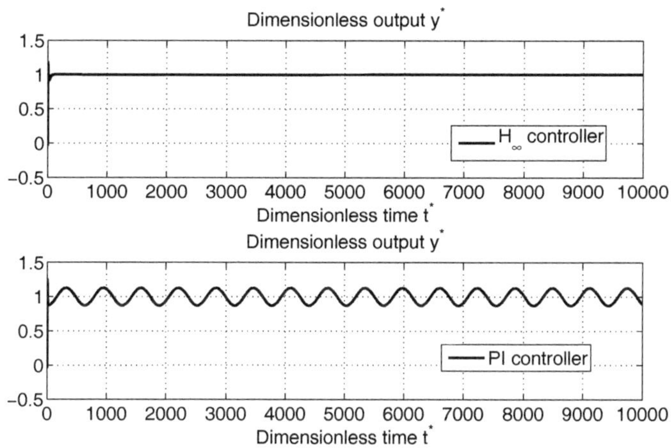

Fig. 7.19 Comparison of the time domain response of the closed-loop system with the PI and the H_∞ controllers (designed with $W_f(s)$) for a unitary sinusoidal perturbation at the frequency $\omega = 0.01$ rad/s

reach the design level after this. The results are first derived for continuous-time control, then extended to discrete-time control. The control action variable can be the upstream discharge or the gate opening. This robust tuning method enables us to:

1. evaluate whether a required performance is possible, and if it is the case,

2. to compute the controller parameters that fulfills the design requirements for all considered hydraulic conditions.

We have used the H_∞ framework to express the control design problem as the minimization of a weighted H_∞ criteria. We first formulated each constraint as a weighted H_∞ constraint on one of the closed-loop transfers. Then, all these constraints were put into a single standard H_∞ problem. The obtained H_∞ controller was shown to be close to a PI controller, except that it had better high frequency roll-off, ensuring a limited control effort and a low noise to input ratio. This method was first applied to an open channel hydrosystem in [4, 5, 7]. Other works on this subject include [8, 1].

References

[1] Cantoni M, Weyer E, Li Y, Ooi SK, Mareels I, Ryan M (2007) Control of large-scale irrigation networks. Proc IEEE 95(1):75–91

[2] Font S (1995) Méthodologie pour prendre en compte la robustesse des systèmes asservis: optimisation H_∞ et approche symbolique de la forme standard. PhD thesis, Université de Paris-Sud Orsay, France

[3] Franklin GF, Powell JD, Emami-Naeini A (2002) Feedback control of dynamic systems, 4th edn. Prentice Hall, Englewood Cliffs, NJ

[4] Litrico X, Fromion V (2002) Real-time management of multi-reservoir hydraulic systems using H_∞ optimization. In: IFAC World Congress, Barcelona

[5] Litrico X, Fromion V (2003) Advanced control politics and optimal performance for an irrigation canal. In: European Control Conference, Cambridge, UK

[6] Litrico X, Fromion V (2004) Simplified modelling of irrigation canals for controller design. J Irrig Drain Eng 130(5):373–383

[7] Litrico X, Fromion V (2006) H_∞ control of an irrigation canal pool with a mixed control politics. IEEE Trans Control Syst Technol 14(1):99–111

[8] Mareels I, Weyer E, Ooi SK, Cantoni M, Li Y, Nair G (2005) Systems engineering for irrigation systems: successes and challenges. Annu Rev Control 29(2):191–204

[9] Pognant-Gros P, Fromion V, Baume J (2001) Canal controller design: a multivariable approach using H_∞. In: European Control Conference, Porto, pp 3398–3403

[10] Safonov M, Athans M (1977) Gain and phase margins for multiloop LQG regulators. IEEE Trans Autom Contr 22(2):173–179

[11] Schuurmans J, Clemmens A, Dijkstra S, Hof A, Brouwer R (1999) Modeling of irrigation and drainage canals for controller design. J Irrig Drain Eng 125(6):338–344

[12] Schuurmans J, Hof A, Dijkstra S, Bosgra OH, Brouwer R (1999) Simple water level controller for irrigation and drainage canals. J Irrig Drain Eng 125(4):189–195

[13] Skogestad S, Postlethwaite I (1998) Multivariable feedback control. Analysis and design. Wiley, New York

[14] Zames G (1966) On the input-output stability of time-varying nonlinear feedback systems - Part I : Conditions derived using concepts of loop gain, conicity, and posivity. IEEE Trans Autom Contr 11:228–238

[15] Zames G (1981) Feedback and optimal sensitivity: Model reference transformations, multiplicative seminorms, and approximate inverses. IEEE Trans Autom Contr 26(2):301–320

[16] Zames G (1981) Feedback and optimal sensitivity: model reference transformations, multiplicative seminorms, and approximate inverses. IEEE Trans Autom Contr 26(2):301–320

[17] Zhou K, Doyle J (1998) Essentials of robust control. Prentice Hall, Upper Saddle River, NJ

Chapter 8
Mixed Control of a Canal Pool

Abstract This chapter develops a mixed control policy for a canal pool. The classical control policies have advantages and drawbacks. The local upstream control policy ensures good performance, but bad water management. The distant downstream control ensures good water management, but relatively bad performance. We propose a mixed local upstream/distant downstream control that ensures good performance and at the same time efficient water management. The controller structure imposes a substitution between the two actuators. Two controller tuning methods are developed, a PI controller and an H_∞ controller.

8.1 Introduction

As already mentioned in Chap. 7, the control structure for a canal pool corresponds to a given trade-off between water resource management and the service to water users. In this section, we show that a multivariable control structure offers enough free parameters to achieve this trade-off.

To this end, we first show that the two classical structures used in canal control (local upstream and distant downstream control) are particular cases of a multivariable controller, each one being specifically in charge of one of the control objectives. This analysis finally leads us to consider the multivariable structure as a means to mix the advantages of each solution.

A naive analysis of the control problem for irrigation canals would be to consider that the control specifications can be reduced to rejecting unmeasured perturbations by controlling the downstream water level. Reducing this problem to this classical control problem leave out an essential aspect of the specifications: the water resource management. Indeed, even if a linear description of the flow is locally licit, it hides the fact that water flows from upstream to downstream. Then, the model is valid for variations around a functioning point, and the sign of the control variables corresponds to the sign of discharge variations upstream or downstream. We can now interpret the control structures based on this remark.

In fact, a control structure using the upstream control variable leads to economic water management, since faced with a decreasing demand, the upstream discharge will be diminished. In other terms, the upstream control variable adapts to the consumed discharge in the pool, and therefore uses only the water necessary to satisfy the effective water demand.

On the contrary, a control structure using the downstream control variable leads to expensive water management. Indeed, faced with a decreasing demand, the only way to maintain the downstream water level is to let the superfluous discharge go downstream.

This chapter first reviews the control objectives for a canal pool, then proposes a rationale to achieve the desired trade-off between performance and water management. We end up with tuning rules for mixed PI controllers. An H_∞ framework to design robust mixed controllers will be presented in Chap. 11 for an experimental canal.

8.2 Control Specifications

8.2.1 Control Objectives

The two classical control structures presented earlier are widely used mainly because they are monovariable control structures, where controller design can be done with classical tools. We now examine the multivariable control problem, where both control action variables u_1 and u_2 are used to control the downstream water level y. The controller design problem leads to finding a controller $K(s)$ that relates the tracking error e to the control vector (u_1, u_2):

$$K(s) = \begin{pmatrix} K_1(s) \\ K_2(s) \end{pmatrix}.$$

Using (7.1), the open-loop transfer matrix is given by:

$$G(s)K(s) = G_1(s)K_1(s) + \tilde{G}_1(s)K_2(s).$$

This controller structure corresponds to the combination of the two classical control structures used in canal control. Indeed, we can rewrite this controller as the sum of a distant downstream controller and a local upstream one:

$$G(s)K(s) = \underbrace{G_1(s)K_1(s)}_{\text{distant downstream controller}} + \underbrace{\tilde{G}_1(s)K_2(s)}_{\text{local upstream controller}}.$$

The two classical control policies for irrigation canals can therefore be interpreted from a control point of view as monovariable cases of a multivariable controller structure.

The control objectives are therefore threefold:

- To maintain the downstream water level by rejecting unmeasured perturbations induced by water users
- To ensure that the effect of perturbations on the downstream discharge are only transient
- To ensure robustness with respect to static gain error of actuators (at least 6 dB of gain margin)

The second objective ensures that, in steady-state, a water demand in the pool can only be satisfied by the upstream discharge.

8.2.2 Mixed Controller Structure

Let us examine the design specifications with respect to the control structure. If the required performance with respect to the user can be satisfied by a distant downstream controller, then there is no need to mix the control methods. However, if the distant downstream controller cannot satisfy the real-time performance specification, it is possible to use the local upstream control for this purpose. It is then necessary to add a constraint on u_2. Since the discharge needs to come from upstream, we would like to use u_2 only for transients, and that in steady-state, only u_1 has an effect on y.

This can be viewed as a problem of substitution of actuators, and can be taken into account with an input cascade framework [2].

The (rapid) upstream control u_2 is used to control the output y_1:

$$u_2 = K_2(s)(r_1 - y_1).$$

And the (slow) distant downstream control u_1 is used to regulate u_2 to a reference r_{u_2} ($r_{u_2} = 0$ in steady-state):

$$u_1 = K_{1b}(s)(r_{u_2} - u_2).$$

r_{u_2} corresponds to the desired flow of the downstream pool. If $r_{u_2} = 0$, one obtains a multivariable controller as noted above with $K_1 = -K_{1b}K_2$.

The controller structure can be schematized as in Fig. 8.1.

This control scheme enables us to mix local upstream and distant downstream control schemes, thus enabling the manager to trade-off between water management requirements and performance with respect to the water user.

Indeed, with the local upstream controller K_2, the relation between the reference r_1, the perturbation p_1, and the control u_2 becomes:

$$u_2 = \frac{K_2}{1+\tilde{G}_1 K_2} r_1 - \frac{K_2 \tilde{G}_1}{1+\tilde{G}_1 K_2} p_1 + \frac{K_2 \tilde{G}_1}{1+\tilde{G}_1 K_2} e^{-\tau_1 s} u_1.$$

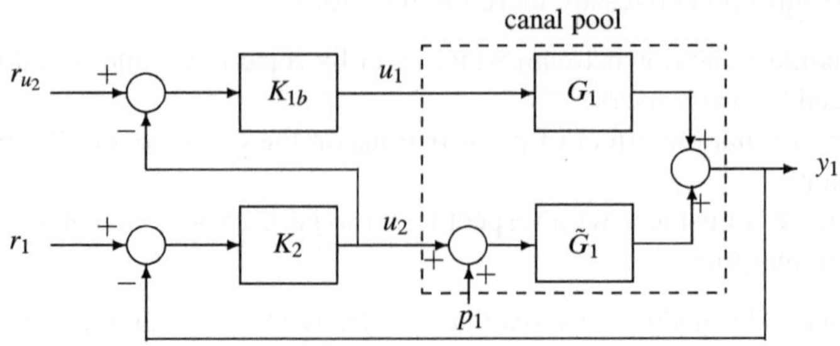

Fig. 8.1 Cascade architecture with two control variables (u_1, u_2) to control one output y_1

Since $\tilde{G}_1$ has an integrator, the transfer function between r_1 and u_2 is zero for low frequencies, and the transfer function between p_1 and u_2 is equal to -1. In this way, the local upstream control gives a "copy" of the perturbation for low frequencies. The proposed mixed controller, which uses u_1 to control u_2, will therefore reject low frequencies perturbations, leading to a substitution between u_2 and u_1.

8.2.3 Analysis of Control Substitution

Let us illustrate the way the substitution works. The water delivered by the upstream or downstream gates can be compared to the one needed by the offtake. Figure 8.2 depicts the dimensionless discharges along time at the downstream end of the pool, in response to an unpredicted withdrawal. The discharge is first compensated by decreasing the downstream flow, i.e., by withdrawing some discharge from the downstream pool. Then, after some delay, the discharge released from upstream reaches the downstream end of the pool, and there is a substitution between the discharge coming from upstream and the one taken from downstream. Finally, the required discharge comes mainly from upstream, as required in order to have a good water management. The resulting discharge at that point is very close to the one that would have been delivered by a pure local upstream controller. However, in steady-state (or for low frequencies), the discharge comes from upstream, since the downstream withdrawal tends asymptotically to zero.

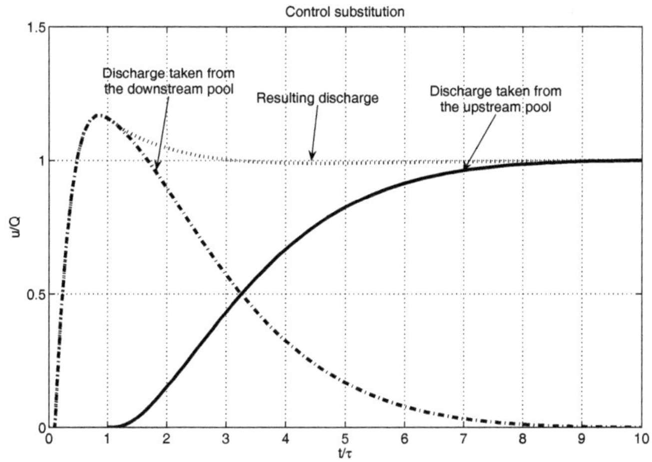

Fig. 8.2 Substitution between upstream and downstream discharges in response to an unpredicted withdrawal

8.3 Tuning of Mixed PI Controllers

8.3.1 Dimensionless System

We use the dimensionless system introduced in Chap. 7, (7.4). We assume that the PI controller K_2^* has already been tuned according to the method proposed in Sect. 7.4. We now want to tune the controller K_{1b}^* in order to ensure that the closed-loop system has the desired properties, such as gain and phase margins.

Let us denote by T_2^* the transfer function from u_1^* to u_2^*. T_2^* is given by:

$$T_2^*(s^*) = \frac{K_2^*(s^*)}{K_2^*(s^*) - s^*} e^{-s^*}.$$

We consider two cases, depending on whether the local upstream controller K_2^* has an integrator or not:

- When K_2^* is a proportional (P) controller, we have

$$T_2^*(s^*) = -\frac{k_{p2}^*}{s^* - k_{p2}^*} e^{-s^*}.$$

  Then T_2^* corresponds to a first-order with delay system, with a static gain equal to 1. Let us remember that the proportional coefficient k_{p2}^* is negative, therefore the pole of $T_2^*(s^*)$ is stable.
- When K_2^* is a PI controller, we have

$$T_2^*(s^*) = -\frac{k_{p2}^*(s^* + 1/T_{i2}^*)}{s^{*2} - k_{p2}^* s^* - k_{p2}^*/T_{i2}^*} e^{-s^*}.$$

Then T_2^* corresponds to a second-order with delay system. The zero in $-1/T_{i2}^*$ can lead to a resonance peak.
We add a first-order filter to remove this resonance peak:

$$f_2^*(s^*) = \frac{1}{1 + T_{i2}^* s^*}.$$

The filtered transfer function is then equal to:

$$T_2^*(s^*) f_2^*(s^*) = \frac{e^{-s^*}}{1 + T_{i2}^* s^* - T_{i2}^*/k_{p2}^* s^{*2}}.$$

In both cases, for low frequencies, the transfer function T_2^* is close to a pure delay e^{-s^*}. We will use this approximation to derive analytical design rules for the controller K_{1b}^*.

8.3.2 *Mixed PI Controller Tuning with Gain and Phase Margins*

We want to design a PI controller $K_{1b}^*(s^*) = k_{pb}^*\left(1 + \frac{1}{T_{ib}^* s^*}\right)$ to control the system represented by the transfer function $T_2^*(s^*)$ (or the filtered transfer function when K_2^* is a PI controller), using the low frequency approximation of $T_2^*(s^*)$ by a pure delay.

To study the frequency response of the controlled system, we evaluate the open-loop transfer function $K_{1b}^*(s^*)T_2^*(s^*)$ on the imaginary axis $s^* = j\omega^*$, where ω^* is the dimensionless frequency and j the imaginary number such that $j^2 = -1$. This leads to:

$$|K_{1b}^*(j\omega^*)e^{-j\omega^*}| = k_{pb}^*\sqrt{1 + \frac{1}{T_{ib}^{*2}\omega^{*2}}}, \tag{8.1a}$$

$$\arg(K_{1b}^*(j\omega^*)e^{-j\omega^*}) = -90 - \frac{180}{\pi}\omega^* + \frac{180}{\pi}\arctan(T_{ib}^*\omega^*). \tag{8.1b}$$

Let us denote by ω_{cb}^* the dimensionless crossover frequency where the gain of the open-loop equals 1, and ω_{180b}^* the frequency where the phase of the open-loop equals $-180°$.

To get a desired gain margin ΔG, the proportional coefficient k_{pb}^* needs to verify:

$$k_{pb}^* \leq 10^{-\frac{\Delta G}{20}}. \tag{8.2}$$

We first use this value to obtain initial values for the integral time T_{ib}^* and the crossover frequency ω_{cb}^*.

At the crossover frequency ω_{cb}^*, one has $|T_2^*(j\omega_{cb}^*)K_{1b}^*(j\omega_{cb}^*)| = 1$, which leads to:

$$T_{ib}^*\omega_{cb}^* = k_{pb}^*\left(1 - k_{pb}^{*2}\right)^{-1/2}. \tag{8.3}$$

In order to get a desired phase margin $\Delta\Phi$, one should choose T_{ib}^* such that:

$$\arctan(T_{ib}^*\omega_{cb}^*) = \frac{\pi}{180}\Delta\Phi - \frac{\pi}{2} + \omega_{cb}^*. \tag{8.4}$$

Using (8.3) and $\arctan(\frac{x}{\sqrt{1-x^2}}) = \arcsin(x)$, we finally get:

$$\omega_{cb}^* = \frac{\pi}{2} - \frac{\pi}{180}\Delta\Phi_b + \arcsin(k_{pb}^*). \tag{8.5}$$

T_{ib} is then given by (8.3).

The inequality $\omega_{cb}^* > 0$ implies that the phase margin is limited:

$$\Delta\Phi_b \leq \Delta\Phi_{\max b} = 90 + \frac{180}{\pi}\arcsin(k_{pb}^*).$$

One should therefore choose a phase margin lower than $\Delta\Phi_{\max b}$.

The frequency ω_{180b}^* can be approximated if we assume that $\arctan(T_{ib}^*\omega_{180b}^*) \approx \pi/4$. Such an approximation is valid if $1/T_{ib}^* \approx \omega_{180b}^*$. Then, using (8.1b), one gets:

$$\omega_{180}^* = \frac{3\pi}{4}. \tag{8.6}$$

Finally, the proportional gain can be chosen as follows:

$$k_{pb}^* = \frac{10^{-\frac{\Delta G}{20}}}{\sqrt{1 + \frac{1}{T_{ib}^{*2}\omega_{180}^{*2}}}}.$$

8.3.3 Robustness Analysis

In the case of one pool, the input robustness margin can be evaluated using the structured singular value with respect to diagonal complex or real uncertainties. Let us denote by $\mu_1 = \max_\omega \mu_\Delta(T_u(j\omega))$ and $\mu_2 = \max_\omega \mu_\Delta(S_u(j\omega))$, where S_u and T_u are, respectively, the input sensitivity and input complementary sensitivity functions. Then, the input complex or real gain margin ΔG verifies:

$$\Delta G \in \left(1 - \frac{1}{\mu_1}, 1 + \frac{1}{\mu_1}\right) \bigcup \left(\frac{\mu_2}{\mu_2 + 1}, \frac{\mu_2}{\mu_2 - 1}\right). \tag{8.7}$$

In our case, due to the specific structure of the controller, the input sensitivity functions have the following expressions:

$$S_u = S_y \begin{pmatrix} 1+\tilde{G}_1K_2 & -\tilde{G}_1K_1 \\ -G_1K_2 & 1+G_1K_1 \end{pmatrix},$$

and

$$T_u = S_y \begin{pmatrix} G_1K_1 & \tilde{G}_1K_1 \\ G_1K_2 & \tilde{G}_1K_2 \end{pmatrix},$$

with S_y the output sensitivity function $S_y = (1+G_1K_1+\tilde{G}_1K_2)^{-1}$.

Let us evaluate the structured singular value of T_u with respect to a diagonal complex uncertainty Δ. In the case of a 2×2 matrix A, the structured singular value μ_Δ is equal to the upper bound [3]:

$$\mu_\Delta(A) = \inf_{D\in\mathscr{D}} \overline{\sigma}(DAD^{-1}).$$

with $\mathscr{D} = \{\text{diag}(d,1) : d \in \mathbb{C}\}$.

After the tedious but straightforward manipulations detailed in Appendix F, the structured singular value of T_u can be computed explicitly:

$$\mu_\Delta(T_u) = |S_y|(|G_1K_1| + |\tilde{G}_1K_2|). \tag{8.8}$$

A similar expression can be obtained for S_u.

In the case of frequency decoupling between both designed monovariable controllers, the multivariable input gain margin is directly related to the input gain margin of both controllers. Indeed, if $|\tilde{G}_1K_2| \ll |G_1K_1|$ in low frequencies and $|G_1K_1| \ll |\tilde{G}_1K_2|$ in high frequencies, there is "frequency decoupling":

- In low frequencies, one has $\mu_\Delta(T_u) \approx (1+G_1K_1)^{-1}|G_1K_1|$, which is the complementary sensitivity function of the distant downstream controller,
- In high frequencies, $\mu_\Delta(T_u) \approx (1+\tilde{G}_1K_2)^{-1}|\tilde{G}_1K_2|$, which is the complementary sensitivity function of the local upstream controller.

Frequency decoupling therefore ensures that SISO robustness margins are recovered in the mixed control setup for the control of a single pool.

8.3.4 Time Domain Analysis

We first design a local upstream PI controller based on Sect. 7.4, assuming that the sampling time period is equal to one fifth of the time-delay. Then, using a gain margin of 10 dB and a phase margin of 43°, this leads to $k_{p2}^* = -4.7$ and $T_{i2}^* = 0.6$. Then, we design a filtered PI controller for the distant downstream part, according to the results developed earlier in this chapter. In the following, we compare the effects of the gain margin and the phase margin on the time domain simulations.

Figure 8.3 depicts the effect of the gain margin on the time domain simulations of the mixed PI controller with filter on a dimensionless ID model for a canal pool.

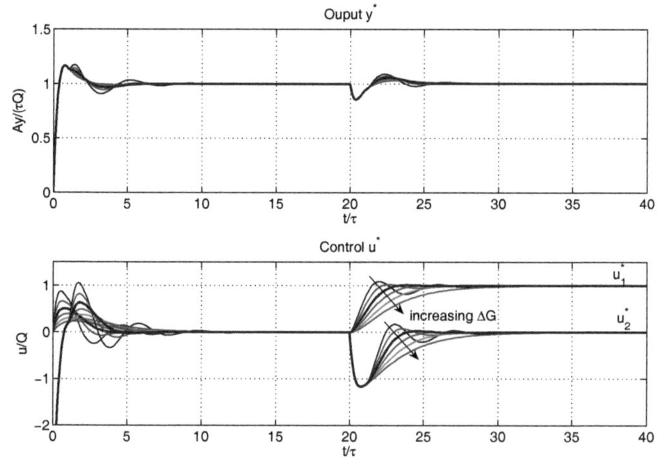

Fig. 8.3 Dimensionless linear simulation of a mixed PI controller for $\Delta\Phi = 0.7\Delta\Phi_{\max}$ and values of the gain margin ΔG from 4 to 19 dB with increments of 3 dB

We observe that increasing the gain margin leads to a longer time of substitution between the upstream discharge u_1^* and the downstream discharge u_2^*. A similar effect occurs when one increases the phase margin ratio. This has only a slight effect on the controlled output, since it is controlled mainly via the local upstream controller.

8.4 Parameterizing the Amount of Local Upstream Control

The previous method allows us to design a mixed controller with a large amount of local upstream control. Now, we extend this mixed control structure to be able to arbitrarily choose the amount of local upstream control, with a parameter $\theta \in [0,1]$.

To do this, we propose to first design a pure distant downstream controller $K_1(s)$, and a mixed controller composed of a proportional local upstream controller K_2 and a controller $K_{1b}(s)$. In that case, it is necessary to remove the integrator in the local upstream controller, otherwise it will always react when the output error is not zero.

The resulting controller is computed as the weighted sum of both controller:

$$u_1 = \theta K_{1b}(s)(r_{u_2} - u_2) + (1-\theta)K_1(s)(r_1 - y_1).$$

The controller structure can be schematized as in Fig. 8.4.

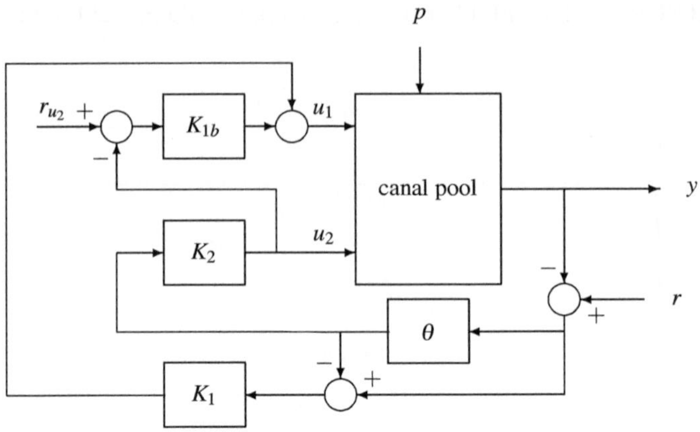

Fig. 8.4 Parameterized mixed control structure

This parameterization enables us to recover a pure distant downstream controller equal to $K_1(s)$ when $\theta = 0$, with no local upstream control. When θ becomes strictly positive, one adds some local upstream control to the multivariable controller. We recover the pure mixed controller when $\theta = 1$. This method provides a way to go continuously from the distant downstream to the mixed controller.

This can only be done when the local upstream controller has no integrator, otherwise even a small value of θ can lead to a large local upstream control, due to the integrator effect.

8.4.1 Controller Tuning

The distant downstream controller $K_1(s)$ is tuned according to the rules specified in Chap. 7. The local upstream controller K_2 is a simple proportional controller, also tuned using the rules specified in Chap. 7. In that case, the tuning rule is very simple, since the desired gain margin directly determines the proportional gain. Finally, the controller $K_{1b}(s)$ is a PI controller tuned according to the rules given in the previous section.

Now, one can choose parameter θ in order to specify the amount of local upstream control. As an example, in Fig. 8.5 we provide the relation between the amount of local upstream control and the value of parameter θ. The percentage of local upstream control is relative to the case of the pure mixed controller, obtained when $\theta = 1$. We observe that the relationship is not linear, and that even small values of θ lead to a large effect in terms of local upstream control. However, one can easily select the value of θ corresponding to a given amount of local control input.

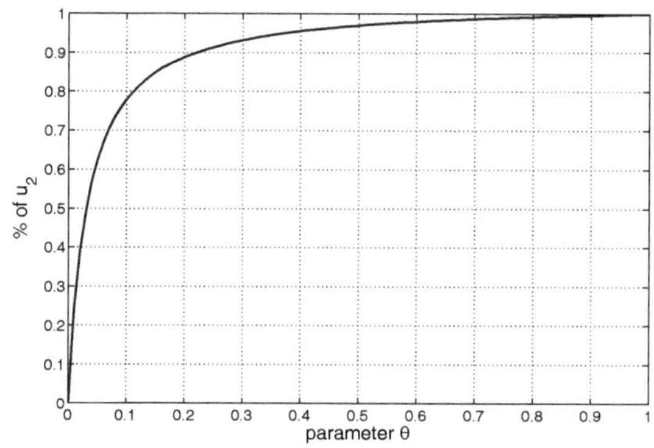

Fig. 8.5 Effect of parameter θ on the control inputs and the controlled output for a canal pool

The proposed parameterization therefore provides a useful tool to the canal manager: it enables the manager to select the amount of desired local upstream control for a given canal pool and directly obtain the controller.

8.4.2 Time Domain Simulations

To illustrate this,in Fig. 8.6 we depict the control actions and the controlled output obtained with various values of parameter θ.

We observe that for $\theta = 0.1$, we already have a very large improvement in the water level control, since the maximum error is reduced by a factor of 2, and the error in terms of volume is about only 25% of the one achieved with the pure distant downstream controller. There is also a decrease in the volume of water released from upstream, since this volume is now taken from the one sent further downstream.

This method paves the way for an optimization procedure, where one can select the controller structure, and then optimize the value of parameter θ in order to fulfill control design objectives.

One could also use this control structure to design a nonlinear controller that can switch from distant downstream to mixed control according to some specifications, such as security issues (local upstream control is selected to prevent overtopping), or water saving issues (local upstream control is only used when the error is negative, in order to save water). One can use the nonlinear μ analysis to study the stability of such control nonlinear schemes [1].

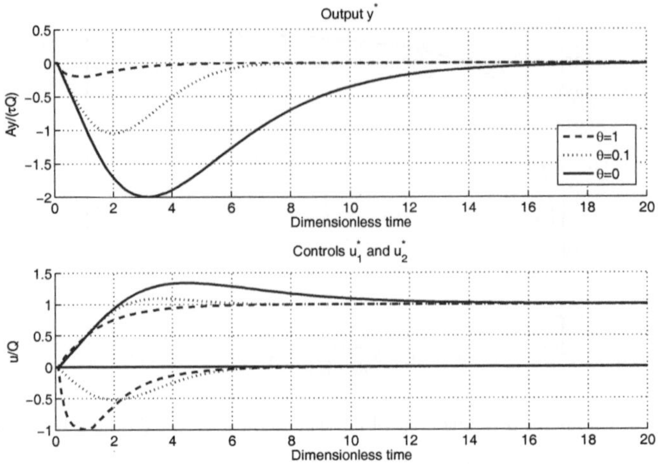

Fig. 8.6 Effect of parameter θ on the control inputs and the controlled output for a canal pool

8.5 Summary

In this chapter, we have presented the mixed control policy for a canal pool, which mixes distant downstream and local upstream control methods. The control structure ensures that the discharge comes mainly from upstream using distant downstream control, and that local upstream control is used only temporarily to respond to rapidly changing demands. This mixed control structure has the ability to trade-off the conflicting objectives: water management and performance with respect to the user. We proposed a tuning method based on PI controllers, leading to a pure mixed controller, and a parameterization enabling to continuously pass from a distant downstream controller to the mixed controller.

References

[1] Fromion V, Scorletti G, Ferreres G (1999) Nonlinear performance of a PID controlled missile: an explanation. Int J Robust Nonlin Contr 9(8):485–518

[2] Skogestad S, Postlethwaite I (1998) Multivariable feedback control. Analysis and design. Wiley, New York

[3] Zhou K, Doyle J (1998) Essentials of robust control. Prentice Hall, Upper Saddle River, NJ

Chapter 9
Open-loop Control of a Canal Pool

Abstract This chapter considers the design of feedforward boundary control laws for a canal pool. We first consider the problem of delivering water downstream according to a prespecified schedule by using the upstream discharge. The controller design is based on frequency domain methods. We then turn towards the similar problem of canceling known disturbances using boundary control. Perfect rejection of measured perturbations at one boundary is obtained by controlling the other boundary. Frequency domain comparisons and time domain simulations illustrate the good performance of the feedforward boundary controller.

9.1 Introduction

Open channels have delayed dynamics, and feedforward or open-loop control is useful in systems subject to large time-delays. The management of open channels requires being able to react to variations in measured variables (feedback) but also to anticipate in order to compensate for the effect of known future variations (feedforward). In this chapter, we deal with the open-loop control of a canal pool, with different problem setups. We first consider the feedforward problem for a canal pool managed with a fixed rotational schedule. In that case, the manager knows in advance the discharge to be delivered at the downstream end of the pool. The open-loop control problem consists of computing the upstream discharge $q(0,t)$ that has to be released in order to meet the downstream demand $q(L,t)$, while keeping the downstream level close to its reference $y(L,t) = 0$. The frequency domain approach developed in this book enables us to find explicit solutions to this problem. We also consider the case where the control action variable is not the upstream discharge, but the upstream gate opening. In that case, the feedforward control incorporates a supplementary term due to the effect of the propagation of the downstream perturbation along the pool.

We then turn towards the similar problem of canceling known disturbances using boundary control. It is shown that a feedforward controller can exactly cancel the disturbances generated at the upstream end using the downstream discharge. We also evaluate the effect of uncertainty in open-loop control.

9.2 Open-loop Boundary Control of a Canal Pool

9.2.1 Problem Statement

We consider a canal pool of length L, which is represented by the linearized Saint-Venant equations:

$$\frac{\partial \xi}{\partial t} + \begin{pmatrix} 0 & 1 \\ \alpha\beta & \alpha - \beta \end{pmatrix} \frac{\partial \xi}{\partial x} + \begin{pmatrix} 0 & 0 \\ -\gamma & \delta \end{pmatrix} \xi = 0. \tag{9.1}$$

This pool is controlled by gates (see Chap. 6) which can be operated to control the discharge. Let us denote by $q_1(t) = q(0,t)$ the upstream discharge, $q_2(t)$ the outgoing discharge and $w_1(t)$ the downstream withdrawal. The downstream discharge is then given by $q(L,t) = q_2(t) + w_1(t)$.

The control problem is to compute $q_1(t)$ such that $q(L,t)$ follows a given trajectory, and that the downstream water level remains constant, i.e., $y(L,t) = 0$. This problem is closely related to the so-called motion planning problem [6].

9.2.1.1 Solution with Upstream Discharge

Let us write these specifications using the frequency domain approach. Using the linearized Saint-Venant input–output transfer matrix (3.83), the condition $y(L,t) = 0$ gives:

$$\hat{y}(L,s) = p_{21}(s)\hat{q}(0,s) + p_{22}(s)\hat{q}(L,s),$$

which implies formally that:

$$\hat{q}(0,s) = -\frac{p_{22}(s)}{p_{21}(s)}\hat{q}(L,s). \tag{9.2}$$

We know that transfer function $p_{21}(s)$ contains a pure time-delay $e^{-\tau_1 s}$, which is not causally invertible. However, if we know the downstream discharge in advance, i.e., if at time t we know $q(L,t+\tau_1)$, the solution is causal, since we know that there are no other noninvertible elements in $\tilde{p}_{21}(s)$.

Equation (9.2) gives the open-loop control in terms of the upstream discharge $\hat{q}(0,s)$ to deliver a given downstream discharge $\hat{q}(L,s)$, in the frequency domain. We will examine below various ways to invert this relation back into the time domain.

9.2.1.2 Solution with an Upstream Gate Opening

Let us now assume that the upstream discharge is delivered by a gate with a constant upstream water level. The linearized gate equation is:

$$q(0,t) = k_u y_0(t) - k_d y(0,t) + k_w u_1(t),$$

with y_0 the gate upstream level, $y(0,t)$ the gate downstream level and u_1 the gate opening. k_u, k_d, and k_w are the gate coefficients, obtained by linearizing the gate equation. Assuming a constant water level upstream of the gate leads to $y_0(t) = 0$, and the evolution of the water level downstream of the gate $y(0,t)$ can be computed with the linearized Saint-Venant model:

$$\hat{y}(0,s) = p_{11}(s)\hat{q}(0,s) + p_{12}(s)\hat{q}(L,s).$$

Then, the open-loop control in terms of the upstream gate opening is obtained as follows:

$$\hat{u}_1(s) = \frac{1}{k_w}\left[\hat{q}(0,s) + k_d\hat{y}(0,s)\right] \tag{9.3a}$$

$$= \frac{1}{k_w}\left[-\frac{p_{22}(s)}{p_{21}(s)} + k_d\left(p_{12}(s) - \frac{p_{11}(s)p_{22}(s)}{p_{21}(s)}\right)\right]\hat{q}(L,s). \tag{9.3b}$$

The open-loop control appears to be the sum of two terms, one linked to the downstream discharge variation, and the other one linked to the change in the upstream discharge due to the variation of the gate downstream level.

Equations (9.2) and (9.3b) provide the general solution for the open-loop control problem in the frequency domain. We now consider various ways to convert this result in the time domain, by examining the different cases.

9.2.2 Horizontal Frictionless Channel Case

In the horizontal frictionless channel case, an analytical solution can be obtained for the problem.

9.2.2.1 Exact Solution

Using the expressions for the input–output transfer functions given by (3.11), the open-loop control input (9.2) can be expressed in terms of the sum of a pure delay and advance of the downstream discharge:

$$q(0,t) = \frac{\alpha}{\alpha+\beta}q(L,t+\tau_1) + \frac{\beta}{\alpha+\beta}q(L,t-\tau_2). \tag{9.4}$$

This expression can be easily understood using the characteristics variables.

Expression with the Characteristics Variables

Using (2.29) giving the change of variable leading to the characteristics variables and (E.13) giving the solution in the time domain, the open-loop control problem can be restated as follows. The boundary condition is imposed at $x = L$:

$$\chi_1(L,t) = \chi_2(L,t) = q(L,t).$$

Therefore, the value of the characteristics variables can be deduced at $x = 0$:

$$\chi_1(0,t) = \chi_1(L,t+\tau_1) = q(L,t+\tau_1)$$
$$\chi_2(0,t) = \chi_2(L,t-\tau_2) = q(L,t-\tau_2).$$

The upstream discharge and water level can be computed using the inverse transform given by (2.30):

$$y(0,t) = \frac{1}{T_0(\alpha+\beta)}[q(L,t+\tau_1) - q(L,t-\tau_2)]$$
$$q(0,t) = \frac{1}{\alpha+\beta}[\alpha q(L,t+\tau_1) + \beta q(L,t-\tau_2)].$$

We recover (9.4) for the open-loop control in terms of the upstream discharge. In the case where the control action variable is the gate opening, we use (9.3a) to compute the gate opening, which leads to:

$$u_1(t) = \frac{1}{k_w(\alpha+\beta)}\left[\left(\alpha+\frac{k_d}{T_0}\right)q(L,t+\tau_1) + \left(\beta-\frac{k_d}{T_0}\right)q(L,t-\tau_2)\right]. \quad (9.5)$$

This provides an explicit solution for the open-loop problem in the horizontal frictionless open channel case.

It is also possible to use the rational approximations developed in Chap. 4 to compute the solution.

9.2.2.2 Rational Approximation

We cannot use directly the modal decomposition of transfer functions $p_{ij}(x,s)$ given by (4.2) to compute the open-loop control. In fact, if one uses the rational approximation in (9.2) directly to compute the open-loop, this will result in an unstable simulation. As we have already seen, the solution given by (9.2) is noncausal. This means that the only invertible part of transfer function $p_{21}(s)$ is the outer part $\tilde{p}_{21}(s)$. The rational approximation of $p_{21}(s)$ therefore contains nonminimum phase zeros, which correspond to unstable poles if one tries to invert the transfer function.

However, one may use the rational approximations of the delay-free system $\tilde{p}_{21}(s)$ to obtain an approximate solution for the open-loop control problem (see the approximation given by (4.3)). In that case, (9.2) and (9.3b) are computed by replacing the infinite dimensional transfer functions $\tilde{p}_{ij}(s)$ by their rational approximations, and the advance part is incorporated into the desired downstream discharge.

9.2.2.3 Simulation Results

We first compare the open-loop control input computed with the exact solution given by (9.4) and the one based on the rational approximations.

Then we use these control inputs to compute the water level deviation using (3.33b), which provides an explicit solution to the linearized Saint-Venant equations in the horizontal frictionless case.

The desired output discharge $q(L,t)$ is specified as a step increase from 0 to 1 at time $t = \tau_1$. The exact solution for the open-loop control input is calculated using (9.4). Rational approximations are computed using the method described above.

Figure 9.1 depicts the open-loop control inputs and the resulting deviations in downstream boundary water levels for various rational approximations of the open-loop controller.

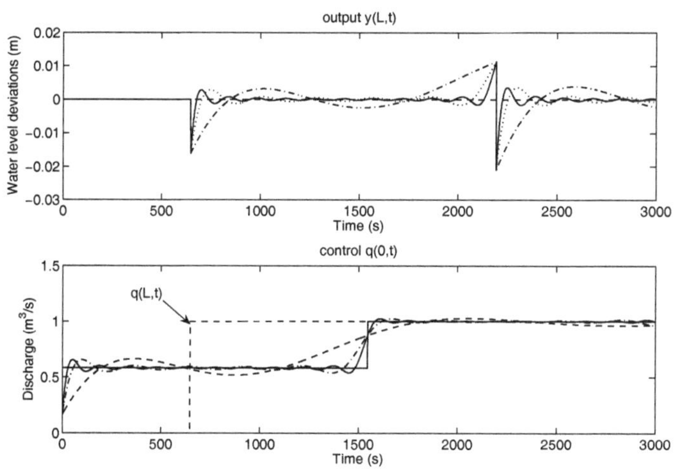

Fig. 9.1 Open-loop control inputs and resulting deviations in downstream water level for a horizontal frictionless channel. Comparison between the exact solution and various rational approximations of the transfer functions with 1 pole, 11 poles, and 21 poles

We observe that the higher the order of the approximation, the lower the error in the water level. However, since this is an open-loop control, all the model errors are visible in the output.

9.2.3 Uniform Flow Case

9.2.3.1 Exact Solution

In the uniform flow case, we can also derive an implicit solution of the feedforward control with the expressions using the characteristics change of variable given by (C.2):

$$\hat{\zeta}_1(x,s) = \hat{q}(x,s) + \frac{T_0 s}{\lambda_2(s)}\hat{y}(x,s),$$

$$\hat{\zeta}_2(x,s) = \hat{q}(x,s) + \frac{T_0 s}{\lambda_1(s)}\hat{y}(x,s).$$

Since we require that $\hat{y}(L,s) = 0$, the boundary condition at $x = L$ can be written as:

$$\hat{\zeta}_1(L,s) = \hat{\zeta}_2(L,s) = \hat{q}(L,s).$$

The first generalized characteristic verifies (3.54):

$$\hat{\zeta}_1(x,s) = e^{\lambda_1(s)x}\hat{\zeta}_1(0,s).$$

As we have seen in (3.72), the first eigenvalue $\lambda_1(s)$ can be decomposed as:

$$\lambda_1(s) = -r_1 - \frac{s}{\alpha} + \theta \hat{F}(s),$$

with $\hat{F}(s)$ the infinite dimensional filter given by (3.68) and $\theta = \frac{(\alpha+\beta)}{2\alpha\beta}$.

This enables us to express the generalized characteristic $\hat{\zeta}_1(L,s)$ as a function of its value at $x = 0$:

$$\hat{\zeta}_1(L,s) = e^{-r_1 L}\left[1 + \hat{H}(\theta L,s)\right]e^{-\tau_1 s}\hat{\zeta}_1(0,s), \tag{9.6}$$

with $\hat{H}(l,s)$ given by (3.69).

Conversely, the generalized characteristic $\hat{\zeta}_2(0,s)$ can be expressed as a function of its value at $x = L$:

$$\hat{\zeta}_2(0,s) = e^{-r_2 L}\left[1 + \hat{H}(\theta L,s)\right]e^{-\tau_2 s}\hat{\zeta}_2(L,s).$$

Finally, using the inverse Laplace transform of $\hat{H}(s)$, we deduce the value of the generalized characteristics variables at $x = 0$ using (3.73):

$$\zeta_1(0,t) = e^{r_1 L} q(L,t+\tau_1) - \int_0^t H(\theta L, t-\tau)\zeta_1(0,\tau)\mathrm{d}\tau$$

$$\zeta_2(0,t) = e^{-r_2 L}\left[q(L,t-\tau_2) + \int_0^t H(\theta L, t-\tau) q(L,\tau-\tau_2)\mathrm{d}\tau\right],$$

with $H(l,t)$ given by (3.71):

$$H(l,t) = \frac{a l e^{-bt}}{\sqrt{t^2-l^2}} I_1\left(a\sqrt{t^2-l^2}\right),$$

and where we have used the fact that $\zeta_1(L,t) = \zeta_2(L,t) = q(L,t)$.

The first equation is implicit, and we need to compute $\zeta_1(0,t)$ as a solution of a fixed point problem. Once this is done, the upstream discharge and water level variations can be computed using (C.4):

$$\begin{aligned} y(0,t) &= \frac{1}{T_0(\alpha+\beta)}[\zeta_1(0,t) - \zeta_2(0,t)] \\ &+ \frac{a}{T_0(\alpha+\beta)} \int_0^t e^{-b(t-\tau)} I_1(a(t-\tau))(\zeta_1(0,\tau) - \zeta_2(0,\tau))\mathrm{d}\tau \\ &+ \frac{c_y}{T_0(\alpha+\beta)} \int_0^t e^{-b(t-\tau)} I_0(a(t-\tau))(\zeta_1(0,\tau) - \zeta_2(0,\tau))\mathrm{d}\tau, \qquad (9.7a) \\ q(0,t) &= \frac{1}{\alpha+\beta}[\alpha\zeta_1(0,t) + \beta\zeta_2(0,t)] \\ &+ \frac{a(\alpha-\beta)}{2(\alpha+\beta)} \int_0^t e^{-b(t-\tau)} I_1(a(t-\tau))(\zeta_1(0,\tau) - \zeta_2(0,\tau))\mathrm{d}\tau \\ &+ \frac{c_q(\alpha-\beta)}{2(\alpha+\beta)} \int_0^t e^{-b(t-\tau)} I_0(a(t-\tau))(\zeta_1(0,\tau) - \zeta_2(0,\tau))\mathrm{d}\tau, \qquad (9.7b) \end{aligned}$$

with

$$\begin{aligned} a^2 &= \frac{4\alpha\beta(\alpha\delta-\gamma)(\gamma+\beta\delta)}{(\alpha+\beta)^4}, \\ c_y &= \frac{(\alpha^2+\beta^2)\delta - (\alpha-\beta)\gamma}{(\alpha+\beta)^2}, \\ c_q &= \frac{2\alpha\beta}{(\alpha+\beta)^2}(2\gamma - (\alpha-\beta)\delta). \end{aligned}$$

When the control input is the upstream gate opening, the open-loop control input can be computed using (9.3a). This provides an implicit solution to the open-loop control problem in the uniform flow case. Explicit rational approximations can be computed based on the rational models developed in Chap. 4.

9.2.3.2 Rational Approximation

As in the horizontal frictionless case, it is also possible to directly compute the open-loop control by using rational approximations for the outer parts of the input–output transfer matrix elements $\tilde{p}_{ij}(s)$, following the results of Chap. 4.

The modal decomposition of transfer functions $\tilde{g}_{ij}(x,s)$ given by (4.10) can be used to compute the open-loop control. In that case, (9.2) and (9.3b) are computed by replacing the infinite dimensional transfer functions $\tilde{p}_{ij}(s)$ by their rational approximation.

9.2.3.3 Simulation Results

In Figs. 9.2 and 9.3, we compare the solutions obtained with rational approximations of various orders. It is clear from the graphs that the higher the order, the smaller the error in the downstream water level. We also observe that the open-loop control for the example canal 1 is rather close to that of the horizontal frictionless channel, while for the example canal 2, the effect of the second characteristics is barely visible. Rather surprisingly, the open-loop control input for the example canal 2 appears to be oscillating, while this is not the case for the example canal 1.

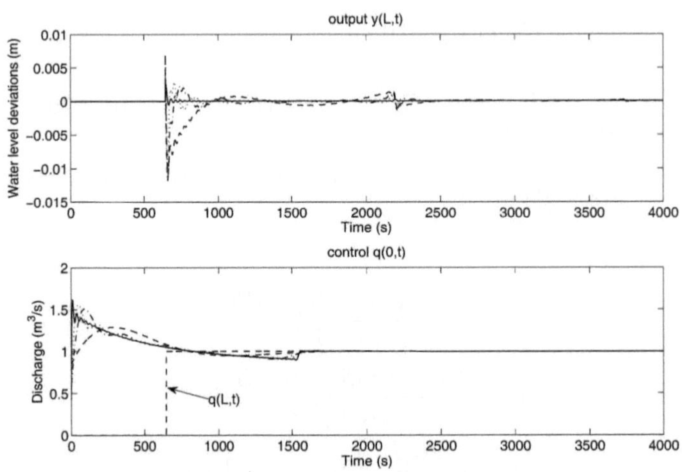

Fig. 9.2 Open-loop control inputs and resulting deviations in downstream water level for canal 1 at uniform flow. Comparison between various rational approximations of the transfer functions with 1 pole, 11 poles, and 21 poles

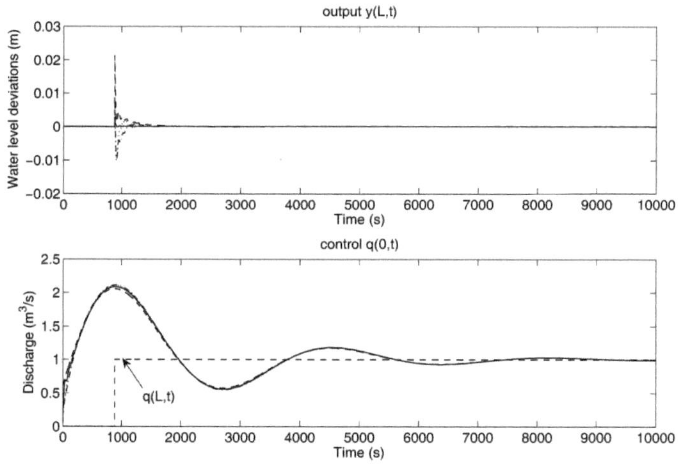

Fig. 9.3 Open-loop control inputs and resulting deviations in downstream water level for canal 2 at uniform flow. Comparison between various rational approximations of the transfer functions with 1 pole, 11 poles, and 21 poles

9.2.4 Nonuniform Flow Case

In the general nonuniform flow case, it is no longer possible to compute an analytical solution for the open-loop control problem. However, we can still use the rational approximations developed in Chap. 4 to solve the problem. Using a rational approximation of the transfer functions $\tilde{p}_{ij}(s)$ based on the modal decomposition, we study the effect of the flow nonuniformity on the open-loop control.

Figures 9.4 and 9.5 depict the open-loop control for a unit step change in downstream discharge for example canals 1 and 2 for nonuniform flow conditions. The accelerating flow corresponds to a downstream water level equal to $0.8Y_n$, and the decelerating flow condition corresponds to a downstream water level equal to $1.2Y_n$.

We can see in both figures that the accelerating flow conditions require larger variations in upstream discharge than the decelerating flow conditions to deliver the same downstream flow. This confirms the fact that nonuniform accelerating flow is more difficult to control.

9.3 Exact Disturbance Cancelation Using Boundary Control

We now focus on a similar problem, the design of a feedforward control law in order to reject measured disturbances using boundary control. We assume that the input

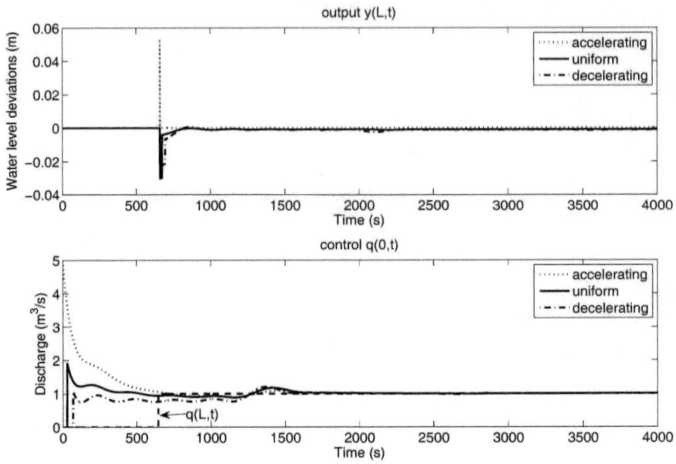

Fig. 9.4 Open-loop control inputs and resulting deviations in downstream water level for canal 1 for nonuniform flow

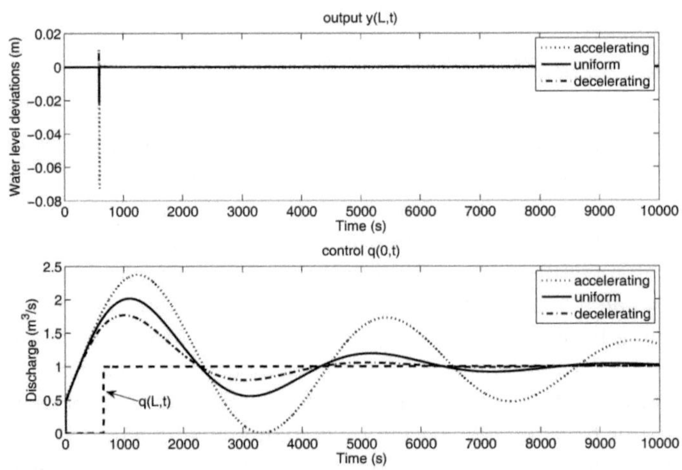

Fig. 9.5 Open-loop control inputs and resulting deviations in downstream water level for canal 2 for nonuniform flow

boundary condition at $x = 0$, $q(0,t)$ is a measured disturbance on the system and that the input boundary condition at $x = L$ is the control variable $q(L,t)$.

The control objective is therefore to design a feedforward boundary controller such that the boundary output variable $y(L,t)$ remains equal to 0.

9.3.1 Feedforward Controller Design

Using the input–output Saint-Venant transfer functions given by (3.50), it is easy to see that a perfect rejection of disturbances provoked by $\hat{q}(s,0)$ can be achieved by specifying $\hat{q}(s,L)$ as follows:

$$\hat{q}(L,s) = K_F(s)\hat{q}(0,s), \tag{9.8}$$

with

$$K_F(s) = -\frac{p_{21}(s)}{p_{22}(s)}. \tag{9.9}$$

In the uniform flow case, we have:

$$K_F(s) = \frac{(\lambda_2(s) - \lambda_1(s))e^{(\lambda_1(s)+\lambda_2(s))L}}{\lambda_2(s)e^{\lambda_2(s)L} - \lambda_1(s)e^{\lambda_1(s)L}}.$$

$K_F(s)$ is an irrational controller that is stable and causal. For its implementation, a rational approximation has to be computed.

In practice, since the system is marginally stable (it contains an integrator), it is not advised to design the open-loop controller represented by (9.9) and implement it with (9.8). Indeed, a constant input perturbation leads to an unbounded output for such an open-loop marginally stable system. Therefore, it is necessary to first *stabilize* the system with a feedback controller and then to design the open-loop controller for the closed-loop system.

For this purpose, let us assume that the system is stabilized with the following proportional boundary control:

$$\hat{q}(L,s) = k_u\hat{y}(L,s) + k_w\hat{w}(s), \tag{9.10}$$

where w is an additional boundary input, and $k_u > 0$ and $k_w \in \mathbb{R}^*$ are real constants[1].

The resulting closed-loop system becomes:

$$\hat{y}(L,s) = \frac{p_{21}(s)}{1-k_u p_{22}(s)}\hat{q}(0,s) + \frac{k_w p_{22}(s)}{1-k_u p_{22}(s)}w(s).$$

Then, for any $k_u > 0$, the closed-loop system is stable (see [3]).

Now, the perfect rejection of the effect of measured upstream boundary perturbations on the downstream boundary is obtained by designing $w(s) = K_{Fw}(s)q(s,0)$ such that $y(L,s) = 0$, i.e.:

$$K_{Fw}(s) = -\frac{1}{k_w}\frac{p_{21}(s)}{p_{22}(s)} = \frac{1}{k_w}K_F(s). \tag{9.11}$$

---

[1] As we have seen in Chap. 6, this can be achieved by a moveable hydraulic structure such as a weir or a gate, and w represents the gate opening or the weir elevation.

Therefore, for a perfect rejection, it is possible to compute $q(L,t)$ using the open-loop controller $K_F(s)$ and to implement it with $w(t) = \frac{q(L,t)}{k_w}$. This is equivalent to inverting (9.10) by assuming that $y(L,s) = 0$.

We focus below on the analysis of $K_F(s)$, keeping in mind that the implementation is done with a stabilizing feedback such as the one in (9.10).

9.3.2 Horizontal Frictionless Channel Case

9.3.2.1 Explicit Solution of the Open-loop Control

When $\gamma = \delta = 0$ the eigenvalues become $\lambda_1(s) = -\frac{s}{\alpha}$ and $\lambda_2(s) = \frac{s}{\beta}$.

Then, the open-loop controller $K_F(s)$ becomes:

$$K_F(s) = \frac{(1+k)e^{-s\frac{L}{\alpha}}}{1+ke^{-\tau s}},$$

where we have used the notations $k = \frac{\beta}{\alpha}$ and $\tau = \frac{L}{\alpha} + \frac{L}{\beta}$.

This infinite dimensional controller can be approximated by a series of delays, using the well-known series expansion:

$$\frac{1}{1+z} = \sum_{n=0}^{\infty}(-1)^n z^n,$$

which is valid for $|z| < 1$.

In our case, $0 < k < 1$ since $\beta < \alpha$, therefore the series converges and we get:

$$K_F(s) = \sum_{n=0}^{\infty}(-1)^n k^n (1+k)e^{-s(\frac{L}{\alpha}+n\tau)}. \tag{9.12}$$

In the time domain, this expression leads to an explicit solution for the open-loop control:

$$q(L,t) = \sum_{n=0}^{\infty}(-1)^n k^n (1+k) q\Big(0, t - \frac{L}{\alpha} - n\tau\Big). \tag{9.13}$$

This expression was first obtained by [1] using a Riemann invariants approach.

Our frequency domain approach allows us to evaluate the approximation error generated by the truncation. We evaluate below the H_∞ norm of the approximation error due to the truncation of the series (9.12).

9.3.2.2 Evaluation of the Truncation Error

When the infinite series (9.12) is truncated, this generates an approximation error, which can easily be evaluated. Indeed, suppose that this series is approximated by the first N elements of the series, and let us denote by $K_{F,N}^{[1]}(s)$ this truncated series. We have:

$$K_{F,N}^{[1]}(s) = \sum_{n=0}^{N} (-1)^n k^n (1+k) e^{-j\omega(\frac{L}{\alpha}+n\tau)}.$$

Then, let us evaluate the norm of the approximation error for $s = j\omega$:

$$\begin{aligned}\left|K_F(j\omega) - K_{F,N}^{[1]}(j\omega)\right| &= (1+k)\left|\sum_{n=N+1}^{\infty}(-1)^n k^n e^{-j\omega(\frac{L}{\alpha}+n\tau)}\right| \\ &= (1+k)k^{N+1}\left|\frac{e^{-j\omega\left(\frac{L}{\alpha}+(N+1)\tau\right)}}{1+ke^{-\tau j\omega}}\right| \\ &= \frac{(1+k)k^{N+1}}{\sqrt{1+k^2+2k\cos(\omega\tau)}}.\end{aligned}$$

Since $|\cos(\omega\tau)| \leq 1$, we have the double inequality:

$$k^{N+1} \leq \left|K_F(j\omega) - K_{F,N}^{[1]}(j\omega)\right| \leq \left(\frac{1+k}{1-k}\right)k^{N+1}.$$

Therefore the worst case approximation error (H_∞ norm) will always be larger than k^{N+1}. Since $k < 1$, this tends towards zero as N tends towards ∞. This convergence can, however, be slow especially for canals where α and β are very close.

9.3.3 Uniform Flow Case

9.3.3.1 Series Decomposition of the Feedforward Controller

In the general case, it is necessary to use a rational approximation of $K_F(s)$ before implementation. One possibility to this end is to expand the transfer function into a series, and then truncate the infinite series.

The poles of the controller are the solutions of the following equation:

$$e^{(\lambda_2(s)-\lambda_1(s))L} = \frac{\lambda_1(s)}{\lambda_2(s)}.$$

They can be shown to be stable, i.e., $\exists\varepsilon > 0$ such that for all k, $\Re(p_k) < -\varepsilon$, if and only if $\gamma \neq 0$ or $\delta \neq 0$.

Let us denote the poles by $(p_n)_{n\in\mathbb{Z}}$, and $p_{-n} = \bar{p}_n$. Since each pole has single multiplicity, the rational approximation problem reduces to finding the residues of $K_F(s)$ such that:

$$K_F(s) = \sum_{n=-\infty}^{\infty} \frac{a_n}{s - p_n}, \tag{9.14}$$

with $a_n = \lim_{s\to p_n}(s - p_n)K_F(s)$, i.e.,

$$a_n = \left[\frac{(\lambda_2 - \lambda_1)e^{(\lambda_1(s)+\lambda_2(s))L}}{(1 + L\lambda_2(s))\lambda_2'(s)e^{\lambda_2(s)L} - \lambda_1'(s)(1 + L\lambda_1(s))e^{\lambda_1(s)L}}\right](p_n),$$

with

$$\lambda_i'(s) = \frac{(\alpha - \beta)}{2\alpha\beta} + (-1)^i \frac{(\alpha + \beta)^2 s + (\alpha - \beta)\gamma + 2\alpha\beta\delta}{2\alpha\beta\sqrt{d(s)}},$$

and $d(s) = (\alpha + \beta)^2 s^2 + 2[(\alpha - \beta)\gamma + 2\alpha\beta\delta]s + \gamma^2$.

9.3.3.2 Rational Approximation

Delay-free Feedforward Controller

There are various possibilities for approximating the feedforward controller (e.g., (9.14) or (9.12)), but the series may not converge very rapidly. It may be necessary to use a large number of terms in order to get a good approximation. As we will show in the application, a better approximation can be achieved by extracting the delay $\exp(-\frac{L}{\alpha}s)$ from the controller and considering the delay-free controller:

$$\tilde{K}_F(s) = K_F(s)e^{\frac{L}{\alpha}s}.$$

In this case, a good low frequency approximate solution of order N is obtained by:

$$\tilde{K}_{F,N}(s) = 1 + \sum_{n=-N}^{N} \left(\frac{\tilde{a}_n}{s - p_n} + \frac{\tilde{a}_n}{p_n}\right), \tag{9.15}$$

with $\tilde{a}_n = \lim_{s\to p_n}(s - p_n)\tilde{K}_F(s)$, i.e.,

$$\tilde{a}_n = \frac{(\lambda_2(p_n) - \lambda_1(p_n))e^{(\lambda_1(p_n)+\lambda_2(p_n)+\frac{p_n}{\alpha})L}}{(1 + L\lambda_2(p_n))\lambda_2'(p_n)e^{\lambda_2(p_n)L} - \lambda_1'(p_n)(1 + L\lambda_1(p_n))e^{\lambda_1(p_n)L}}.$$

Finite Bandwidth Approximation

The above approaches have considered series decomposition of the feedforward controller, with a given order, leading to a given truncation error. Another interesting possibility is to consider a finite bandwidth approximation of the feedforward controller, with a bounded error for higher frequencies. This is coherent with the fact that the control is implemented by limited bandwidth actuators. In this case, the problem can be stated as an H_∞-like minimization one, where one tries to minimize the norm of the difference between the feedforward controller $K_F(s)$ and its rational approximation $K_{FA}(s)$:

$$\min |K_F(j\omega) - K_{FA}(j\omega)| \text{ for } \omega \in [0, \omega_0],$$

and

$$|K_F(j\omega) - K_{FA}(j\omega)| < K \text{ for } \omega > \omega_0.$$

Under some conditions, this problem is a convex one and has already been solved [4]. But this approach is not so simple to implement, since the frequency bandwidth of interest remains to be determined, and the system is uncertain in high frequencies. In order to take these uncertainties into account, it is more realistic to directly consider the robust feedforward control problem, as in [5].

9.3.4 Application

9.3.4.1 Closed-loop Controller

The stabilizing proportional boundary control is performed with a moveable weir, defined by the linearized equation:

$$\hat{q}(s, L) = k_u \hat{y}(s, L) + k_w \hat{w}(s),$$

with $k_u = 0.6\sqrt{2g}L_w$, where L_w is the length of the weir and $k_w = -k_u$. In the simulations, the weir length is equal to $0.5T_0$.

9.3.4.2 Feedforward Controller

Bode Diagrams

The feedforward controller is obtained following (9.9). Figure 9.6 depicts the Bode diagram of controller $K_F(s)$ for the example canal 1. Figure 9.6 compares the feedforward controllers $K_F(s)$ for three canals with the same length, same α and β, but with different slope and friction, leading to different γ and δ. It is clear from Fig. 9.6 that even small changes of the slope and the friction dramatically change the Bode diagram of the feedforward controller. This remark justifies our approach, which explicitly takes into account nonzero γ and δ parameters in the feedforward control design.

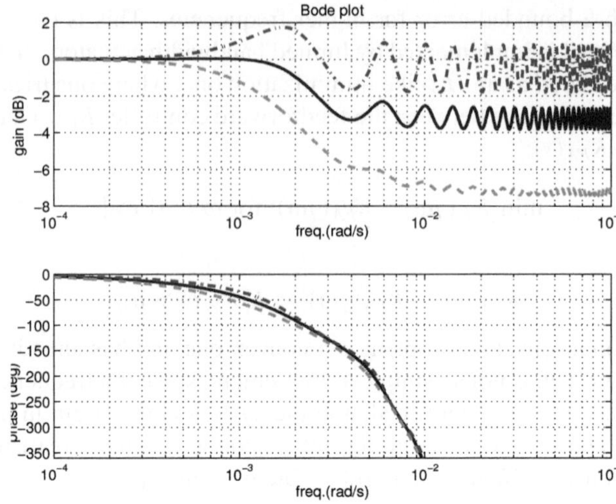

Fig. 9.6 Bode plot of $K_F(s)$ with $(\alpha, \beta, \gamma, \delta) = (4.63, 3.33, 2.7 \times 10^{-3}, 3 \times 10^{-3})$ (—). The two others plots are obtained with parameters γ and δ changed by $\pm 50\%$: $\gamma_1 = \gamma/1.5$, $\delta_1 = \delta/1.5$ (– – · – –), and $\gamma_2 = 1.5\gamma$, $\delta_2 = 1.5\delta$ (– –)

Rational Approximations

The rational approximations obtained with (9.14) are compared to $K_F(s)$ in Fig. 9.7. The Bode diagram shows that a higher number of poles gives a better approximation of the controller.

We compared the rational approximations obtained with (9.15) to $K_F(s)$. The Bode plot in Fig. 9.8 shows that a higher number of poles gives a better approximation of the controller. In this case, a good approximation is obtained with only two pairs of poles, leading to a delay-free controller of dimension 4.

Simulation Results

Figure 9.9 compares the water level $y(t, L)$ and the feedforward discharge $q(t, L)$ computed with controller $\tilde{K}_F(s)$ of (9.15) with five poles and the controller of [1] obtained with (9.13). It is clear from Fig. 9.9 that a controller designed by assuming $\delta = \gamma = 0$ gives large poorly damped oscillations of the control $q(t, L)$, leading to a large error in the downstream water level. On the contrary, when the damping linked to the nonzero slope and friction is taken into account, the control $q(t, L)$ is much smoother and the output error much smaller.

This is also clear from the Bode plot of Fig. 9.10, which compares the Bode diagram of the controller $K_F(s)$ to that of the controller of [1].

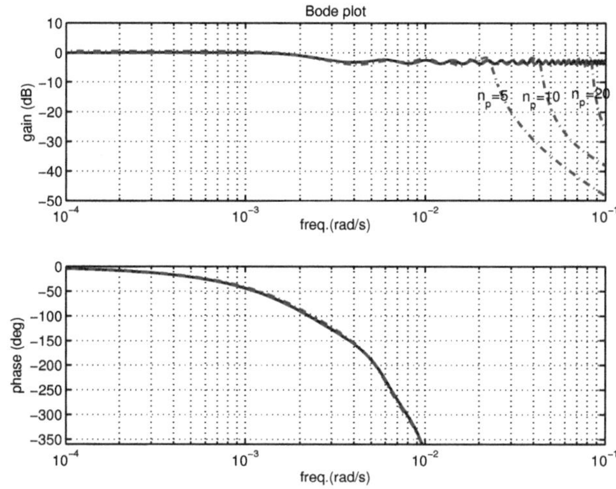

Fig. 9.7 Bode plot of $K_F(s)$ (—) and rational approximations obtained with (9.14), using 5, 10 and 20 poles (– – · – –)

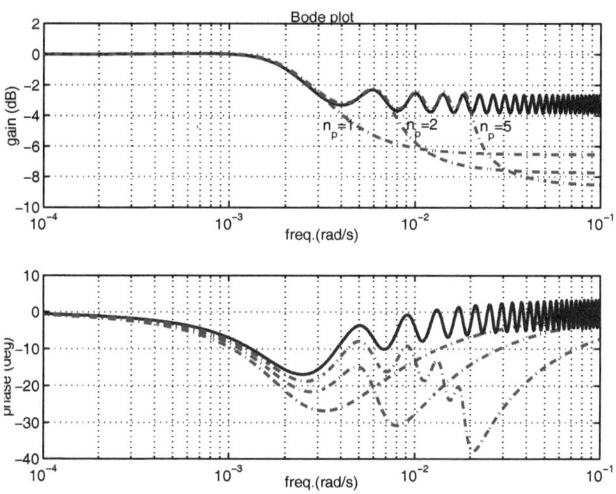

Fig. 9.8 Bode plot of $K_F(s)e^{\frac{L}{\alpha}s}$ (—) and rational approximations obtained with (9.15), using 1, 2 and 5 poles (– – · – –)

9.4 On Feedforward Control with Model Uncertainty

We first discuss the control design for rejecting output system perturbation. Let us focus on the case when the hydraulic system is modeled by the following equation:

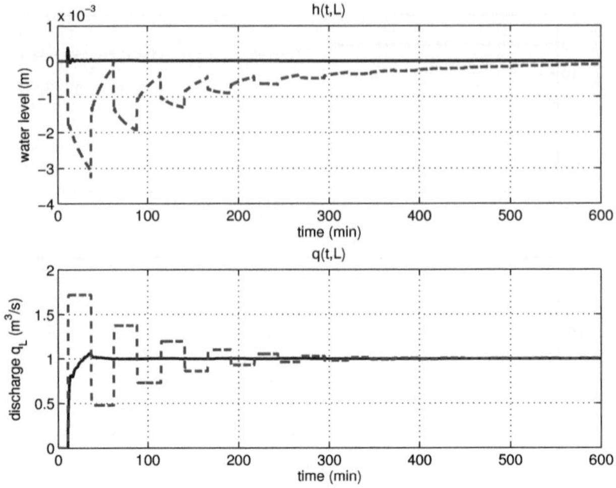

Fig. 9.9 Downstream water level error $y(t,L)$ and feedforward control $q(t,L)$ for a rational approximations of $\tilde{K}_F$ using five poles (—) and the controller of [1] (– –)

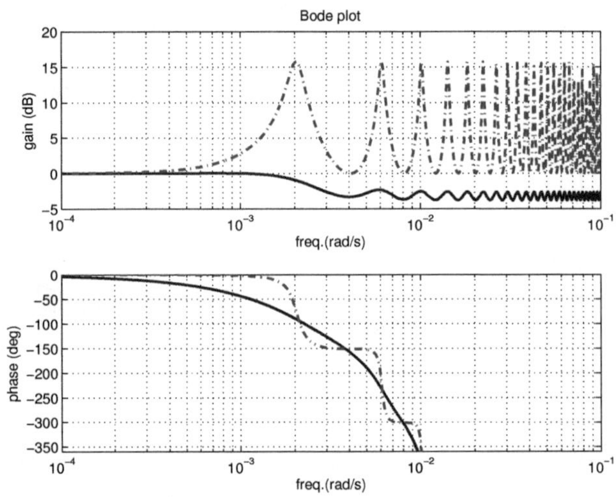

Fig. 9.10 Bode diagram of $K_F(s)$ (—) and of the controller of [1] (– – · – –)

$$y = \frac{e^{-\tau s}}{s}u - \frac{1}{s}p, \tag{9.16}$$

where p is the perturbation input and u the control input. Assume that we want to achieve the best rejection for any perturbation input p with a feedforward control law $u = F(s)y$.

This problem can be formulated as minimizing the H_∞ norm of the transfer function between p and y, that is:

$$\inf_{F(s)\in\mathscr{H}_\infty} \left\|1 - F(s)e^{-\tau s}\right\|_\infty = \gamma.$$

Following [7] we know that $\gamma = 1$, which implies that perturbation rejection can be achieved only on a finite interval of frequencies. This first claim is a theoretical point of view, and does not indicate any bound on the size of the interval. As a matter of fact, the interval is directly limited by the model uncertainty. To illustrate this fact, let us consider that the delay of system (9.16) belongs to:

$$\tau \in \left[(1-\Delta_\tau)\tau_0, (1+\Delta_\tau)\tau_0\right],$$

and let us assume that the measured perturbation is a sinusoid, i.e., $p(t) = a\sin(\omega_p t)$. Let us furthermore assume that the perturbation is known in the future, say τ_0 seconds in advance. Thus it is possible to perfectly reject it with the feedforward control $F(s)$ such that $F(j\omega_p) = e^{\tau_0 j\omega_p}$, which, with respect to the perturbation signal, introduces a lag effect given by $\phi = \tau_0\omega_p$.

Due to the fact that $\tau \in \left[(1-\Delta_\tau)\tau_0, (1+\Delta_\tau)\tau_0\right]$, the lag effect is given by

$$\Delta\phi \in \left[-\Delta_\tau\tau_0\omega_p,\ \Delta_\tau\tau_0\omega_p\right].$$

For the frequency ω_p^* such that $\Delta_\tau\tau_0\omega_p^* = \pi$, the feedforward law can increase the perturbation effect by 2. Therefore, the uncertainty level on the delay constrains the maximal bandwidth where feedforward allows us to improve performance.

The above point is also valid in closed-loop cases. On the one hand, following [2], the closed-loop system explicitly contains the time-delay, on the other hand, the frequency interval where the sensitivity function is less than 1 is also constrained by the delay. Thus, the frequency where the previous argument is used is in "open-loop".

When the sensitivity function is less than 1, the feedback "shrinks" the uncertainty, which allows us to improve the quality of the feedforward loop. It is therefore better to use a controller structure with a feedback loop and a feedforward loop that uses the perturbation measurement to improve its rejection.

9.5 Summary

In this chapter, we have developed various methods for feedforward boundary control of an open channel pool.

We first considered the design of an open-loop controller to compensate for perfectly known future discharge perturbations occurring at the downstream end of a pool. The solution has been obtained using various methods, with an explicit analytical solution in the horizontal frictionless case, and an implicit one in the uniform flow case. Approximate solutions have also been obtained in the uniform and nonuniform cases. Then, we considered the case of boundary cancelation of upstream discharge perturbation. We proposed a series of feedforward controller designs: analytically computed irrational controller, its rational approximation, and a delay-free controller. In addition, for the case of a horizontal frictionless channel, we recovered the time domain result obtained by [1] and derived bounds on approximation error due to infinite series truncation.

The use of a frequency domain approach has been shown to provide efficient methods for open-loop boundary control of an open channel.

References

[1] Bastin G, Coron J, D'Andréa-Novel B, Moens L (2005) Boundary control for exact cancellation of boundary disturbances in hyperbolic systems of conservation laws. In: Proc. 44th IEEE Conf on Decision and Control, Sevilla, pp 1086–1089

[2] Khargonekar P, Poolla K (1986) Robust stabilization of distributed systems. Automatica 22(1):77–84

[3] Litrico X, Fromion V (2006) Boundary control of linearized Saint-Venant equations oscillating modes. Automatica 42(6):967–972

[4] Litrico X, Fromion V (2006) H_∞ control of an irrigation canal pool with a mixed control politics. IEEE Trans Control Syst Technol 14(1):99–111

[5] Litrico X, Fromion V, Scorletti G (2007) Robust feedforward boundary control of hyperbolic conservation laws. Networks Heterogeneous Media 2(4):715–729

[6] Rudolph J (2004) Planning trajectories for a class of linear partial differential equations: an introduction. In: Sciences et Technologies de l'Automatique (e-sta), (http://www.e-sta.see.asso.fr), vol 1

[7] Tannenbaum A (1992) Frequency domain methods for the H_∞ optimization of distributed systems, in Curtain RF, Bensoussan A, Lions JL (eds.), *Analysis and optimization of systems: State and frequency domain approaches for infinite dimensional systems*, Lecture Notes in control and information sciences, vol 185, Springer, New York

Part III
Control of a Multiple-pool Canal

We now consider the case of a multiple-pool canal, where multivariable controllers can be used. In Chap. 10, we extend the control methods of a canal pool by considering decentralized control schemes: distant downstream, local upstream, and mixed control policies.

Then, in Chap. 11, we present the application of the modeling and control methods developed in the book to the case of an experimental canal. The experimental results show the effectiveness of our approach, since simulations and experiments are very close.

Finally, in Chap. 12, we consider another hydrosystem where the modeling and control approaches may be useful: regulated rivers where a dam or multiple dams enable us to control the upstream discharges, and intermediate measurements can be used to estimate water intake or offtake.

Chapter 10
Decentralized Control of a Multiple-pool Canal

Abstract In this chapter, we study the extension of the control policies developed in Chaps. 7 and 8. We first study the case of a two-pool canal, controlled with distant downstream, local upstream, and mixed control policies. We study the stability and performance of decentralized controllers designed separately for each pool. We then extend the results to the case of a multiple-pool canal.

10.1 Introduction

Irrigation canals have a series structure, which is classically used to propose decentralized control structures. Many decentralized control techniques have already been proposed in research papers (see [4] and references therein, [8, 1, 11, 3], or applied in real situations [5, 7]). They usually implement the classical distant downstream control policy, which is parsimonious from the resource management point of view, but has a low performance with respect to water users. The other classical control policy is the local upstream control, which is very performing from the user point of view, but consumes a lot of water. These decentralized control schemes have the advantage of being easily tuned and implemented. Their structure also facilitates fault diagnosis and localization, leading to easy maintenance. However, each of them has limitations, linked to the fact that they use only one control action variable in each pool to control the output. Recent methods have been developed to design controllers with a specific structure, but these methods are computationally demanding and may be difficult to implement for large-scale systems such as irrigation canals.

The main issue at stake for irrigation canal control is to provide a design method to tune simple controllers that achieve a desired trade-off between water resource management and performance with respect to water users. Local upstream and distant downstream control can be viewed as solutions to one of the two design specifications. In Chap. 8, we have shown how to design a mixed controller for one canal pool in order to recover the real-time performance of the local upstream controller, while keeping the water management of the distant downstream controller.

This mixed controller ensures that the mean value of the discharge comes from upstream, as in distant downstream control, but uses some local upstream control in order to dramatically increase the performance with respect to the user.

In this chapter, we extend the mixed control strategy to a multiple pool irrigation canal, with guaranteed stability. In this way, we are able to design a multivariable controller for an irrigation canal with any required performance level (high frequency control performed by local upstream controllers) and ensure that the average value of the resource comes from upstream (low frequency control performed by distant downstream controllers). The proposed solution encompasses both classical control policies, which are based on monovariable control design. A multivariable mixed control design enables us to achieve any desired level of performance, while guaranteeing good water management. This mixed controller can be designed and implemented in a structured semi-decentralized way, with no increase in complexity compared to classical control methods. It therefore meets the design requirements elegantly.

10.1.1 Structure of a Canal

An irrigation canal can be represented as a series of pools (see Fig. 10.1). For pool i we denote by u_i the control variable (discharge) at the upstream end, u_{i+1} the control variable at the downstream end, y_i the controlled variable (water depth at the downstream of pool i), and p_i the load disturbances (water offtake).

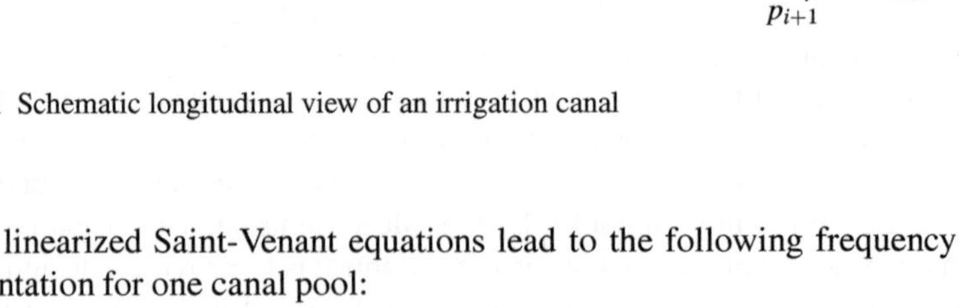

Fig. 10.1 Schematic longitudinal view of an irrigation canal

The linearized Saint-Venant equations lead to the following frequency domain representation for one canal pool:

$$y_i(s) = G_i(s)u_i(s) + \tilde{G}_i(s)(u_{i+1}(s) + p_i(s)), \tag{10.1}$$

where the disturbance $p_i(s)$ (corresponding to the unknown withdrawal) is supposed to act additively with the downstream discharge $u_{i+1}(s)$.

Remark 10.1. Let us recall that transfer functions $G_i(s)$ and $\tilde{G}_i(s)$ can be approximated by an integrator delay model, leading to:

$$G_i(s) = \frac{e^{-\tau_i s}}{A_i s}, \tag{10.2a}$$

$$\tilde{G}_i(s) = -\frac{1}{A_i s}, \tag{10.2b}$$

with τ_i the time-delay for downstream propagation and A_i the backwater area. The delay τ_i and the integrator gain can be obtained analytically from the hydraulic parameters of the pool (see Chaps. 4 and 5).

Based on (10.1), a multiple-pool canal is represented by the following model:

$$y(s) = G(s)u(s) + \tilde{G}(s)p(s),$$

with G a bidiagonal matrix:

$$G(s) = \begin{pmatrix} G_1(s) & \tilde{G}_1(s) & 0 & 0 & 0 & 0 \\ 0 & \ddots & \ddots & 0 & 0 & 0 \\ 0 & 0 & G_i(s) & \tilde{G}_i(s) & 0 & 0 \\ 0 & 0 & 0 & \ddots & \ddots & 0 \\ 0 & 0 & 0 & 0 & G_n(s) & \tilde{G}_n(s) \end{pmatrix},$$

and

$$\tilde{G}(s) = \begin{pmatrix} \vdots & \vdots & \vdots & \vdots \\ 0 & \tilde{G}_i(s) & 0 & 0 \\ 0 & 0 & \tilde{G}_{i+1}(s) & 0 \\ \vdots & \vdots & \vdots & \vdots \end{pmatrix}.$$

10.1.2 The Case of a Two-pool Canal

Let us now consider the generalization of the three classical control schemes to a multiple-pool canal system. For simplicity, but with no loss of generality, we focus in the following on the analysis of a two-pool canal. The structure and results obtained in this specific case also apply in the general case, as will be demonstrated below.

In this case, the canal is represented by the following equation, where the dependence in s has been omitted:

$$\begin{pmatrix} y_1 \\ y_2 \end{pmatrix} = \underbrace{\begin{bmatrix} G_1 & \tilde{G}_1 & 0 \\ 0 & G_2 & \tilde{G}_2 \end{bmatrix}}_{G} \begin{pmatrix} u_1 \\ u_2 \\ u_3 \end{pmatrix} + \underbrace{\begin{bmatrix} \tilde{G}_1 & 0 \\ 0 & \tilde{G}_2 \end{bmatrix}}_{\tilde{G}} \begin{pmatrix} p_1 \\ p_2 \end{pmatrix}.$$

Let r_1 and r_2 denote the reference values for the outputs y_1 and y_2, and $e_1 = r_1 - y_1$ and $e_2 = r_2 - y_2$ denote the output errors.

We first recall how both classical control policies can be directly extended to the case of multiple-pool irrigation canals. These control policies are monovariable in essence and need only be slightly modified to enhance their performance, by adding feedforward decoupling terms. We then show that the mixed control scheme, although multivariable in essence, can also be tuned using monovariable design methods.

10.2 Distant Downstream Control of a Two-pool Canal

Let us study the stability and performance of a canal controlled with decentralized distant downstream controllers. In this case, the structure of the controller is given by

$$K^{dd} = \begin{pmatrix} K_{11} & 0 \\ 0 & K_{22} \\ 0 & 0 \end{pmatrix},$$

with K_{11} and K_{22} monovariable distant downstream controllers for each pool. The open-loop transfer matrix is then given by

$$L = GK^{dd} = \begin{pmatrix} G_1K_{11} & G_1K_{12} + \tilde{G}_1K_{22} \\ 0 & G_2K_{22} \end{pmatrix}.$$

With this controller, the closed-loop system is represented by the following set of equations:

$$\begin{cases} u_1 = K_{11}(r_1 - y_1) = K_{11}(s)e_1 \\ u_2 = K_{22}(r_2 - y_2) = K_{22}(s)e_2 \\ u_3 = 0 \\ y_1 = G_1(s)u_1 + \tilde{G}_1(s)(u_2 + p_1) \\ y_2 = G_2(s)u_2 + \tilde{G}_2(s)(u_3 + p_2) \end{cases}.$$

10.2.1 Stability

This leads to the following closed-loop transfer matrix $M(s)$ relating the tracking errors e_1, e_2 to the perturbations p_1, p_2:

$$\begin{pmatrix} e_1 \\ e_2 \end{pmatrix} = \begin{pmatrix} M_1 & M_1K_{22}M_2 \\ 0 & M_2 \end{pmatrix} \begin{pmatrix} p_1 \\ p_2 \end{pmatrix}, \qquad (10.3)$$

with $M_1 = -\tilde{G}_1(1 + G_1K_{11})^{-1}$ and $M_2 = -\tilde{G}_2(1 + G_2K_{22})^{-1}$.

Following the results presented in Appendix E and assuming that $L(s)$ contains no unstable hidden modes, the system is closed-loop stable if the Nyquist plot of $\det(I+L(j\omega))$ makes a number of anti-clockwise encirclements of the origin equal to the number of unstable poles of $L(s)$ and does not pass through the origin. Since $L(s)$ is upper triangular, we thus have

$$\det(I+L(s)) = M_1(s)M_2(s),$$

and the multivariable system stability naturally derives from the monovariable controllers design.

Actually, robustness is also inherited from monovariable design, since the system is *structurally* upper triangular (the m_{21} term of matrix M is null, even for model mismatch). Indeed let us consider that G is subject to model uncertainties. Let us denote by $\check{G}$ the modified model, we then have:

$$\begin{pmatrix} y_1 \\ y_2 \end{pmatrix} = \underbrace{\begin{bmatrix} \check{G}_1 & \check{\tilde{G}}_1 & 0 \\ 0 & \check{G}_2 & \check{\tilde{G}}_2 \end{bmatrix}}_{\check{G}} \begin{pmatrix} u_1 \\ u_2 \\ u_3 \end{pmatrix} + \underbrace{\begin{bmatrix} \check{\tilde{G}}_1 & 0 \\ 0 & \check{\tilde{G}}_2 \end{bmatrix}}_{\check{\tilde{G}}} \begin{pmatrix} p_1 \\ p_2 \end{pmatrix}.$$

We then have the equality:

$$\det(I+\check{G}(s)K^{dd}(s)) = \check{M}_1(s)\check{M}_2(s),$$

with $\check{M}_1(s) = -\check{\tilde{G}}_1(1+\check{G}_1K_{11})^{-1}$ and $\check{M}_2(s) = -\check{\tilde{G}}_2(1+\check{G}_2K_{22})^{-1}$. Therefore, the multivariable uncertain system is stable *if and only if* the monovariable uncertain systems are stable.

As a consequence, we can link the multivariable input margins to the monovariable ones. To this end, let us consider structured multiplicative uncertainties acting on the system input:

$$\check{G}_1 = G_1(1+\Delta_1) \text{ and } \check{G}_2 = G_2(1+\Delta_2),$$

where Δ_1 and Δ_2 are two stable linear systems with a finite H_∞ norm less than or equal to $\alpha \geq 0$. The input module margin then corresponds to the greatest value of α such that the closed-loop system remains stable despite the structured dynamical uncertainties. In the same vein, the real input margin corresponds to the robustness margin assuming that the diagonal uncertainties are now two real gains with a norm less than α. In our case, we then have

$$\det(I+\check{G}(s)K^{dd}(s)) = -\tilde{G}_1(1+G_1(1+\Delta_1)K_{11})^{-1}\tilde{G}_2(1+G_2(1+\Delta_2)K_{22})^{-1},$$

and thus the input margin associated to the multivariable system is then given by

$$\alpha = \min(\alpha_1, \alpha_2),$$

where α_i is the greatest value such that $(1+G_i(1+\Delta_i)K_{ii})^{-1}$ remains stable for any $\|\Delta_i\|_\infty \leq \alpha_i$. Consequently, the multivariable input margin, which corresponds to the robustness margins with respect to structured diagonal input uncertainties [9] is equal to the worst margin associated to the monovariable controller. This important result can directly be extended to a multiple-pool canal: the decentralized controller obtained by the concatenation of monovariable distant downstream controllers tuned separately for each pool is stable and has a robustness margin equal to the worst input margin of monovariable controllers.

10.2.2 Performance Analysis

Let us now examine the performance of such a decentralized controller, by studying the gain of $M(j\omega)$ as a function of ω. In the multivariable case, the classical module of a transfer function is extended through the use of a matrix norm as, for example, the largest singular value (see [9] and references therein).

The singular values of a matrix $M \in \mathbb{C}^{n\times p}$, denoted $\sigma(M)$ are defined as the square roots of the eigenvalues of M^*M, where M^* stands for the complex conjugate transpose of M. One may show that the largest singular value, denoted $\bar{\sigma}$ is a matrix norm, and we have:

$$\|e(j\omega)\| = \|M(j\omega)p(j\omega)\| \leq \bar{\sigma}(M(j\omega))\|p(j\omega)\|.$$

By definition of the largest singular value, we know that there exists a couple of perturbations $(\bar{p_1}(j\omega), \bar{p_2}(j\omega))$ such that the error norm reaches the upper bound of the inequality. The couple $(\bar{p_1}(j\omega), \bar{p_2}(j\omega))$ is called the worst case perturbation. Since in our case the perturbations are unknown a priori, the largest singular value $\bar{\sigma}(M(j\omega))$ provides a good estimate of the performance of a controlled system.

In the present case, we have $\bar{\sigma}(M(j\omega)) = \sqrt{\lambda_{\max}(\mathscr{M}(j\omega))}$, with $\mathscr{M}(j\omega) = M(j\omega)M(j\omega)^*$, i.e.,

$$\mathscr{M} = \begin{pmatrix} |M_1|^2 + |M_1K_{22}M_2|^2 & M_1K_{22}|M_2|^2 \\ |M_2|^2K_{22}^*M_1^* & |M_2|^2 \end{pmatrix},$$

where we have omitted the argument $j\omega$ for readability.

Let us compute the singular value. For this purpose, we denote by λ_1 and λ_2 the two eigenvalues of $\mathscr{M}$, which are solutions of the second-order polynomial equation:

$$\lambda^2 - (|M_1|^2 + |M_2|^2 + |M_1K_{22}M_2|^2)\lambda + |M_1M_2|^2 = 0.$$

Their product is given by:

$$\lambda_1\lambda_2 = |M_1(j\omega)|^2|M_2(j\omega)|^2,$$

and is independent of the coupling, while their sum

$$\lambda_1+\lambda_2=|M_1(j\omega)|^2+|M_2(j\omega)|^2+|M_1(j\omega)|^2|K_{22}(j\omega)|^2|M_2(j\omega)|^2,$$

is always larger than $|M_1(j\omega)|^2+|M_2(j\omega)|^2$ when $K_{22}\neq 0$. So generally, we have

$$\bar{\sigma}(M(j\omega))\geq \max(|M_1(j\omega)|,|M_2(j\omega)|),$$

for all the frequencies, and thus the coupling between pools necessarily decreases the overall performance in terms of perturbation rejection. Let us now study how this performance loss impacts the pools, by analyzing the case of dissymmetric performance for each pool.

10.2.2.1 Case Where $|M_2(j\omega)|\ll|M_1(j\omega)|$ and $|M_2(j\omega)|\ll 1$

By assuming that $M_2(j\omega)\ll 1$, we have $u_2(j\omega)\approx p_2(j\omega)$ since in this case the controller of the second pool is able to "perfectly" reject the perturbation acting on it. The feature of the control structure allows to deduce the announced result. The closed-loop system is then given by:

$$\begin{cases} e_1(j\omega)\approx M_1(j\omega)(p_1(j\omega)+p_2(j\omega)) \\ e_2(j\omega)=M_2(j\omega)p_2(j\omega)\end{cases}. \tag{10.4}$$

This means that when the performance of the downstream pool is better than that of the upstream pool, the perturbation $p_2(j\omega)$ is added to the perturbation $p_1(j\omega)$. Therefore the first pool is perturbed not only by its perturbation p_1, but also by that of the second pool p_2. This leads to a strong interaction between pools, and to a strong inequity between water users: the upstream ones have to reject all the perturbations of the downstream ones.

Remark 10.2. This result can also be obtained directly from the singular values of M. Indeed, using the above approximations, the eigenvalues of $\mathscr{M}$ verify

$$\lambda_1+\lambda_2\approx 2|M_1(j\omega)|^2+|M_2(j\omega)|^2.$$

Since the eigenvalues verify $\lambda_1\lambda_2=|M_1(j\omega)|^2M_2(j\omega)|^2$ and $|M_2(j\omega)|\ll|M_1(j\omega)|$, the largest singular value of $M(j\omega)$ can be approximated by

$$\bar{\sigma}(M(j\omega))\approx\sqrt{2}|M_1(j\omega)|.$$

This worst case gain corresponds to the amplification factor of the system when two identical perturbations act on the pools. In that case the error can be computed using (10.4) by taking $p_1(j\omega)=p_2(j\omega)=p(j\omega)$. This leads to $e_1(j\omega)\approx 2M_1(j\omega)p(j\omega)$, and finally the ratio of the norm of the tracking error and the perturbation is:

$$\frac{\sqrt{\|e_1(j\omega)\|^2+\|e_2(j\omega)\|^2}}{\sqrt{\|p_1(j\omega)\|^2+\|p_2(j\omega)\|^2}}\approx\frac{\|2M_1(j\omega)p(j\omega)\|}{\sqrt{\|p(j\omega)\|^2+\|p(j\omega)\|^2}}=\sqrt{2}|M_1(j\omega)|.$$

This remark shows the importance of studying the singular value to evaluate the performance of a feedback loop for unknown perturbations.

10.2.2.2 Case Where $|M_1(j\omega)| \ll |M_2(j\omega)|$ and $|M_1(j\omega)| \ll 1$

Let us now consider the case where the first pool is much more efficient than the second one in rejecting perturbations occurring at frequency ω.

Then, using similar arguments as above, we can show that the largest singular value of M can be approximated by

$$\bar{\sigma}(M(j\omega)) \approx |M_2(j\omega)|,$$

since the coupling term appearing in the sum $\lambda_1 + \lambda_2$ can be neglected. Then, the overall performance is only linked to that of the second pool. This is due to the fact that the perturbation generated at the downstream end of pool 1 by the control input for pool 2 is very quickly attenuated.

10.2.2.3 About Equity and Decoupling

Requiring equity between water users corresponds to having a diagonal matrix M with identical transfer functions on the diagonal. Then, the controlled system would be treating each user in the same way. However, as we have seen before, such an objective is not compatible with the decentralized distant downstream controller structure. Indeed, for such a control structure, to have a quasi-diagonal matrix M, one needs to have the coupling term $M_1(j\omega)K_{22}(j\omega)M_2(j\omega) \approx 0$. This is only possible when the performance of one pool is larger than the other one. The decoupling can therefore be obtained, but at the expense of equity.

As we will now show, it is possible to decrease the coupling term. A perfect decoupling is actually possible in downstream control, but leads to an unrealistic solution (see below).

10.2.3 Feedforward Decoupler

10.2.3.1 Adding a Feedforward Term

The perturbation generated by the control input u_2 on the first pool is perfectly known. One can therefore use this knowledge to improve the performance of the control system, without impacting its stability and robustness properties.

To this end, we use an additional term K_{12} in the controller:

$$K^{dd} = \begin{pmatrix} K_{11} & K_{12} \\ 0 & K_{22} \\ 0 & 0 \end{pmatrix}.$$

With this new term, the closed-loop system becomes

$$\begin{pmatrix} e_1 \\ e_2 \end{pmatrix} = \begin{pmatrix} M_1 & M_1 K_{22} M_2 \left(1 + K_F \frac{G_1}{\tilde{G}_1}\right) \\ 0 & M_2 \end{pmatrix} \begin{pmatrix} p_1 \\ p_2 \end{pmatrix}, \tag{10.5}$$

with $K_F = K_{12} K_{22}^{-1}$ a feedforward term.

It is clear that the additional term does not modify the stability and robustness properties of the closed-loop, since the transfer matrix remains triangular, and that only the inverses of $(1 + G_1 K_{11})$ and $(1 + G_2 K_{22})$ appear in matrix M. This implies that the stability and robustness properties of the multivariable controller are identical to those of the monovariable controllers.

The feedforward term $K_F(s)$ can now be chosen in order to reduce the coupling effect highlighted above.

10.2.3.2 Perfect Decoupling: A Noncausal Solution

A perfect decoupling is achieved if the nondiagonal term of matrix M is zero. An algebraic solution to this problem would be to choose K_F such that:

$$K_F(s) = -\frac{\tilde{G}_1(s)}{G_1(s)}. \tag{10.6}$$

However, one may show that this solution generally leads to a noncausal term. Assuming that transfers $\tilde{G}_1$ and G_1 are given by (10.2a–10.2b), $K_F(s) = e^{\tau_1 s}$ is a predictor, and depending on the term K_{22}, $K_{12} = K_F K_{22}$ may not be causal.

When K_{22} is a classical PI controller, K_{12} is noncausal. However, if K_{22} includes a time-delay, which is larger than τ_1, then the product $K_F(s)K_{22}(s)$ becomes causal. This solution is equivalent to designing a distant downstream controller K_{22} for a system where the delays of the first and the second pools are added. This enables us to perfectly decouple the pools, but at the expense of the performance of the second one, which is now controlled as if its delay incorporates that of the first pool τ_1 in addition to its own delay τ_2. In practice, such a controller would "wait" a time τ_1 for the water to come from upstream before beginning to act on u_2 to compensate for the perturbation p_2. Extended to a canal, this solution leads to very poor performance. Indeed, each monovariable controller has to control a "delayed system" with a delay equal to the sum of all the canal pool delay. So clearly, this solution is unrealistic, and we study below a practical solution to this decoupling problem.

10.2.3.3 Decoupling: A Practical Solution

In practice, one would like to keep the distant downstream controller K_{22} initially designed for the second pool without modifying it. In that case, a perfect decoupling is no longer possible. We therefore want to find K_{12} that minimizes the coupling effect. We formulate this using the largest singular value $\bar{\sigma}(M(j\omega))$ i.e., the square root of the largest eigenvalue of:

$$\mathcal{M} = \begin{pmatrix} |M_1|^2 + |M_1 K_{22} M_2 Q|^2 & -M_1 K_{22} Q |M_2|^2 \\ -|M_2|^2 Q^* K_{22}^* M_1^* & |M_2|^2 \end{pmatrix},$$

with $Q(s) = 1 - K_F(s)\frac{G_1(s)}{\tilde{G}_1(s)}$, and where the dependence in $j\omega$ has been omitted for readability.

We know that the product of the eigenvalues of $\mathcal{M}$ is always given by:

$$\lambda_1 \lambda_2 = |M_1(j\omega)|^2 |M_2(j\omega)|^2,$$

and that their sum is given by:

$$\lambda_1 + \lambda_2 = |M_1(j\omega)|^2 + |M_2(j\omega)|^2 + |M_1(j\omega)|^2 |K_{22}(j\omega)|^2 |M_2(j\omega)|^2 |Q(j\omega)|^2.$$

A naive solution would be to try to find the transfer function $K_F(s) \in H_\infty$ that minimizes the coupling for all the frequencies

$$\gamma_1 = \inf_{K_F(s) \in H_\infty} \max_{\omega \in [0,\infty)} \left\| 1 - K_F(j\omega) \frac{G_1(j\omega)}{\tilde{G}_1(j\omega)} \right\|_\infty,$$

which is equivalent to this H_∞ norm problem:

$$\gamma_1 = \inf_{K_F(s) \in H_\infty} \left\| 1 - K_F(s) \frac{G_1(s)}{\tilde{G}_1(s)} \right\|_\infty.$$

However, this problem is equivalent to the one presented in the previous section and we are then not surprised that it does not yield an interesting result. Indeed, using the simple approximation (10.2a–10.2b), one may show that:

$$\inf_{K_F(s) \in H_\infty} \left\| 1 - K_F(s) e^{-\tau_1 s} \right\|_\infty = 1.$$

This theoretical result provided in [10] shows that a causal decoupler cannot improve the performance evaluated over all the frequency range. However, this does not hold when one considers a limited frequency bandwidth. Indeed, for the low frequencies, we know that with $K_F(s) = 1$, we have $\|1 - e^{-\tau_1 j\omega}\| \approx 0$ for $\omega \approx 0$ and thus a low pass filter allows us to ensure decoupling for low frequency perturbations. Often, a constant gain is used in practice by hydraulic engineers who use a decoupler with constant gain between 0.5 and 1 (see the dynamic regulation of the Canal de Provence [5]). However, there exist frequencies where $e^{-\tau_1 j\omega} \approx -1$, which lead to $|1-e^{-\tau_1 j\omega}| \approx 2$ and thus explain why the use of a low pass filter is better for this feedforward decoupler.

10.2.3.4 Summary

Finally, the multivariable system appears to be directly linked in terms of performance and robustness to the monovariable systems. However, in this case, once the distant downstream controllers are tuned, there is no way to increase the performance with respect to water users. In particular, we have shown that the coupling between pools cannot be perfectly removed and only low frequency decoupling can be achieved.

10.2.4 Simulation Results

The simulation results are obtained on a simplified integrator delay model of the canal pools, in order to facilitate the analysis. So, the considered system consists of a two-pool canal represented by a series of two identical dimensionless ID models, that is for $i \in \{1,2\}$ we have:

$$G_i(s) = \frac{e^{-s}}{s} \text{ and } \tilde{G}_i(s) = -\frac{1}{s}.$$

As we have shown earlier in the book, this simple model gives a good low frequency representation of the dynamics of an open channel. We compare the simulation results for a decentralized distant downstream controller with and without decoupler.

Figure 10.2 depicts the response of a decentralized distant downstream controller with and without decoupler to a perturbation in pool 2. We observe that the water withdrawal in pool 2 is compensated by the increase in discharge u_2. When there is no decoupler, the water level y_1 drops due to the increase of u_2. When there is a decoupler, the input u_1 reacts more quickly, and as a consequence, the water level y_1 goes back to its reference more quickly. The decoupler has a clear effect on the performance.

10.3 Local Upstream Control of a Two-pool Canal

Let us now study the stability and performance of a canal controlled with decentralized upstream control. In this case, the structure of the controller is given by

$$K^{lu} = \begin{pmatrix} 0 & 0 \\ K_{21} & 0 \\ 0 & K_{32} \end{pmatrix},$$

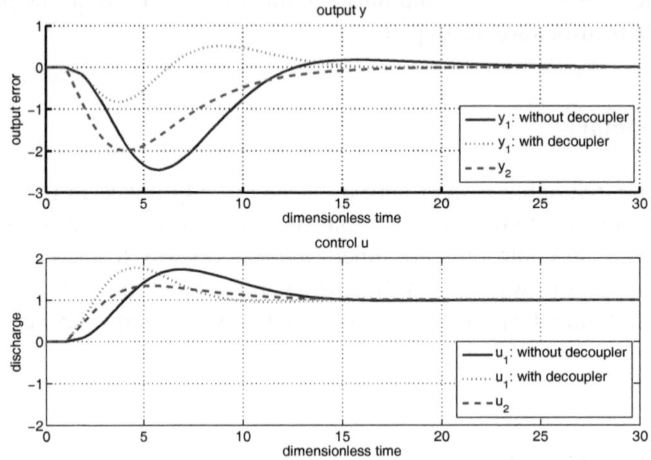

Fig. 10.2 Distant downstream control of a canal with two pools. Response to a perturbation in pool 2

with K_{21} and K_{32} monovariable local upstream controllers for each pool.

The open-loop of the system is then

$$L = GK^{lu} = \begin{pmatrix} \tilde{G}_1 K_{21} & 0 \\ G_2 K_{21} + \tilde{G}_2 K_{31} & \tilde{G}_2 K_{32} \end{pmatrix},$$

and the closed-loop is then given by the following set of equations:

$$\begin{cases} u_1 &= u_{\max} \\ u_2 &= K_{21}(r_1 - y_1) = K_{21}(s)e_1 \\ u_3 &= K_{32}(r_2 - y_2) = K_{32}(s)e_2 \\ y_1 &= G_1(s)u_1 + \tilde{G}_1(s)(p_1 - u_2) \\ y_2 &= G_2(s)u_2 + \tilde{G}_2(s)(p_2 - u_3) \end{cases}.$$

where the first input u_1 delivers the maximum discharge into the canal.

10.3.1 Stability

The closed-loop system gives the relation between tracking errors e_1, e_2 and disturbances p_1, p_2 as follows:

$$\begin{pmatrix} e_1 \\ e_2 \end{pmatrix} = \begin{pmatrix} \tilde{M}_1 & 0 \\ \tilde{M}_1 K_{21} \frac{G_2}{\tilde{G}_2} \tilde{M}_2 & \tilde{M}_2 \end{pmatrix} \begin{pmatrix} p_1 \\ p_2 \end{pmatrix}, \tag{10.7}$$

with $\tilde{M}_1 = -\tilde{G}_1(1 + \tilde{G}_1 K_{21})^{-1}$, $\tilde{M}_2 = -\tilde{G}_2(1 + \tilde{G}_2 K_{32})^{-1}$.

Then, the matrix $\tilde{M}$ is *structurally* lower triangular and the results obtained in the case of distant downstream control also apply in this case: the multivariable control system is stable if and only if the monovariable systems are stable, and the robustness properties of the SISO case are recovered in the MIMO case for structured diagonal input uncertainties.

10.3.2 Performance Analysis

As in the distant downstream case, the performance of the closed-loop system can be studied by analyzing the singular values of matrix $\tilde{M}(s)$. Thus the product of the eigenvalues of $\mathscr{M}(j\omega) = \tilde{M}(j\omega)\tilde{M}(j\omega)^*$

$$\lambda_1\lambda_2 = |\tilde{M}_1(j\omega)|^2|\tilde{M}_2(j\omega)|^2,$$

are independent of the coupling and their sum

$$\lambda_1 + \lambda_2 = |\tilde{M}_1(j\omega)|^2 + |\tilde{M}_2(j\omega)|^2 + |\tilde{M}_1(j\omega)|^2|K_{21}(j\omega)|^2\left|\frac{G_2(j\omega)}{\tilde{G}_2(j\omega)}\right|^2|\tilde{M}_2(j\omega)|^2$$

is always larger than $|\tilde{M}_1(j\omega)|^2 + |\tilde{M}_2(j\omega)|^2$, and thus we necessarily have:

$$\bar{\sigma}(\tilde{M}(j\omega)) > \max(|\tilde{M}_1(j\omega)|, |\tilde{M}_2(j\omega)|),$$

which means that the coupling deteriorates the overall performance compared to that obtained in the monovariable case. This can be examined in two different cases.

10.3.2.1 Case Where $|\tilde{M}_2(j\omega)| \ll |\tilde{M}_1(j\omega)|$ and $|\tilde{M}_2(j\omega)| \ll 1$

Let us first consider that the second pool has a better performance than the first one, which can be expressed as follows:

$$|\tilde{M}_2(j\omega)| \ll 1 \text{ and } |\tilde{M}_1(j\omega)| \gg |\tilde{M}_2(j\omega)|.$$

Following the lines of Sect. 10.2, we can show that the product $\tilde{M}_1(j\omega)K_{21}(j\omega)$ is bounded by a constant:

$$|\tilde{M}_1(j\omega)K_{21}(j\omega)| \leq 2,$$

and that the gain of the ratio $G_2/\tilde{G}_2$ is also constant, since it is close to the time-delay of pool 2:

$$\left|\frac{G_2(j\omega)}{\tilde{G}_2(j\omega)}\right|^2 \approx 1.$$

Finally, we conclude that the coupling term appearing in the sum $\lambda_1+\lambda_2$ can be neglected, which leads to

$$\lambda_1+\lambda_2 \approx |\tilde{M}_1(j\omega)|^2.$$

This means that the multivariable system is fully decoupled, and has the performance of the monovariable ones. Intuitively, due to its efficiency, the controller of the second pool rejects the perturbation generated by the first pool.

10.3.2.2 Case Where $|\tilde{M}_1(j\omega)| \ll |\tilde{M}_2(j\omega)|$ and $|M_1(j\omega)| \ll 1$

Let us now consider the symmetric case, where the upstream pool has a better performance than the downstream one, i.e.:

$$|\tilde{M}_1(j\omega)| \ll 1 \text{ and } |\tilde{M}_2(j\omega)| \gg |\tilde{M}_1(j\omega)|.$$

It an easily be shown using similar arguments that the largest singular value is given by

$$\bar{\sigma}(\tilde{M}(j\omega)) \approx 2|\tilde{M}_2(j\omega)|.$$

In that case, the control input u_2 will generate a discharge perturbation for the downstream pool that will be the copy of the perturbation p_1. Then, the second pool will have to reject a perturbation equal to the sum p_1+p_2. By contrast to downstream decentralized controller, we will show that the feedforward term allows us to fully remove the strong coupling.

10.3.3 Feedforward Decoupler

The perturbation generated by the first pool on the second one is perfectly known, since it corresponds to the control input u_2, and we will show that it is possible to compensate its effect by adding a feedforward term to the controller, without modifying the stability properties of the closed-loop system. This feedforward term is similar to that in the distant downstream case, except that it reports the control input of the first pool u_2 towards the one of the second pool.

This corresponds to the following set of equations:

$$\begin{cases} u_1 = u_{\max} \\ u_2 = K_{21}(r_1-y_1) = K_{21}(s)e_1 \\ u_3 = K_{32}(r_2-y_2) = K_{32}(s)e_2 + K_F(s)u_2 \\ y_1 = G_1(s)u_1 + \tilde{G}_1(s)(p_1-u_2) \\ y_2 = G_2(s)u_2 + \tilde{G}_2(s)(p_2-u_3) \end{cases} .$$

In that case, the controller becomes

$$K^{lu} = \begin{pmatrix} 0 & 0 \\ K_{21} & 0 \\ K_{31} & K_{32} \end{pmatrix}.$$

The closed-loop system gives the relation between tracking errors e_1, e_2, and disturbances p_1, p_2 as follows:

$$\begin{pmatrix} e_1 \\ e_2 \end{pmatrix} = \begin{pmatrix} \tilde{M}_1 & 0 \\ \tilde{M}_1 K_{21} \left(\frac{G_2}{\tilde{G}_2} + K_F \right) \tilde{M}_2 & \tilde{M}_2 \end{pmatrix} \begin{pmatrix} p_1 \\ p_2 \end{pmatrix}, \tag{10.8}$$

with $K_F = K_{31} K_{21}^{-1}$ a feedforward term.

It is important to note that the addition of the feedforward term does not modify the stability and robustness properties of the system, since only the inverses $(1 + \tilde{G}_1 K_{21})^{-1}$ and $(1 + \tilde{G}_2 K_{32})^{-1}$ appear in the equations. As we have seen above, the interactions between pools decrease the performance of the overall closed-loop system. For an "exact" decoupling the off-diagonal elements of matrix $\tilde{M}(s)$ must be zero, i.e.:

$$K_F(s) = -\frac{G_2(s)}{\tilde{G}_2(s)}.$$

Assuming that transfers $\tilde{G}_2$ and G_2 are given by (10.2a–10.2b), one gets $K_F(s) = e^{-\tau_2 s}$. In this case "exact" decoupling is possible, without changing the performance of each pool, since the transfer function is causal. This solution is consistent with our physical intuition, which requires u_3 to compensate for the perturbation generated by the control input u_2 when it reaches the end of the second pool, after a time-delay τ_2.

This solution assumes perfect knowledge of the delay, and in practice, it is necessary to consider model uncertainties in the design of this feedforward controller, which will limit the performance. It is then necessary to use also a low pass filter avoiding mismatch compensation due to delay uncertainty.

10.3.4 Simulation Results

Figure 10.3 depicts the response of a decentralized local upstream controller with and without decoupler to a perturbation in pool 1. We observe that the water withdrawal in pool 1 is compensated by the decrease in discharge u_2. When there is no decoupler, the water level y_2 drops as this perturbation propagates along pool 2. Then u_3 reacts in order to bring back y_2 to its reference level. When there is a decoupler, the input u_3 reacts exactly as needed for this known perturbation and as a consequence, the water level y_2 is not affected by the perturbation occurring in pool 1. There is exact decoupling in that case.

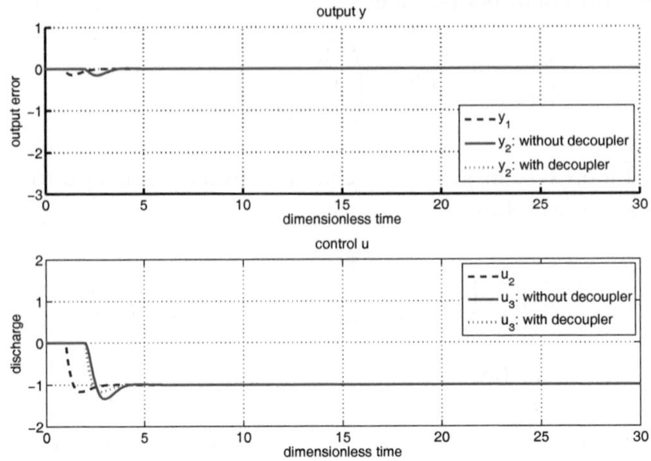

Fig. 10.3 Local upstream control of a canal with two pools. Response to a perturbation in pool 1

One should note that in reality, with uncertain models, it is difficult to completely cancel a known perturbation. In that case, one should rather filter the high frequency component of the feedforward term.

10.4 Mixed Control of a Two-pool Canal

We now present the multivariable mixed control scheme, which aims at combining the advantages of both classical control policies for a multiple-pool canal.

10.4.1 Control Structure

We now consider a fully multivariable controller K, which gives the control vector (u_1, u_2, u_3) from the error vector (e_1, e_2):

$$K = \begin{pmatrix} K_{11} & K_{12} \\ K_{21} & K_{22} \\ K_{31} & K_{32} \end{pmatrix}.$$

The open-loop transfer matrix is then given by

$$L = \begin{pmatrix} G_1 K_{11} + \tilde{G}_1 K_{21} & G_1 K_{12} + \tilde{G}_1 K_{22} \\ G_2 K_{21} + \tilde{G}_2 K_{31} & G_2 K_{22} + \tilde{G}_2 K_{32} \end{pmatrix},$$

which can be restated as the sum of the open-loops obtained with a local upstream and a distant downstream controller:

$$L = \underbrace{\begin{pmatrix} G_1K_{11} & G_1K_{12}+\tilde{G}_1K_{22} \\ 0 & G_2K_{22} \end{pmatrix}}_{\text{distant downstream}} + \underbrace{\begin{pmatrix} \tilde{G}_1K_{21} & 0 \\ G_2K_{21}+\tilde{G}_2K_{31} & \tilde{G}_2K_{32} \end{pmatrix}}_{\text{local upstream}}.$$

By combining these two controllers, one aims at satisfying the two following objectives:

- Reject low frequency perturbations using the upstream discharge
- Increase the performance with respect to water users by using the downstream discharge to reject perturbations occurring in higher frequencies

By contrast to the two previous multivariable structured controllers, the robustness (and even the stability) of the mixed multivariable controller remains a complicated issue during the design. This drawback is not due to the mixed controller, but it is a general issue for the design of multivariable controllers. This problem has been deeply investigated during the last 30 years [6, 9, 12] and has lead to the development of powerful synthesis and analysis tools. For the stability and robustness analysis, currently we have quite a complete panel of powerful tools that allow us to easily handle the robustness analysis of multivariable systems. By contrast, even if lot of effort has been made to develop the multivariable synthesis tools, the tools are not as efficient as the one associated to the analysis tool. Since the purpose of this book is not to present a complete panel of possible synthesis tools, in the sequel we will develop a design method allowing us to easily show the importance of the mixed multivariable policies for canal control. Even if the proposed solution has some limitations, it paves the way to the use of more advanced design methods such as the H_∞ approach (see Chap. 11).

As shown above, it is possible in the local upstream control scheme to perfectly decouple the system. Therefore, it is possible to design a controller K_{31} such that

$$G_2K_{21}+\tilde{G}_2K_{31}=0. \tag{10.9}$$

In this way, the open-loop control matrix $G(s)K(s)$ becomes upper triangular:

$$G(s)K(s) = \begin{pmatrix} G_1K_{11}+\tilde{G}_1K_{21} & G_1K_{12}+\tilde{G}_1K_{22} \\ 0 & G_2K_{22}+\tilde{G}_2K_{32} \end{pmatrix}.$$

This last expression can be considered as the series of the following fictive controller:

$$\tilde{K}(s) = \begin{pmatrix} \frac{G_1}{\tilde{G}_1}K_{11}+K_{21} & \frac{G_1}{\tilde{G}_1}K_{12}+K_{22} \\ 0 & \frac{G_2}{\tilde{G}_2}K_{22}+K_{32} \end{pmatrix},$$

and the following fictive system:

$$\tilde{G} = \begin{pmatrix} \tilde{G}_1 & 0 \\ 0 & \tilde{G}_2 \end{pmatrix}.$$

With these notations, the closed-loop system is given by:

$$\begin{aligned} y(s) &= G(s)K(s)(r(s)-y(s)) + \tilde{G}(s)p(s) \\ &= \tilde{G}(s)\tilde{K}(s)(r(s)-y(s)) + \tilde{G}(s)p(s) \end{aligned}.$$

Therefore, the perturbations and the errors are linked by the relation

$$\begin{pmatrix} e_1 \\ e_2 \end{pmatrix} = \begin{pmatrix} \tilde{\tilde{M}}_1(s) & \tilde{\tilde{M}}_1(s)\tilde{K}_{12}(s)\tilde{\tilde{M}}_2(s) \\ 0 & \tilde{\tilde{M}}_2(s) \end{pmatrix} \begin{pmatrix} p_1 \\ p_2 \end{pmatrix}, \tag{10.10}$$

where $\tilde{\tilde{M}}_i(s)$ is given by, for $i = 1,2$:

$$\tilde{\tilde{M}}_i(s) = (I + \tilde{G}_i\tilde{K}_{ii})^{-1}\tilde{G}_i,$$

and where

$$\tilde{K}_{ij}(s) = \frac{G_i}{\tilde{G}_i}K_{ij} + K_{i+1j}.$$

So through the design of K_{31} such that (10.9) holds true, we recover an upper triangular closed-loop system, whose nominal stability are directly linked to the ones of the mixed controllers of each pool. But obviously, this is not a *structural* upper triangular matrix, and thus it is necessary to take care of the model uncertainties acting on the m_{21} term, since it is generally not null in the case of model mismatch. In this context, a way to robustify the control with respect to off-diagonal model uncertainties is to add a low-pass filter in K_{21} and K_{31} in order to avoid the robustness problem induced by high frequencies uncertainties. In practice, the controller robustness analysis has to be done a posteriori by computing the structured singular value μ (see [9]).

10.4.2 Performance Analysis

The mixed control structure is therefore similar to the classical local upstream and distant downstream control structures: it is possible to tune independently a mixed controller for each pool, and then design decoupling elements to improve performance. In the nominal case, matrix $\tilde{M}$ is similar to the matrix M obtained in the decentralized distant downstream case. However, contrarily to this latter case, the performance is no longer limited by the time-delay since $\tilde{G}$ has no delay.

We know that the product of the minimal and maximal singular values of $\tilde{\tilde{M}}(s)$ is independent of the coupling, i.e.:

$$\overline{\sigma}(\tilde{\tilde{M}}(j\omega))\underline{\sigma}(\tilde{\tilde{M}}(j\omega)) = |\tilde{\tilde{M}}_1(j\omega)||\tilde{\tilde{M}}_2(j\omega)|,$$

and that the sum is given by

$$\overline{\sigma}(\tilde{\tilde{M}}(j\omega))^2 + \underline{\sigma}(\tilde{\tilde{M}}(j\omega))^2 = |\tilde{\tilde{M}}_1(j\omega)|^2 + |\tilde{\tilde{M}}_2(j\omega)|^2 + |\tilde{\tilde{M}}_1(j\omega)|^2|\tilde{K}_{12}(j\omega)|^2|\tilde{\tilde{M}}_2(j\omega)|^2.$$

The use of a decoupler for the distant downstream part $K_{12} = K_{22}$ implies that:

$$\tilde{K}_{12} = \frac{G_1}{\tilde{G}_1} K_{12} + K_{22}$$

is zero for low frequencies. This ensures a decoupling for low frequencies, since in that case, the system (10.10) becomes diagonal for low frequencies:

$$\begin{pmatrix} e_1 \\ e_2 \end{pmatrix} \approx \begin{pmatrix} \tilde{\tilde{M}}_1(s) & 0 \\ 0 & \tilde{\tilde{M}}_2(s) \end{pmatrix} \begin{pmatrix} p_1 \\ p_2 \end{pmatrix}.$$

This ensures that for low frequencies, the performance of the two-pools canal is close to that of each pool considered independently.

The mixed control scheme therefore provides an efficient solution to the problem of designing a multivariable controller for multiple-pool irrigation canals. The main point is that it enables us to efficiently trade-off performance versus water management.

10.4.3 Simulation Results

The response of the multivariable mixed controller is depicted in Fig. 10.4. The real-time performance is very similar to that of the decentralized local upstream controller, but the water comes from upstream. As a result, the water levels are maintained close to their reference values, but the local upstream controls are only used temporarily, while the effect of the upstream discharge reaches the downstream end of the pool.

To illustrate the ability of the mixed control scheme to achieve control objectives for irrigation canals, the following experiment has been simulated: pool 1 is perturbed by high frequency but low amplitude water withdrawals, while pool 2 is subject to low frequency but high amplitude water withdrawals (see Fig. 10.5).

For comparison purposes, the simulation of the decentralized distant downstream controller is first depicted in Fig. 10.6. The performance in this case is limited by the time-delay, and the controller cannot efficiently reject the high frequency perturbation p_1. However, the low frequency perturbation p_2 is slowly but efficiently rejected.

The results of the mixed controller are depicted in Fig. 10.7. The local upstream controller of pool 1 efficiently rejects the high frequency perturbation, but at the same time is able to follow the low frequency water demand of pool 2. This is visible when the demand in pool 2 changes, corresponding to a change in the downstream discharge setpoint. In terms of water level control, the performance is more

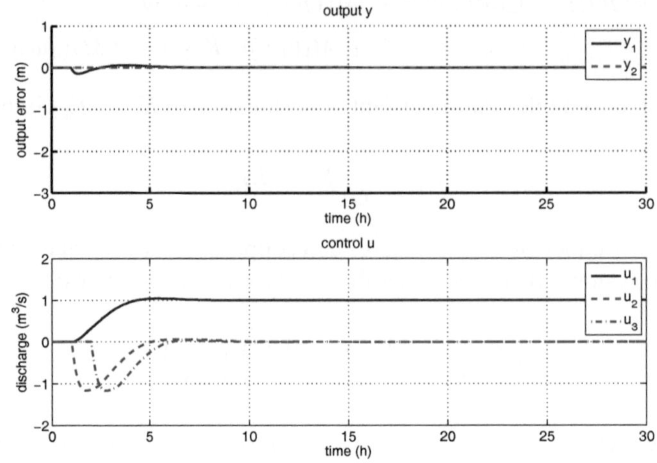

Fig. 10.4 Mixed control of a canal with two pools. Response to a perturbation in pool 1

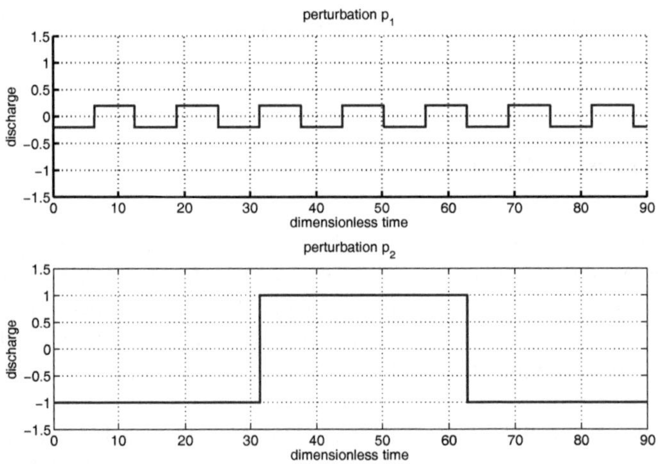

Fig. 10.5 Perturbations in pool 1 (high frequency low amplitude) and pool 2 (low frequency high amplitude)

than ten times better with the mixed controller than with the distant downstream controller. Due to the structure of the controller, this result is obtained while guaranteeing that the perturbations occurring in low frequencies are rejected with the upstream discharge, as in the distant downstream control case.

We conclude this section by noting that the controllers presented were successfully tested on the experimental canal of Évora (see Chap. 11).

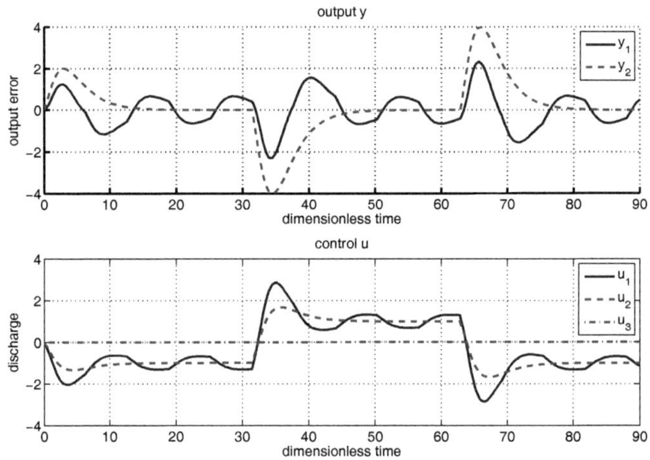

Fig. 10.6 Distant downstream control of a canal with two pools. Response to a high frequency perturbation in pool 1 and a low frequency perturbation in pool 2

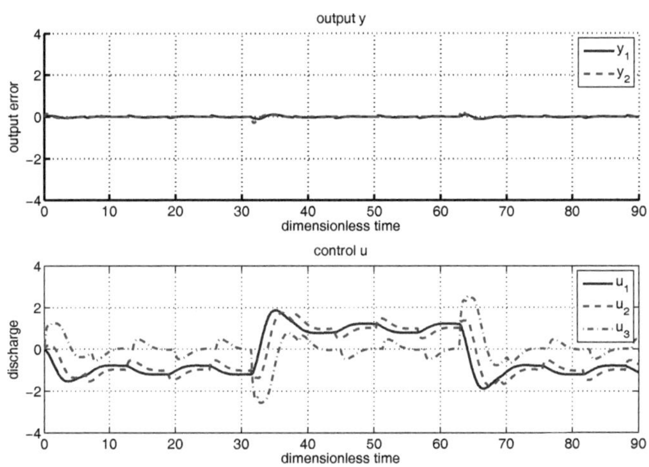

Fig. 10.7 Mixed local upstream/distant downstream control of a canal with two pools. Response to a high frequency perturbation in pool 1 and a low frequency perturbation in pool 2

10.5 Control of a Multiple-pool Canal

10.5.1 Mixed Control of a Multiple-pool Canal

Let us now consider the extension of the above results to the case of a multiple-pool canal.

The obtained mixed controlled structure can be implemented as a fully multivariable centralized controller. But one may use its particular structure to implement it in a structured semi-decentralized fashion, which will enable easier maintenance.

In this way, it is clear that each local controller only communicates with its closest neighbors located upstream and downstream. There is no need to implement other communication lines.

As shown above, each control action variable u_i can be decomposed as the sum of a local upstream control u_i^{lu} and a distant downstream control u_i^{dd}:

$$u_i = u_i^{lu} + u_i^{dd},$$

with u_i^{lu} acting mainly in high frequencies and u_i^{dd} in low frequencies. These control actions are given by:

$$\begin{aligned} u_i^{lu} &= K_{ii-1} e_{i-1}, \\ u_i^{dd} &= K_{ii} e_i, \end{aligned}$$

and by construction $|K_{ii-1}(j\omega)| \ll |K_{ii}(j\omega)|$ in low frequencies and $|K_{ii-1}(j\omega)| \gg |K_{ii}(j\omega)|$ in high frequencies.

The off-diagonal elements of the controller matrix K are then obtained by the decoupler's rules. To simplify the exposition, choosing a constant distant downstream feedforward decoupler equal to 1 leads to:

$$K_{ij}(s) = K_{jj}(s) \quad \forall i < j,$$

and choosing a local upstream feedforward decoupler equal to a pure delay leads to:

$$K_{ij}(s) = e^{-(\sum_{k=i-1}^{j+1} \tau_k)s} K_{j+1j}(s) \quad \forall i > j+1.$$

Such a control scheme can be schematized as in Figs. 10.8 and 10.9.

Such a bidiagonal control structure has the advantage of easy implementation, and easy fault diagnosis and fault recovery. Maintenance is also facilitated, while controller tuning is only done locally. This structured mixed control scheme therefore has many appealing aspects.

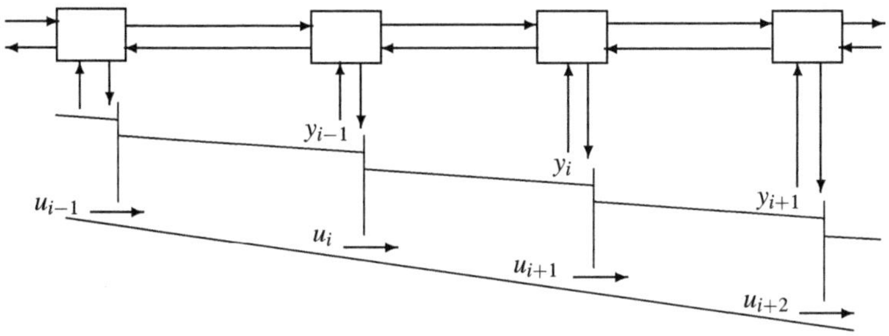

Fig. 10.8 General layout of the structured mixed control scheme

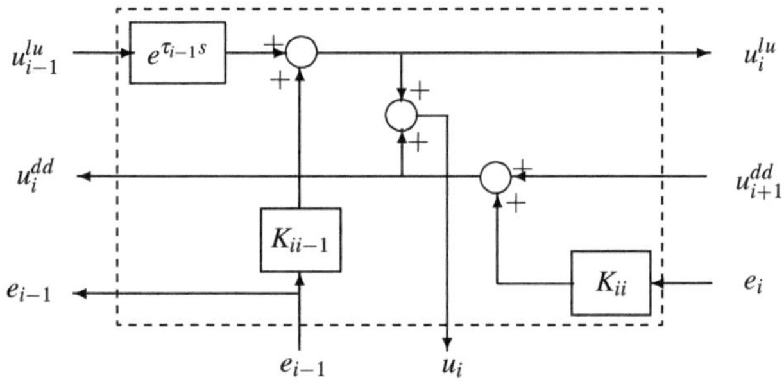

Fig. 10.9 Local control structure for the structured mixed control scheme

10.5.2 Mixed Controller Design According to Water Delivery Priorities

When dealing with multiple-pool canals, the performance appears to be more complex than the mere deviation of water levels from reference values. Indeed, there may be priorities between pools, and one should not necessarily evaluate the performance of a controller in terms of absolute water level deviations.

To illustrate this point, various simulations are performed on a three-pool canal. The scenario consists of unpredicted withdrawals occurring successively in each pool. The first withdrawal occurs in the first pool at time $t^* = 1$, the second occurs in the second pool at time $t^* = 50$, and the third one occurs in the third pool at time $t^* = 100$. We assume that the pools are identical, and we compare various possible control policies, according to various priorities on water level control.

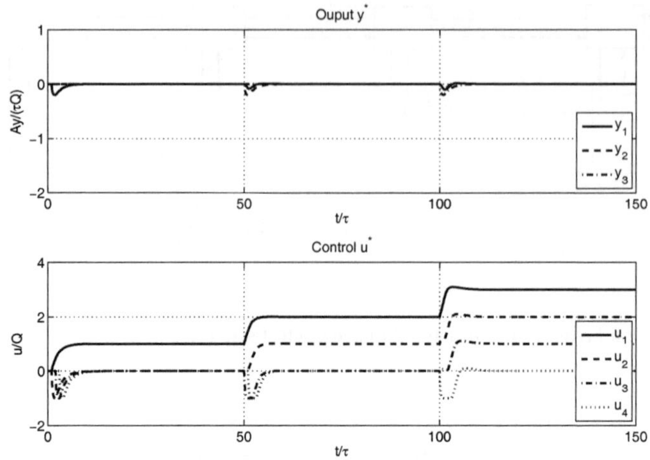

Fig. 10.10 Mixed multivariable control of a three-pool canal

10.5.2.1 Pure Mixed Control

We assume here that the downstream end of the canal can be used to control the level. In that case, one can use a pure mixed controller on each pool, i.e., the parameter θ is chosen equal to 1 for each pool. The local upstream controller K_2^* is a proportional controller, the mixed controller K_{1b}^* is tuned with a gain margin of 13 dB and a phase margin ratio equal to 0.7, leading to $k_{p1b}^* = 0.16$ and $T_{i1b}^* = 0.43$.

We use local upstream decouplers for each pool, and the distant downstream decouplers are chosen equal to 0.8, 0.9, and 1 from upstream to downstream, in order to reduce the amplification effect of an upstream propagating perturbation (see [2]).

Simulation results are depicted in Fig. 10.10, where we see that each pool responds in the same way to the perturbation. This corresponds to the maximum achievable performance on this canal. However, this is achieved to the extent of the water that is released downstream.

10.5.2.2 Pure Distant Downstream Control

If the last pool cannot be controlled with a local upstream controller, we can use a pure distant downstream controller for each pool, and add distant downstream decouplers to reduce the coupling.

This leads to the simulation results depicted in Fig. 10.11. Each pool has the same performance, which is much lower than in the case where local upstream control is used.

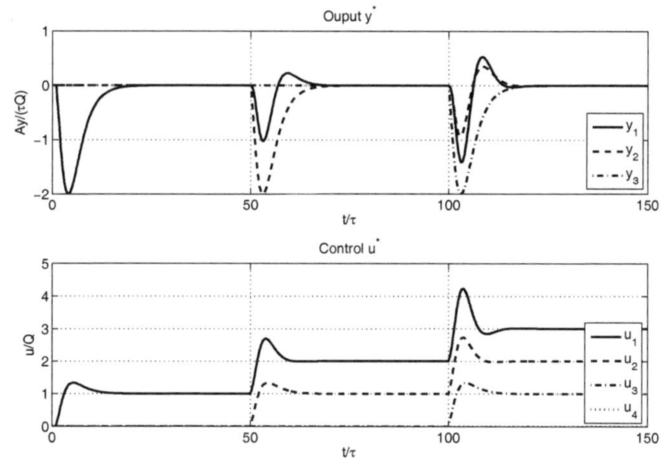

Fig. 10.11 Distant downstream multivariable control of a three-pool canal

This decentralized distant downstream controller ensures that the water comes from upstream and that no water is lost at the downstream end. It uses the assumption that each pool has the same priority in terms of water level control, which is not always the case.

10.5.2.3 Intermediate Solutions

Let us now assume that the last pool has a low priority with respect to the first two, and that it is essential to have a good performance in terms of water level control for the first two pools. Then, one can choose to design a distant downstream controller for pool 3, and mixed controllers for pools 1 and 2. This can be easily done using the proposed control structure.

Simulation results are depicted in Fig. 10.12. We can see that the water levels in pools 1 and 2 are controlled very efficiently, and that the perturbations are transferred to the third pool, the performance of which is much lower. In fact, this pool has a performance similar to that of a canal with a delay equal to the sum of the delays of the canal (in that case, a dimensionless delay of 3). All the perturbations generated by the upstream pools appear in the output of the third pool, since one cannot cancel them using the local upstream control action.

An intermediate solution can be obtained by choosing $\theta = 0.1$ for the two first pools. The water levels in the first two pools deviate a little bit more from their setpoint, and the perturbations transferred to the last pool are less important (see Fig. 10.13).

If we now assume that the second pool has a low priority, and that the local upstream control can be used in the third pool, then we obtain the simulation results

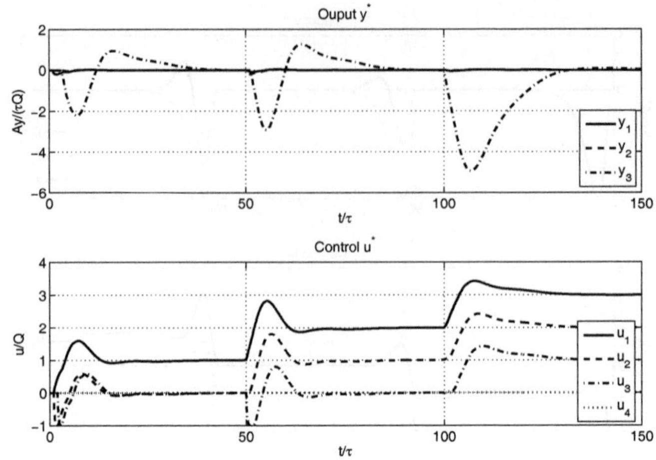

Fig. 10.12 Multivariable control of a three-pool canal with perturbations transferred to the last pool. $\theta = [1,1,0]$

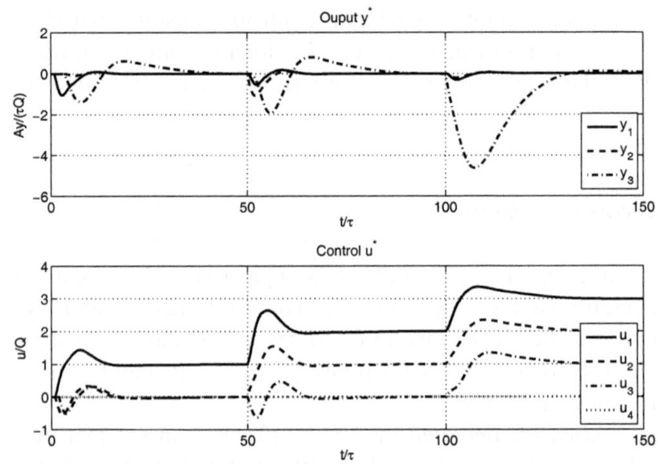

Fig. 10.13 Multivariable control of a three-pool canal with perturbations transferred to the last pool. $\theta = [0.1,0.1,0]$

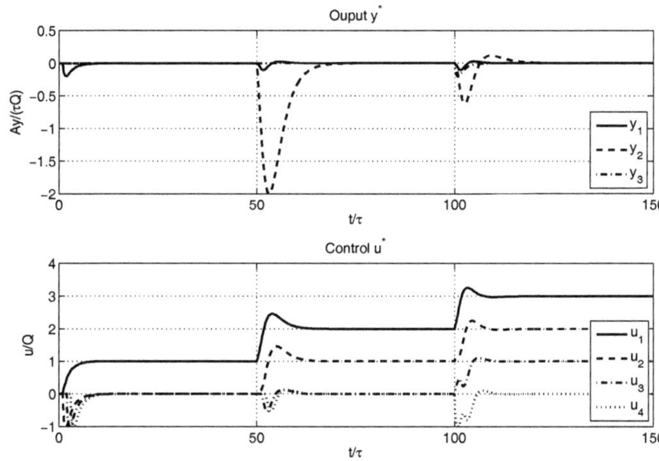

Fig. 10.14 Multivariable control of a three-pool canal, with perturbations transferred to the second pool. $\theta = [1, 0, 1]$

depicted in Fig. 10.14. The perturbations are transferred to the second pool, and partly to the downstream end.

Finally, let us assume that the second pool has a large capacity, i.e., that the integrator area is large compared to the other pools. We consider a simulation where the coefficient $A_d = 5$ for the second pool, and the delay remains equal to 1. In that case, using a controller similar to the one in Fig. 10.14, which transfers the perturbations to this pool, leads to the results depicted in Fig. 10.15. The deviation in water level from the setpoint in pool 2 is almost equal to those of the other pools. This shows that one can design multivariable controllers that effectively transfer the perturbations to pools with larger capacity or lower water delivery priority.

These simulations show that the multivariable design method enables us to effectively achieve the desired trade-off between water management and real-time performance. In the case of a multiple-pool canal, the main issue lies in the definition of the control objectives, which need to include the priorities between pools, and the possible control actions.

10.6 Summary

This chapter has reviewed classical multivariable decentralized control policies for an irrigation canal (distant downstream control and local upstream control) in terms of stability, robustness, and performance. We have proposed a new multivariable structured control policy that combines both classical con-

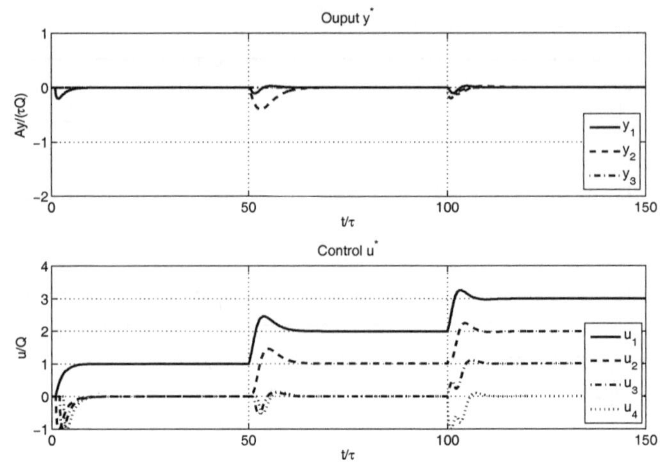

Fig. 10.15 Multivariable control of a three-pool canal, with perturbations transferred to the second pool. $\theta = [1, 0, 1]$. The second pool has a capacity five times larger than the other pools

trol policies, in order to trade-off between resource management and performance with respect to the water user. This multivariable mixed control scheme has been shown to be stable if and only if the mixed controllers for each pool are stable. Its robustness has been evaluated and shown to be close to that computed separately for each pool. A simple design method has been provided to design cascade input PI controllers for each pool.

The three control structures have been evaluated in simulation to show their relative performance. The mixed controller structure appears to be an efficient solution to the problem of designing local controllers for an irrigation canal.

In any case, the mixed control structure enables us to increase the water management efficiency of local upstream control and to increase the performance with respect to water users for distant downstream control. It enables us to efficiently mix the water management of distant downstream control and the time domain performance of local upstream control. The proposed method has the same complexity as the classical control methods, while dramatically improving their performance.

References

[1] Baume JP, Malaterre PO, Sau J (1999) Tuning of PI to control an irrigation canal using optimization tools. In: Workshop on modernization of irrigation water delivery systems, Phoenix, pp 483–500

[2] Cantoni M, Weyer E, Li Y, Ooi SK, Mareels I, Ryan M (2007) Control of large-scale irrigation networks. Proc IEEE 95(1):75–91

[3] Clemmens A, Schuurmans J (2004) Simple optimal downstream feedback canal controllers: theory. J Irrig Drain Eng 130(1):26–34
[4] Malaterre PO, Rogers DC, Schuurmans J (1998) Classification of canal control algorithms. J Irrig Drain Eng 124(1):3–10
[5] Rogier D, Coeuret C, Brémond J (1987) Dynamic regulation on the Canal de Provence. Proceedings of a ASCE symposium, Portland pp 180–200
[6] Safonov MG (1980) Stability and robustness of multivariable feedback systems. Series in signal processing, optimization, and control; 3, MIT Press, Cambridge
[7] Schuurmans J, Hof A, Dijkstra S, Bosgra OH, Brouwer R (1999) Simple water level controller for irrigation and drainage canals. J Irrig Drain Eng 125(4):189–195
[8] Seatzu C (1999) Design and robustness analysis of decentralized constant volume-control for open-channels. Appl Math Mod 23(6):479–500
[9] Skogestad S, Postlethwaite I (1998) Multivariable feedback control. Analysis and design. Wiley, New York
[10] Tannenbaum A (1992) Analysis and optimization of systems: State and frequency domain approaches for infinite dimensional systems, Lecture Notes in control and information sciences, vol 185, Springer, New York, chap Frequency domain methods for the H_∞ optimization of distributed systems
[11] Weyer E (2002) Decentralised PI controller of an open water channel. In: 15th IFAC World Congress, Barcelona, Spain
[12] Zhou K, Doyle J (1998) Essentials of robust control. Prentice Hall, Upper Saddle River, NJ

Chapter 11
Experimental Results on a Small-scale Canal

Abstract In this chapter, we present experimental results obtained on a small-scale canal located in Portugal. Most of the results presented in this book have been tested on this canal, from the modeling part to the control schemes, for one and multiple pools. The results show that the Saint-Venant equations for open channels can very accurately represent the system's dynamics, and that the proposed control strategies are effective in practice.

11.1 Description of the Experimental Facility

The methods developed in the book have been tested on a small-scale canal. This experimental facility belongs to the Hydraulics and Canal Control Center (NuHCC) of Professor Manuel Rijo[1] at the University of Évora (Portugal). This canal possesses four pools separated by three undershot gates (see Fig. 11.1).

The experimental canal is a lined canal, with a general trapezoidal cross-section of bottom width 0.15 m, sides slope 6.6, and depth 0.90 m (see Fig. 11.2). The last downstream 7 m of the canal also have a rectangular cross-section of width 0.7 m. The overall canal is 145.5 m long and the average longitudinal bottom slope is about 1.5×10^{-3}. The maximum flow is 90 l/s. Figure 11.3 provides a general overview of the canal.

There is an offtake p at the downstream end of the pool of the orifice type (a lateral hole close to the bottom of the canal) with an additional external pipe, equipped with an electromagnetic flowmeter and a motorized butterfly valve.

For real-time control purposes, the canal is equipped with water level sensors installed in off-line stilling wells along the canal. The sensor is of the float and

[1] We are thankful to Professor Manuel Rijo and his students Carina Arranja and Miguel Nunes de Almeida for their help during the experiments. This work was supported by the French Ministry of Foreign Affairs and GRICES (Gabinete de Relações Internacionais da Ciência e do Ensino Superior), via the French-Portuguese scientific collaboration projects PAI no. 547-B4 in 2000-2003 and PESSOA no. 09685UE in 2005-2007.

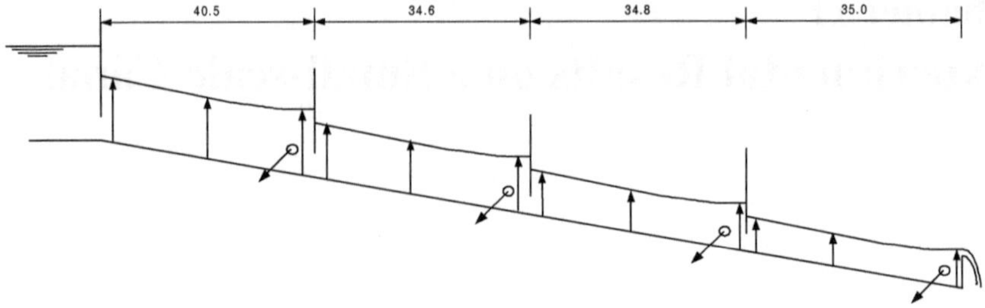

Fig. 11.1 Longitudinal view of the experimental canal

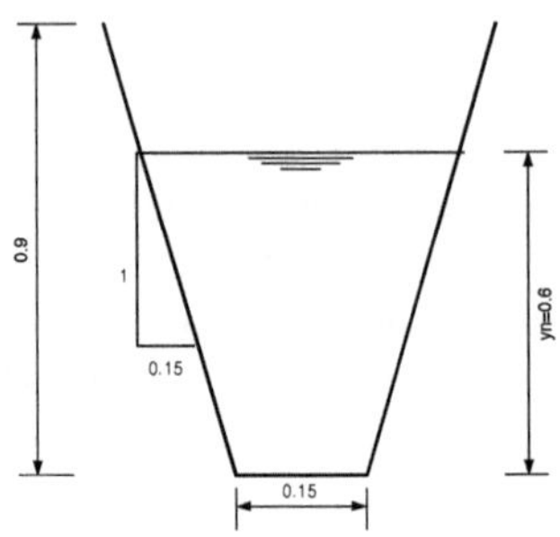

Fig. 11.2 Section of the experimental canal

counter-weight type attached by a stainless steel tape; this tape runs over a sprocket wheel. The wheel movements are transmitted to a potentiometer that transmits the analogical inputs corresponding to the water surface to the controller.

The canal inlet is equipped with a motorized flow control valve that delivers a given discharge. The downstream end is controlled with a motorized rectangular sluice gate (overshot gate). The other cross-structures are motorized undershot sluice gates. All the structures can be remotely controlled in real-time using a SCADA (supervisory control and data acquisition) system.

11.2 Modeling

11.2.1 Nonlinear Model Calibration

11.2.1.1 Steady Flow Model Calibration

For a given constant upstream discharge, the water levels and gate openings are monitored in each pool. This steady flow period enables us to identify the Manning coefficient n for each pool and the discharge coefficient at each structure. The hy-

Fig. 11.3 General view of the experimental canal with a motorized gate and two water level sensors located in off-line stilling wells

draulic calibration is done classically by finding the value of the Manning coefficient that minimizes the error between the computed water levels given by the solution of (2.16a–2.16b) and the measured ones. The discharge coefficient for the overshot gate is also computed by inverting the gate equation (11.2) with the measured discharge and water levels.

In our case, the calibration was performed for a steady flow regime of 45 l/s and a downstream water elevation $Y_L = 0.6$ m. This led to a discharge coefficient of 0.4 for the overshot gate and a Manning coefficient of 0.017 for the canal.

11.2.1.2 Modeling of Actuators and Sensors

Actuators dynamics are identified using a step response. The mechanical system is represented by a linear model of first-order with transmission delay. The upstream actuator u_1 delivering a discharge is modeled with a transfer function $F_1(s)$ identified on a step response of the actuator:

$$F_1(s) = \frac{e^{-4s}}{8s+1}. \tag{11.1}$$

In this case, the input is the desired upstream discharge and the output is the upstream discharge effectively delivered by the gate.

Because the sensors' dynamics are very rapid compared to the canal dynamics, they are neglected in the following (i.e., their transfer function is equal to 1).

11.2.1.3 Hydraulic Structure Equation

The overshot gate hydraulic structure is modeled using the classical equation (free flow case) [3]:

$$Q(L) = C_d L_w \sqrt{2g}(Y(L) - Z_w)^{3/2}, \tag{11.2}$$

with $Q(L)$ the discharge through the structure, $Y(L)$ the water level upstream of the structure, Z_w the weir elevation and L_w the weir width. C_d is a discharge coefficient, close to 0.4 in this case, which takes into account the flow contraction.

11.2.1.4 Unsteady Flow Model Validation

The steady flow calibration is sufficient to obtain a good dynamic model of a canal. For illustration purposes, we performed various experiments to validate the model in unsteady flow for different flow configurations. The canal is simulated using SIC, a computer program developed by Cemagref [9]. This mathematical model solves the full nonlinear Saint-Venant equations using a finite difference scheme (the Preissmann scheme). The model appeared to be very accurate, even for different flow conditions. Figure 11.4 corresponds to the same regime as the one used for the steady flow calibration.

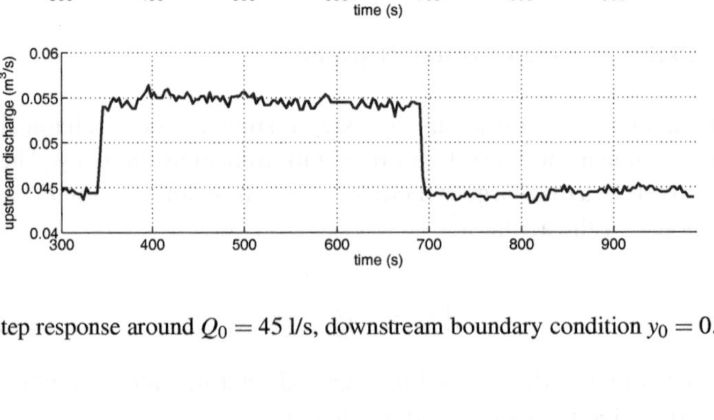

Fig. 11.4 Step response around $Q_0 = 45$ l/s, downstream boundary condition $y_0 = 0.6$ m

We also verified that the model is still able to accurately reproduce the level variations when the downstream boundary condition and the upstream discharge have changed from the values used for identification (see Figs. 11.5 and 11.6).

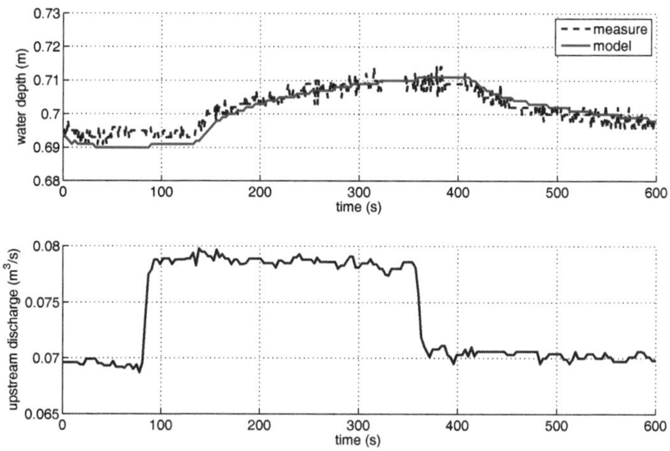

Fig. 11.5 Step response around $Q_0 = 70$ l/s, downstream boundary condition $y_0 = 0.7$ m

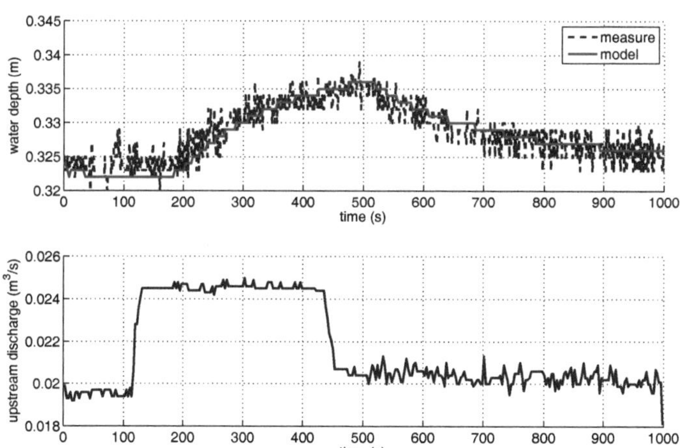

Fig. 11.6 Step response around $Q_0 = 20$ l/s, downstream boundary condition $y_0 = 0.3$ m

The prediction ability of the Saint-Venant equations is striking, in spite of their relatively simple assumptions (hydrostatic pressure distribution, one-dimensional flow, motion by slice, etc.). This validates the full hydraulic model for this canal, i.e., the Saint-Venant equations and hydraulic structures.

We now focus on the design approach, which uses a set of linear models to design and evaluate the robustness of the controller.

11.2.2 Control Oriented Linear Model

The linear model used for control design is obtained following Chap. 3, using the steady-state conditions computed using Saint-Venant equations.

Figure 11.7 presents the Bode plots of the continuous model $p_{21}(s)$ of the Évora canal relating the upstream discharge to the downstream water elevation for a set of discharges. The frequency response of the canal shows that it behaves as an integrator for low frequencies. This is coherent with the physical interpretation: the canal is similar to a reservoir whose level varies according to the input discharge. The transfer $p_{21}(s)$ includes a time-delay, which makes the phase decrease towards $-\infty$. Finally, the system has a nonzero gain for high frequencies. This is explained by the fact that there are oscillating modes in high frequencies corresponding to the interactions of waves propagating upstream at speed $C_0 - V_0$ and downstream at speed $V_0 + C_0$.

It is clear that the model is very sensitive to the flow rate: the delay changes from 55 s for $Q_0 = 80$ l/s to 77 s for $Q_0 = 1$ l/s, and the oscillating modes also change (especially their dampening factor). The model corresponding to the lower discharge (in this case 1 l/s) has the largest delay and the least dampened oscillating modes.

This model needs to be completed with a model of the dynamics of the actuators and sensors. The equations describing hydraulic structures' interactions with the flow are linearized and added to the model. The hydraulic structure (overshot gate) is modeled by linearizing (11.2):

$$q = k_u y, \tag{11.3}$$

with $q = q(L)$ the deviation from stationary discharge through the structure and $y = y(L)$ the deviation from stationary upstream water level. Coefficient k_u is given by:

$$k_u = \frac{3}{2} C_d L_w \sqrt{2g} (Y_0 - Z_w)^{1/2}.$$

A linear model of the canal pool is then given by:

$$y = \frac{F_1(s) p_{21}(s)}{1 - k_u p_{22}(s)} u_1 + \frac{p_{22}(s)}{1 - k_u p_{22}(s)} p, \tag{11.4}$$

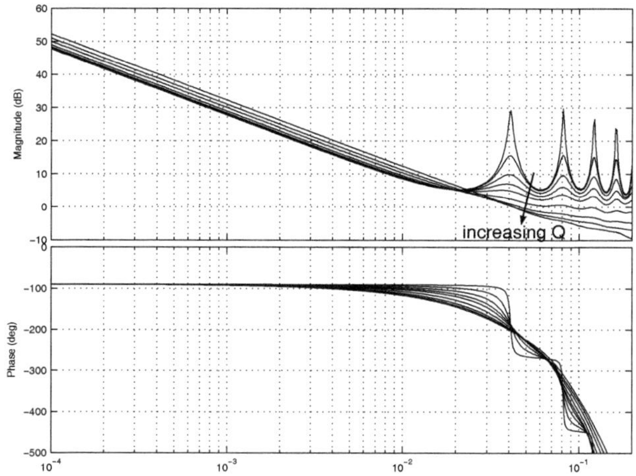

Fig. 11.7 Bode plots of the continuous model $p_{21}(s)$ for $Q_0 \in \{1, 5, 10, 15, 20, 30, 45, 60, 80\}$ l/s, $Y_L = 0.6$ m

where y is the downstream water level, u_1 the upstream control (the actuator control input, equivalent to the upstream discharge $q(0,t)$), and p is the downstream perturbation (corresponding to the unknown water withdrawal).

These transfer functions are obtained by combining the transfer matrix (4.1) representing the flow dynamics and the hydraulic structure equation (11.3) representing the local feedback induced by the hydraulic structure.

The Bode plots of the models obtained for the Évora canal are depicted in Fig. 11.8. One can observe that the static gain also varies with the discharge. These models will be used to design simple controllers (filtered PI).

11.3 Robust Distant Downstream PI Control of a Canal Pool

We adapt here the method developed in Chap. 7 to the case of a canal with a fixed downstream cross-structure.

11.3.1 PI Controller Design

Let us first consider the model corresponding to 45 l/s. As mentioned above, and validated in previous works [5, 10], this model appears to be similar to a first-order with delay model. There are many ways to tune a PI controller for such a system [1]. The method used is adapted from that proposed in Chap. 7.

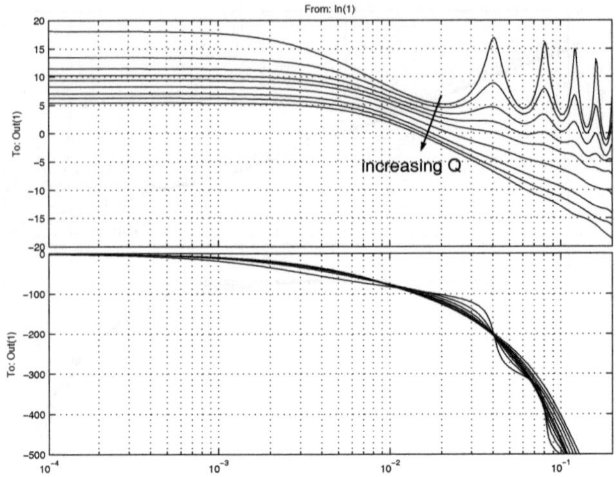

Fig. 11.8 Bode plots of the complete model for $Q \in \{1,5,10,15,20,30,45,60,80\}$ l/s, $Y_L = 0.6$ m

We used a simple approximation: the gain margin ΔG is supposed to be obtained at the frequency ω_{180} where the system's phase equals $-180°$. This is valid if the phase added by the controller's zero is negligible at this frequency. Then, the controller proportional gain k_p is directly linked to the desired gain margin by the relation:

$$k_p = \frac{10^{-\Delta G/20}}{|G(j\omega_{180})|}.$$

The integrator gain T_i is then tuned in order to obtain a good performance in rejecting unmeasured perturbations. Reasonable values range from 2 to 5 times τ (the time-delay of the system).

The presented approach is simple, but effective, and leads to a good approximation of the controller. Fine manual tuning is sometimes necessary to obtain a required level of performance. A low-pass filter is added in order to filter the high frequency noise occurring on the output y. This leads to a filtered PI controller.

Our objective it not to obtain a controller that stabilizes a single model, but rather a family of linear models. The design approach is therefore iterative: the obtained controller margins are evaluated over the different models, and the controller is modified until the minimum gain margin is about 2 dB. This can be done by increasing (or decreasing) the gain margin for the nominal case.

We designed two controllers using this design method, a robust controller with a gain margin of about 9 dB in the nominal case and a rapid controller with a gain margin of about 7 dB in the nominal case.

The robust distant downstream controller (a filtered PI) is obtained as:

$$K_1(s) = \frac{0.324s + 2.7 \times 10^{-3}}{s(25^2 s^2 + 25s + 1)}.$$

The gain and phase margins evaluated on the nominal plant are, respectively, 8.8 dB and 62.6°. The robust controller is stable for a set of linear models corresponding to discharges $Q \in \{1,5,10,15,20,30,45,60,80\}$ l/s (see Fig. 11.9). The open-loop is depicted on a Nichols chart, which allows to evaluate on the same graphic the gain, phase and complementary modulus margins. We observe that the smallest margins are obtained for the model corresponding to the lowest discharge (1 l/s).

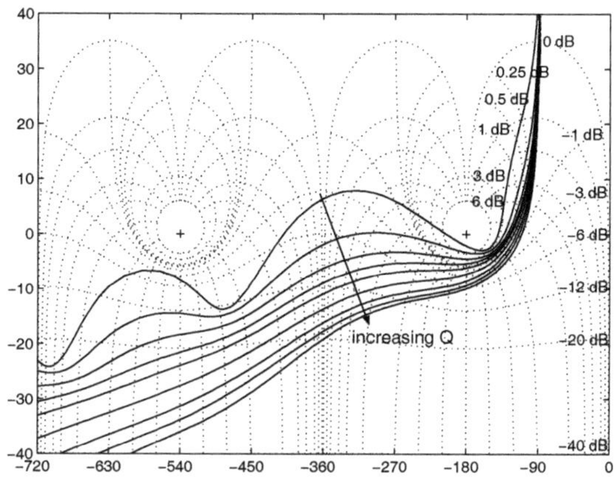

Fig. 11.9 Nichols chart for the robust distant downstream controller, for $Q \in \{1,5,10,15,20,30,45,60,80\}$ l/s, $Y_L = 0.6$ m

The rapid distant downstream controller is obtained as:

$$K_2(s) = \frac{0.5s + 4 \times 10^{-3}}{s(10^2 s^2 + 10s + 1)}$$

The gain and phase margins evaluated on the nominal plant are, respectively, 7.3 dB and 59°.

This controller has a better real-time performance, but is less robust, since according to the model, the closed-loop is unstable for reference discharges lower than 10 l/s (see Fig. 11.10).

This can also be checked from Fig. 11.11, where the gain margin of both open-loop systems (with the robust and rapid distant downstream controller) are plotted with respect to the reference discharge. It is clear that the robust controller has a sufficient gain margin for all linear models obtained for discharges ranging from 1 to 80 l/s, whereas the rapid controller leads to an unstable loop for linear models obtained for discharges lower than 10 l/s. Therefore, such a controller should not perform correctly when tested around 10 l/s.

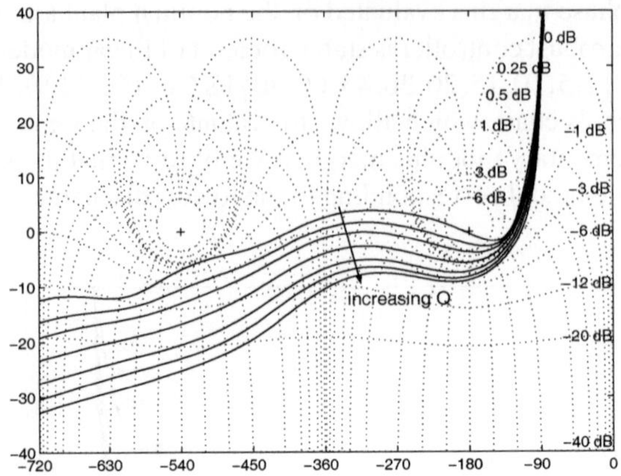

Fig. 11.10 Nichols chart of the rapid distant downstream controller, for $Q \in \{10,15,20,30,45,60,80\}$ l/s, $Y_L = 0.6$ m

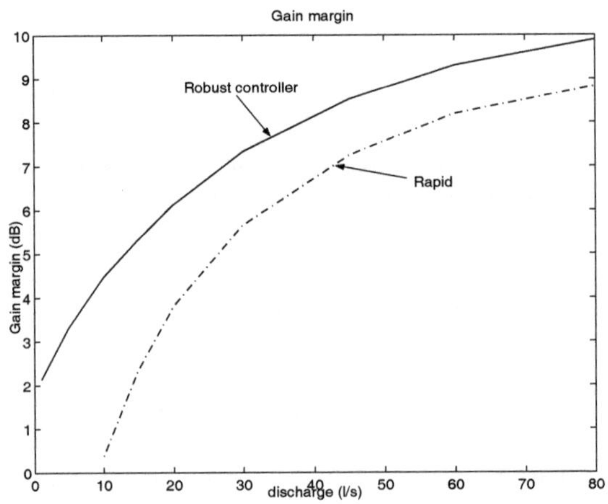

Fig. 11.11 Gain margins of the two distant downstream controller, for $Q \in \{1,5,10,15,20,30,45,60,80\}$ l/s, $Y_L = 0.6$ m

11.3.2 Experimental Validation

We first evaluate the robust controller performance for different reference discharges, before trying to push our approach towards its limits, by using the model to predict instability, with the rapid controller.

11.3.2.1 Experiments with the Robust Controller

Figure 11.12 gives the experimental results obtained with a robust distant downstream filtered PI controller around 45 l/s. A downstream withdrawal of 10 l/s (0.01 m^3/s) is done at time $t = 40$ s. The controller is able to reject this unknown perturbation in about 400 s (time for the output y to reject 90% of the perturbation). The outlet is closed at time $t = 750$ s, and the response appears to be symmetrical. Figure 11.12 also depicts the simulation results obtained with the linear model deduced from Saint-Venant and hydraulic structures equations. The comparison shows that the linear simulation is very close to the experimental results.

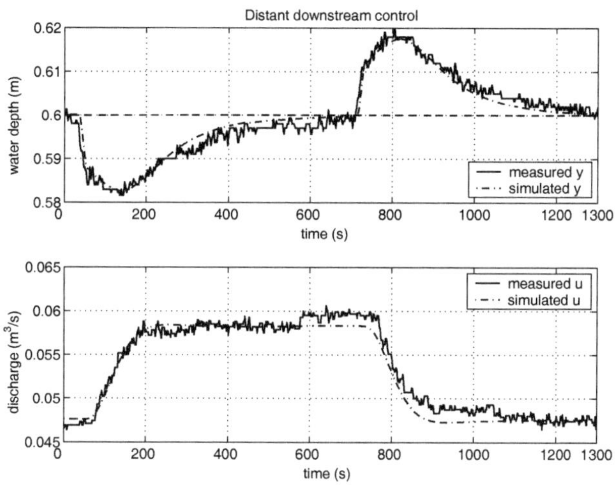

Fig. 11.12 Experimental response of the robust distant downstream filtered PI controller to a downstream withdrawal around $Q = 45$ l/s. Comparison with a linear simulation ($- \cdot -$)

Figure 11.13 shows the same experiment repeated for a reference flow $Q = 60$ l/s and compared with a linear simulation on the model corresponding to 60 l/s. The controller still behaves correctly, and the linear simulation is very accurate, as in the nominal case. Other experiments were carried out for different reference discharges, which confirmed these results.

Our model-based approach is therefore validated for "normal" functioning conditions. However, is this model able to predict instability?

11.3.2.2 Experiments with the Rapid Controller

According to our model, the rapid controller K_2 should not perform correctly when tested around 10 l/s. Is this theoretical result representative of the reality? Experiments were carried out to test this hypothesis. Figures 11.14 and 11.15 show the

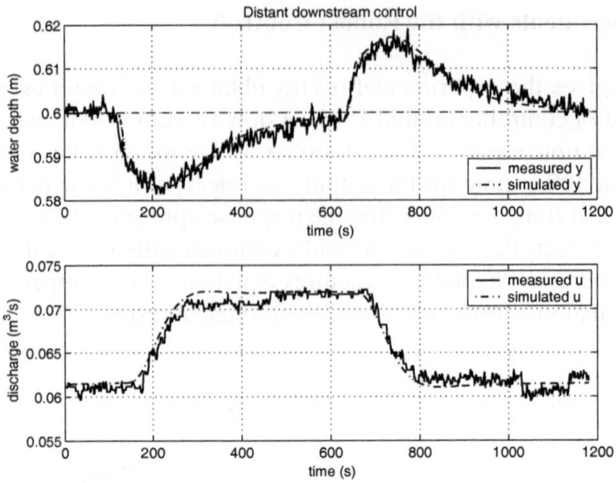

Fig. 11.13 Experimental response of the robust distant downstream filtered PI controller to a downstream withdrawal around $Q = 60$ l/s. Comparison with a linear simulation (— · —)

response of the controlled system to a downstream withdrawal around 45 and 60 l/s. The experimental system behavior is once again very close to the linear simulations.

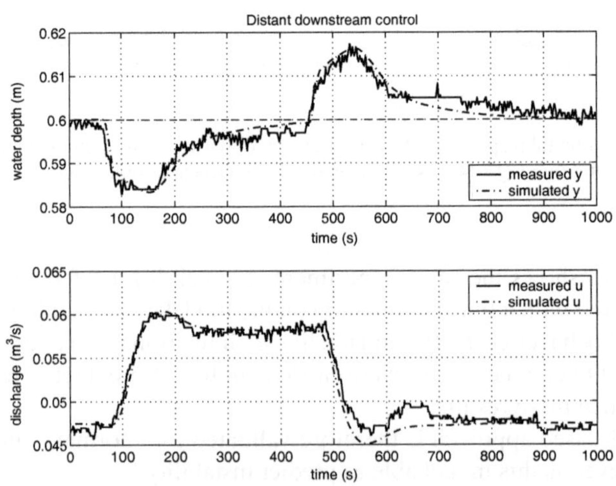

Fig. 11.14 Experimental response of the rapid distant downstream filtered PI controller to a downstream withdrawal around $Q = 45$ l/s. Comparison with a linear simulation (— · —)

An experiment was carried out around 15 l/s to see the prediction ability of the model for low discharges (see Fig. 11.16). The oscillations predicted by the linear

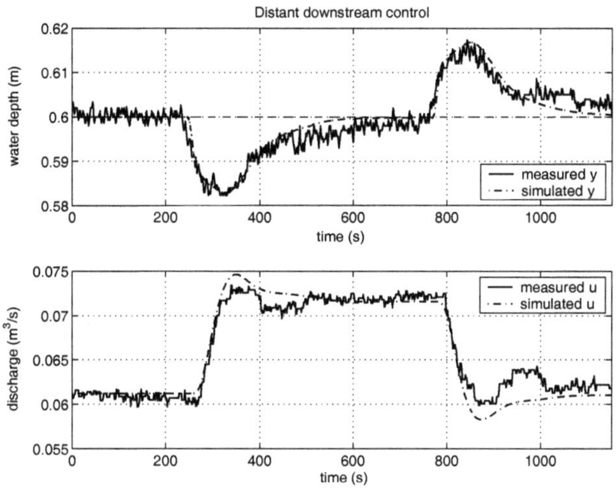

Fig. 11.15 Experimental response of the rapid distant downstream filtered PI controller to a downstream withdrawal around $Q = 60$ l/s. Comparison with a linear simulation $(- \cdot -)$

model are not so important in reality, but this may be due to limitations of the actuators. However, the overall dynamics are well represented by the model, which maybe underestimates the damping in the canal for low discharges.

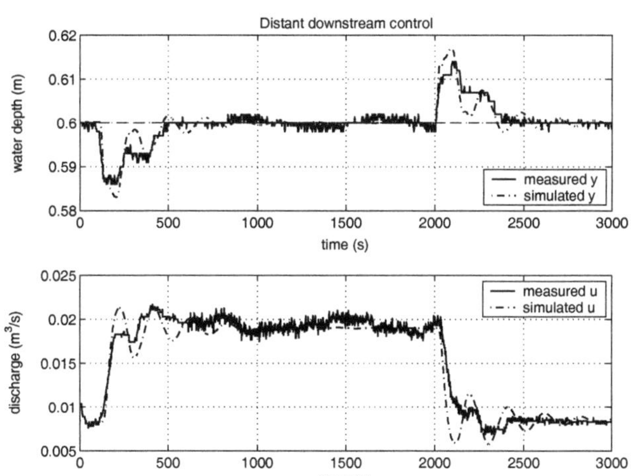

Fig. 11.16 Experimental response of the rapid distant downstream filtered PI controller to a downstream withdrawal around $Q = 15$ l/s. Comparison with a linear simulation $(- \cdot -)$

A step response of the controlled system is tested around 15 l/s. The water level setpoint is changed from 0.6 m to 0.65 at time $t = 10$ s. Figure 11.17 shows that the system is oscillating, and the linear simulation is still able to rather correctly estimate the system's behavior.

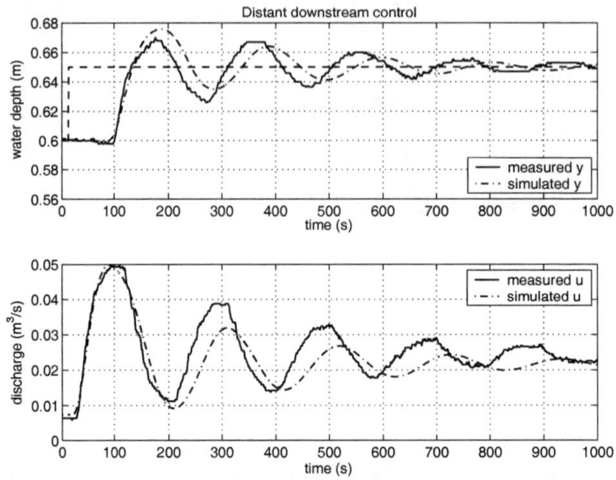

Fig. 11.17 Step response of the rapid distant downstream filtered PI controller: Experimental measurement around $Q = 20$ l/s. Comparison with a linear simulation $(- \cdot -)$

These experimental results show the usefulness of our model-based approach: such an approach is able to accurately predict the system behavior around steady-state regimes corresponding to various discharges. This is exactly what is needed in our framework, which uses the gain-scheduling heuristic rule: the stability of stationary linearizations allows us to guarantee the stability of the nonlinear system. Therefore, our methodology can be used to design efficient controllers for irrigation canals.

11.4 Mixed H_∞ Control of a Canal Pool

We now recast the mixed controller design problem as an H_∞ optimization problem, which can naturally take into account performance and robustness issues. In our case, we need to ensure robustness with respect to static gain actuators uncertainties, which can be done by ensuring sufficient real input gain margins. In the following, we compare H_∞ controllers with distant downstream and mixed control policies.

11.4.1 H_∞ Design Specifications

Following the approach already developed in Chap. 7, design specifications are formulated using an H_∞ four-block type criterion. Let the system be described by:

$$\begin{bmatrix} y \\ u_2 \end{bmatrix} = G_a \begin{bmatrix} u_1 \\ u_2 \end{bmatrix} + \widetilde{G}_a p,$$

with $G_a = \begin{pmatrix} G_1 & \tilde{G}_1 \\ 0 & 1 \end{pmatrix}$ and $\widetilde{G}_a = \begin{pmatrix} \tilde{G}_1 \\ 0 \end{pmatrix}$.

For the mixed controller, in order to constrain the downstream control u_2 to go asymptotically to zero, $r_{u_2} - u_2$ is fed back into the controller as a tracking error (see Fig. 8.1). For usual functioning, r_{u_2} is zero. Therefore, imposing a low gain in low frequencies for the transfer function that links the reference r_{u_2} to the error $r_{u_2} - u_2$ will impose a low value for u_2 at these frequencies.

The closed-loop system that links the reference $\bar{r} = [r, r_{u_2}]^T$ and the perturbation p to the tracking error $\bar{e} = [r - y, r_{u_2} - u_2]^T$ and the controlled input $u = [u_1, u_2]^T$ is given by

$$\begin{bmatrix} \bar{e} \\ u \end{bmatrix} = \begin{pmatrix} S_a & S_a\widetilde{G}_a \\ K_aS_a & K_aS_a\widetilde{G}_a \end{pmatrix} \begin{bmatrix} \bar{r} \\ p \end{bmatrix}, \tag{11.5}$$

where K_a is the controller for the augmented system G_a and $S_a = (I + G_aK_a)^{-1}$ is the sensitivity function for system G_a.

11.4.2 A Solution as a Four-block Criterion

The design specifications are then formulated using the following criteria, where the goal is to find the smallest $\gamma_1 > 0$ and the stabilizing controller K_a such that

$$\left\| \begin{pmatrix} W_1S_a & W_1S_a\widetilde{G}_aW_3 \\ W_2K_aS_a & W_2K_aS_a\widetilde{G}_aW_3 \end{pmatrix} \right\|_\infty \leq \gamma_1, \tag{11.6}$$

with $W_1 = \text{diag}(W_{11}, W_{12})$, $W_2 = \text{diag}(W_{21}, W_{22})$, $W_{ik}, W_{ik}^{-1} \in RH_\infty$, and $W_3 \in \mathbb{R}$ is a scaling factor acting on the perturbation. The system augmented with weighting functions is described in Fig. 11.18.

As in the distant downstream control case, $W_1, W_1^{-1} \in RH_\infty$ is used to specify tracking performances, perturbation rejection and modulus margin. Here, W_1 is chosen diagonal $W_1 = \text{diag}(W_{11}, W_{12})$. $W_2, W_2^{-1} \in RH_\infty$ is used to specify high frequency constraints on the controlled inputs. W_2 is also chosen diagonal $W_2 = \text{diag}(W_{21}, W_{22})$, allowing us to constrain command effort and effects of sensor noise command. W_3 is a scaling factor acting on the perturbation. The weighting functions W_{ik} are chosen of the first-order, as in (7.27).

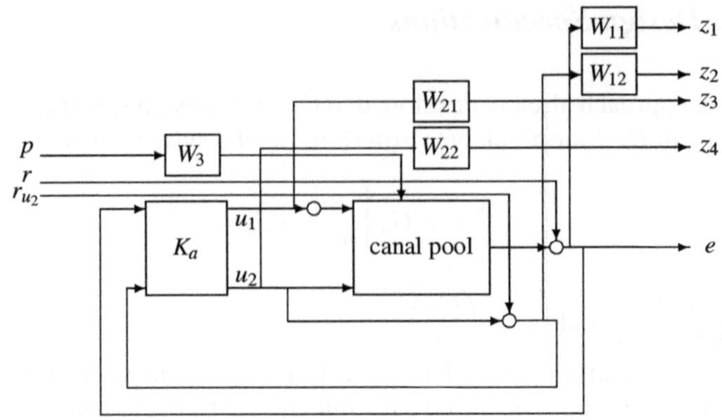

Fig. 11.18 Augmented system for H_∞ optimization

The performance requirements are easy to specify with these weighting functions:

- W_{11} specifies the overall performance of the system with respect to the users requirement. This specification has to be compatible with bandwidth constraints associated to the upstream control actuator. The system bandwidth is closely related to the frequency $\omega_{11,c}$ where the gain of W_{11} equals 1.
- W_{12} corresponds to the substitution specification between downstream and upstream actuators. It has to be chosen in order to be compatible with the time-delay, i.e., $\omega_{12,c} \leq 1/\tau_1$.
- W_{21} allows us to impose roll-off on the upstream control actuator and it has to be compatible with the substitution specification associated to W_{12}. Following the chosen parameterization of the weighting function, $\omega_{21,c} \approx 10 \times \omega_{12,c}$.
- W_{22} allows us to impose strong roll-off on the downstream control actuator and it has to be compatible with the performance specification associated to W_{11}. Following the chosen parameterization of the weighting function, $\omega_{22,c} \approx 10 \times \omega_{11,c}$. Let us finally note that this roll-off is necessary in order to avoid any active control of canal oscillating modes by the downstream control actuator.

It appears more difficult to directly specify robustness requirements, since the chosen criterion does not explicitly consider robustness constraints. However, an input margin constraint can be specified by a careful choice of weighting functions. Indeed, using (8.8), we see that by constraining S_y, K_1, and K_2, one can constrain the input gain margin. This can be done via the parameter $G_{11,\infty}$ of W_{11}, which allows us to constrain S_y, and the parameters $G_{21,\infty}$ of W_{21} and $G_{22,\infty}$ of W_{22}, which allow us to constrain K_1 and K_2.

The weighting functions are tuned sequentially: first W_{22} is tuned to specify the maximum bandwidth for local upstream control, then W_{11} is tuned to specify the global real-time performance, then W_{12} is used to specify the actuators substitution.

W_{21} and the scaling factor W_3 are used to specify control effort and robustness requirements.

11.4.3 Design of H_∞ Controllers

Three different H_∞ controllers are designed and tested on the canal, to show the ability of the proposed control scheme to provide a compromise between real-time performance and water management issues (see Chap. 8 and also [7]). The first controller is a pure distant downstream controller, where the control action u_2 is not operated, therefore its real-time performance with respect to the user is rather low, as it is limited by the time-delay. The second controller is a mixed controller, using u_2 to increase the real-time performance. The third one is a mixed controller with a high real-time performance, close to that obtained by a pure local upstream controller. For the three controllers, the optimization resulted in $\gamma \approx 1$. The parameters for the weighting functions of the three controllers are given in Tables 11.1–11.3. The parameter ω_c for weighting function W_{11} gives a good approximation of the controlled system bandwidth. This parameter is increased from 5×10^{-3} rad/s for the first controller to 2.5×10^{-2} rad/s for the second, and 5×10^{-2} rad/s for the third controller, i.e., 10 times the bandwidth of the first one. This illustrates the ability of the control scheme to "tune" the real-time performance with respect to the user.

Table 11.1 Parameters of the first-order weighting functions given by (7.27), controller 1

	W_{11}	W_{12}	W_{21}	W_{22}	W_3
G_0	1000	1	0.9	0.5	0.48
G_∞	0.4	0.9	1000	1000	
ω_c	5×10^{-3}	5×10^{-5}	10^{-2}	10^{-5}	

Table 11.2 Parameters of the first-order weighting functions given by (7.27), controller 2

	W_{11}	W_{12}	W_{21}	W_{22}	W_3
G_0	1000	10000	0.9	0.2	0.69
G_∞	0.8	0.2	1000	1000	
ω_c	2.5×10^{-2}	5×10^{-3}	10^{-2}	10^{-1}	

Figure 11.19 presents the transfer functions appearing in the H_∞ criteria with associated constraints for controller 2. Each transfer is indeed below the associated constraint, showing that the obtained γ is smaller than 1.

Figure 11.20 depicts the singular values of the three controllers. It is clear that the first controller uses only one control input (corresponding to the distant downstream

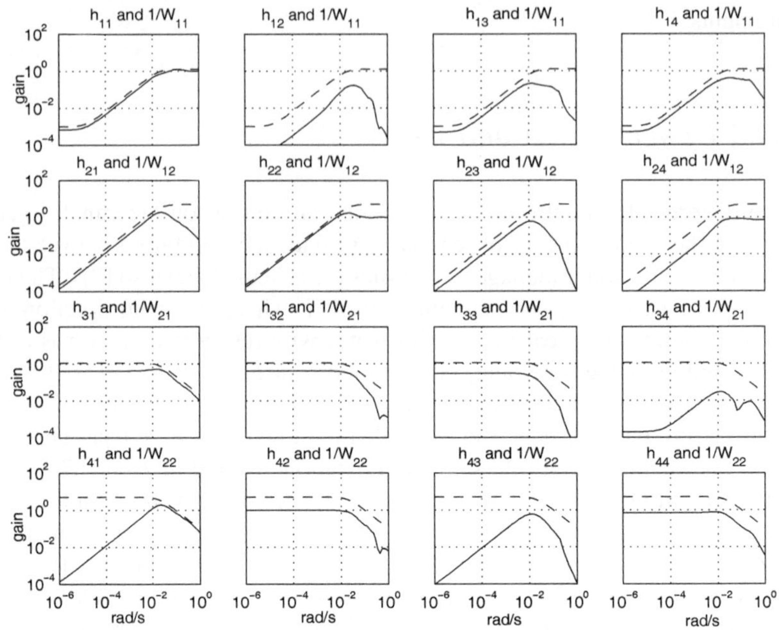

Fig. 11.19 Closed loop transfer functions (—) and associated inverse frequency weighting functions (– –) for controller 2. h_{ij} denotes the transfer between the jth input and the ith output of the augmented system used in the H_∞ criteria (11.6). © IEEE

Table 11.3 Parameters of the first-order weighting functions given by (7.27), controller 3

	W_{11}	W_{12}	W_{21}	W_{22}	W_3
G_0	1000	10000	0.9	0.2	0.34
G_∞	0.8	0.2	1000	1000	
ω_c	5×10^{-2}	5×10^{-3}	2×10^{-2}	2×10^{-1}	

control policies) and controller 3 has a larger bandwidth than controller 2. This better real-time performance is obtained at the expense of some robustness. Indeed, let ΔG denote the real multivariable gain margin (static gain margin). Using (8.7), we obtain:

$$\Delta G_{\min} \leq \Delta G \leq \Delta G_{\max},$$

with $\Delta G_{\min}$ and $\Delta G_{\max}$ given in Table 11.4 for the three controllers.

The controllers were obtained with the Robust Control Toolbox™ of MATLAB®. The order of the controllers is 15, i.e., the order of the system augmented by the order of the weighting functions. The controllers have been discretized at a sampling time of 0.125 s, and we used a balanced realization for implementation.

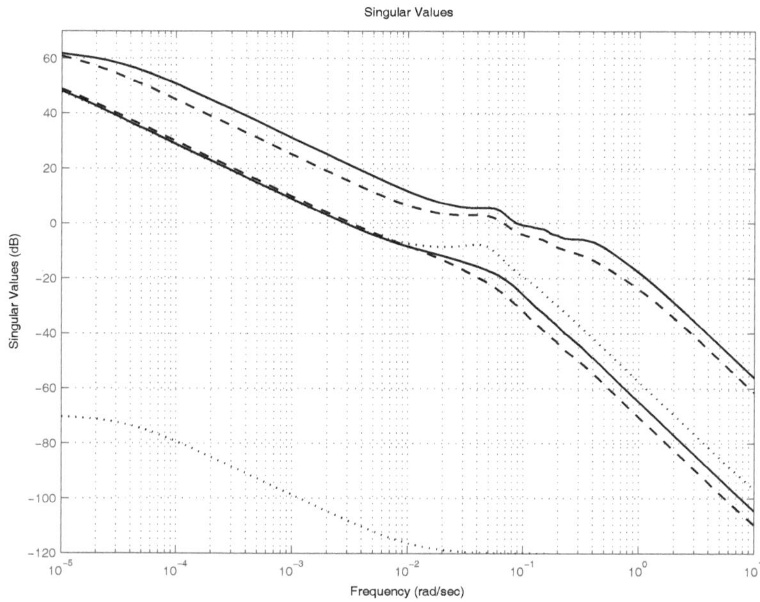

Fig. 11.20 Singular values of controller 1 (...), controller 2 (– –), and controller 3 (—). © IEEE

Table 11.4 Robustness margin of the three controllers

Controllers	1	2	3
$\Delta G_{\min}$ (dB)	$-\infty$	-5	-4.2
$\Delta G_{\max}$ (dB)	10.2	8.8	8

11.4.4 Experimental Results

We now compare the three controllers tested on the real canal, with the same scenario: an unpredicted downstream withdrawal of 10 l/s stopped after stabilization.

Figure 11.21 depicts the experimental results obtained with the distant downstream H_∞ controller 1. A downstream withdrawal of 10 l/s (0.01 m^3/s) is done at time $t = 20$ s and stopped at time $t = 700$ s. In that case, the downstream action variable is not operated, and only the upstream discharge is used to compensate for the water withdrawal.

Figure 11.22 depicts the experimental results obtained with the mixed H_∞ controller 2. The downstream outlet is opened at time $t = 110$ s and stopped at time $t = 650$ s. The controller reacts as expected: first the downstream overshot gate is closed in order to maintain the output y at the target $y_c = 0.6$ m, then the substitution with the upstream control occurs; the downstream gate opens gradually while the upstream discharge increases in order to compensate for the withdrawal. In steady-state (between 300 s and 900 s), the upstream discharge is 10 l/s higher than the

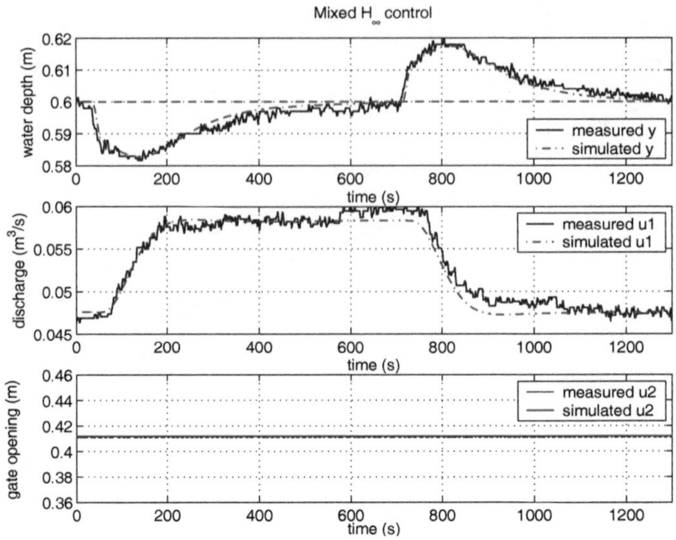

Fig. 11.21 Experimental response of the distant downstream H_∞ controller 1 to a downstream withdrawal, comparison with a linear simulation. © IEEE

initial one, which corresponds exactly to the withdrawal. This ensures that the water needed downstream comes from the resource, located upstream, and is not taken from the water needed downstream.

Figure 11.23 depicts the experimental results obtained with the mixed H_∞ controller 3. The downstream outlet is opened at time $t = 150$ s and stopped at time $t = 850$ s. Here, the controller also reacts as expected and more quickly than the mixed controller 2.

The multivariable H_∞ controllers enable us to recover the real-time performance of a pure local upstream controller while ensuring that in steady-state, the discharge is delivered by the upstream control (as for a distant downstream controller). In addition, one observes that the linear simulation reproduces rather accurately the dynamic behavior of the closed-loop system. The main discrepancy concerns the downstream actuator u_2, which has a 2 mm dead band, which is not modeled here. This, together with the important measurement noise explains the differences between the experiment and the simulation. The response times are, however, very close and the controller reacts as expected.

Finally, the proposed solution appears to stabilize a large set of stationary linearizations of the canal. This is not surprising, since the change of functioning point has a slow dynamics in the case of a canal. Therefore, the gain-scheduling heuristic is fulfilled: the parameters defining the linearizations vary very slowly compared to the system's dynamics [2]. This is coherent with the widespread engineering practice of using gain-scheduling heuristics to design efficient controllers.

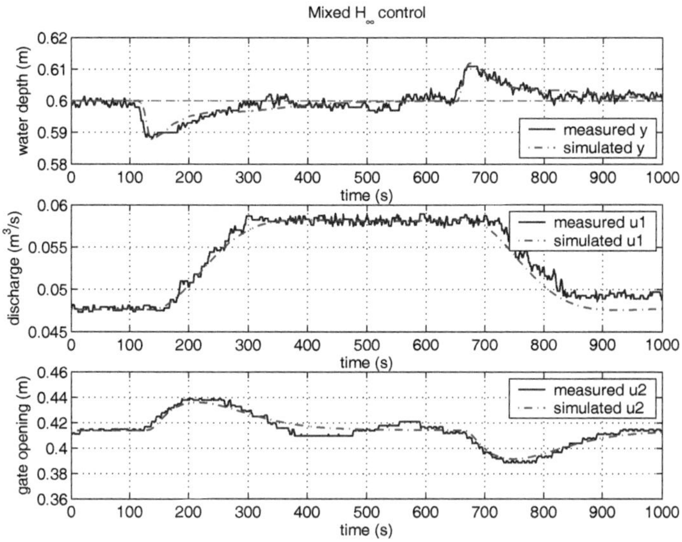

Fig. 11.22 Experimental response of the mixed H_∞ controller 2 to a downstream withdrawal, comparison with a linear simulation. © IEEE

11.5 Boundary Control of Oscillating Modes

In this section, we apply the proportional controller obtained in Sect. 6.4.2 to a portion of the experimental canal of Évora. We present an illustrative result, showing the effect of boundary control on the first oscillating mode.

11.5.1 Experimental Setup

The canal inlet is equipped with a motorized flow control valve that delivers a discharge u_1 in l/s. During the experiment, the canal pool is excited with a sinusoidal input discharge $u_1 = 10 + 5\sin\omega t$ l/s, with $\omega = 0.082$ rad/s, corresponding to the first oscillating mode frequency. Around 10 l/s, the delays are given by $\tau_1 = 34.6$ s, and $\tau_2 = 41.8$ s. The velocity and celerity are, respectively, $V_0 = 0.19$ m/s and $C_0 = 1.98$ m/s, while the top width is $T_0 = 0.36$ m. The optimal proportional controller is therefore equal to $k_u^* = 0.78$ m^2/s [6].

Five water level sensors are installed in off-line stilling wells along the canal pool: one at the upstream end of the pool ($x = 0$), three at $x = L/4$, $x = L/2$ and $x = 3L/4$, and one at the downstream end of the pool ($x = L$). The boundary controller only uses the data from the sensor located at the downstream end of the canal pool. The other sensors are used to monitor the water level along the pool. The boundary controller is implemented on the offtake p, located at the downstream end of the

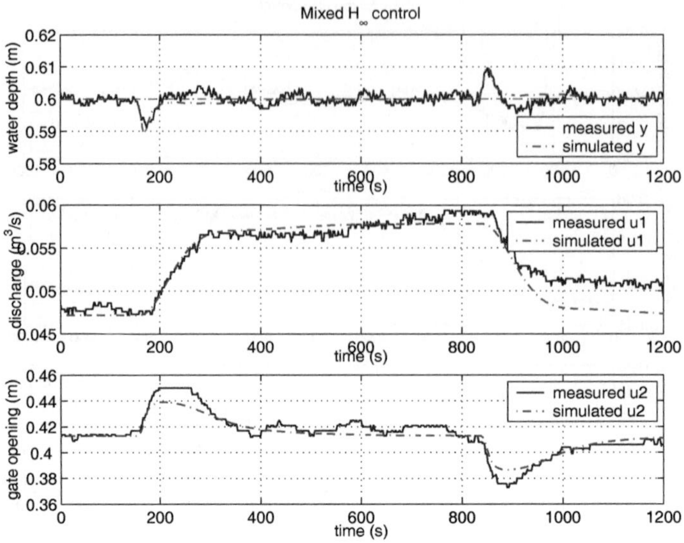

Fig. 11.23 Experimental response of the mixed H_∞ controller 3 to a downstream withdrawal, comparison with a linear simulation. © IEEE

pool. This offtake is equipped with an electromagnetic flowmeter and a motorized butterfly valve that can react quickly.

11.5.2 Experimental Results

At the beginning of the experiment, no control is performed, i.e., the outlet discharge is imposed to a constant value. Then, at $t = 1250$ s, the proportional boundary controller is put on (see Fig. 11.24). The water level oscillations are dramatically reduced, and this is also the case along the whole canal pool (see Figs. 11.25 and 11.26).

These experimental results are in very good accordance with the theoretical linear predictions. Indeed, the dotted lines in Figs. 11.24–11.26 were obtained with a linear simulation using the measured input sinusoidal excitation. The linear model is obtained from a rational approximation of the Saint-Venant transfer matrix following [4], and a linear transfer function to account for the actuator dynamics (see [8] for details).

The amplitude and the frequency of the oscillations are very well predicted by the linear model. The prediction at $x = L/2$ is the one that presents the largest deviation from measured values (see Fig. 11.25). This is due to the fact that this point is an oscillation node where the measuring noise becomes predominant with respect to the dynamics.

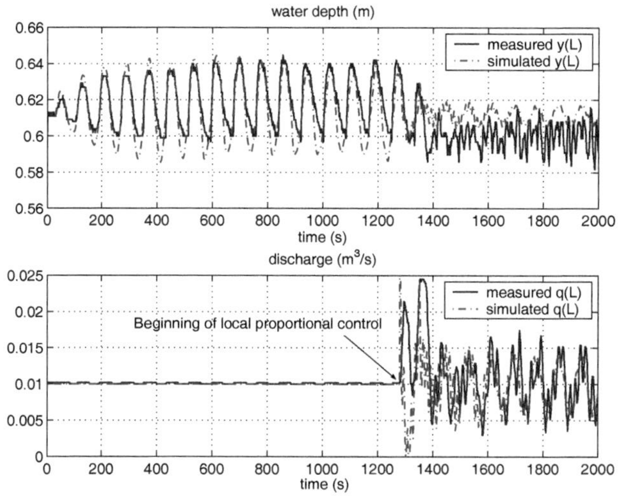

Fig. 11.24 Water level $y(L)$ and outlet discharge $q(L)$ without and with the proportional controller k_u^*, comparison with a linear simulation

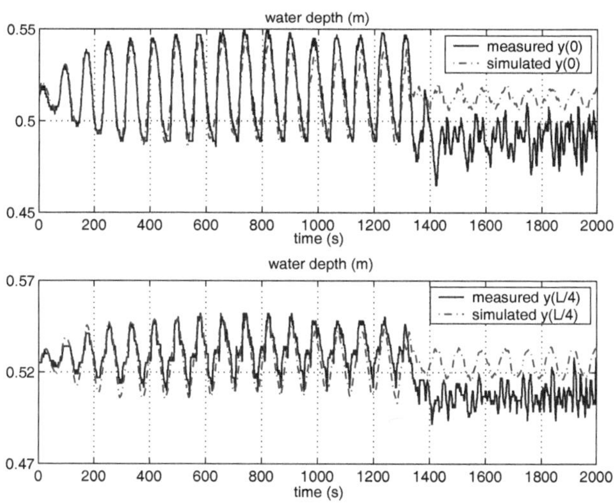

Fig. 11.25 Water levels at $x = 0$ and $x = L/4$ without and with the proportional controller k_u^*, comparison with a linear simulation

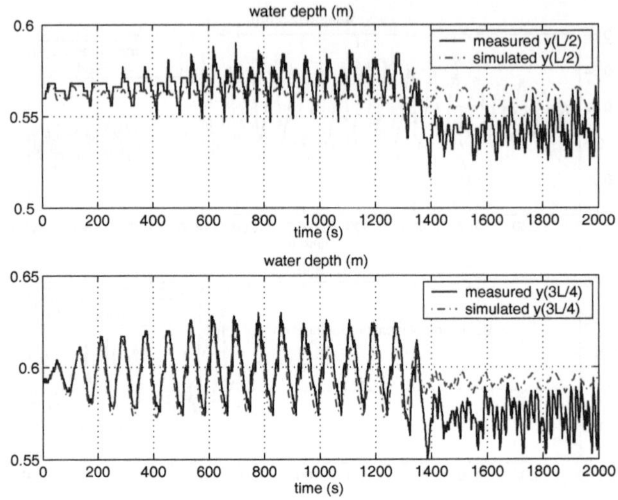

Fig. 11.26 Water levels at $x = L/2$ and $x = 3L/4$ without and with the proportional controller k_u^*, comparison with a linear simulation

There is a small mismatch when the controller is put on, due to a change in functioning point. This nonlinear effect cannot be taken into account by the linear model. But the linear model appears to capture the dynamics of the canal very well, not only at the boundaries, but over *the entire length* of the canal pool. This is a supplementary validation of our approach which is based on linearized equations. One should notice that the input variations represent $\pm 50\%$ of the input discharge, which shows the ability of the linear model to predict the evolution of the system even relatively far away from the linearization regime.

These experimental results provide a validation of our approach for the boundary control of a canal pool. As we have already seen earlier, the linear model appears to capture the system's dynamic behavior very well.

11.6 Decentralized Control of a Two-pool Canal

We now use the intermediate gate to control two pools of the laboratory canal. Each pool is 73 m long, with an average longitudinal bottom slope of 1.5×10^{-3}, a trapezoidal geometry, with bottom width 0.15 m, and side slope 6.6. The average discharge is 45 l/s. This leads to the following ID model parameters: $A_1 = A_2 = 21$ m^2 and $\tau_1 = \tau_2 = 30$ s.

The canal is equipped with water level sensors at the downstream end of each pool. The canal inlet is equipped with a motorized flow control valve that delivers a discharge u_1. The intermediate actuator is an undershot gate u_2, and the downstream end is controlled with an overshot gate u_3.

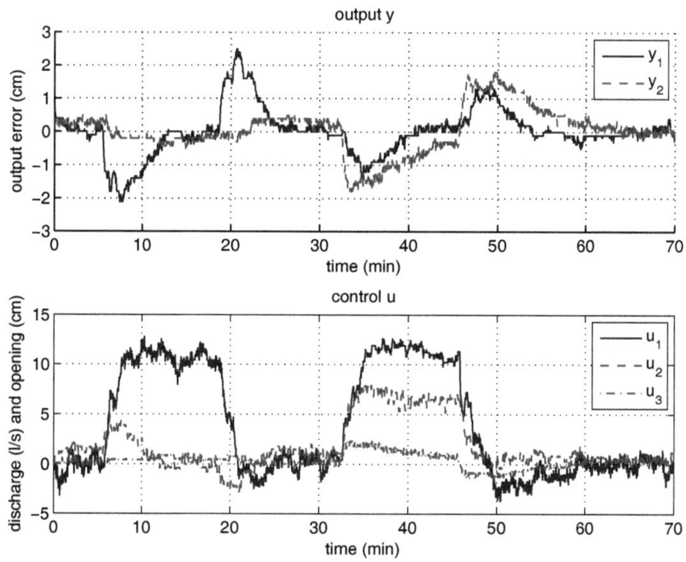

Fig. 11.27 Distant downstream control of a canal with two pools. Response to an unpredicted withdrawal in pool 2

11.6.1 Distant Downstream Control

Figure 11.27 depicts the closed-loop reaction to an unpredicted withdrawal in pool 2 at time $t = 200$ s. Due to the delay, this perturbation causes a large output error of 2 cm, and the output is brought back to the set-point in about 300 s, which is 20 times the time-delay. This time domain performance is coherent with the gain and phase margins of the controller. But, as mentioned above, once the distant downstream SISO controllers are tuned, there is no possible way to increase performance with respect to water users.

11.6.2 Local Upstream Control

For the same scenario used in the simulation of the distant downstream controller, the local upstream controller performs very well (see Fig. 11.28). The maximum error is 1 cm, and the perturbation is rejected in less than 150 s.

The difference in the time domain performance with the distant downstream controller is striking. However, such a controller cannot be used in irrigation canal management, since it assumes that enough water is flowing in the canal to satisfy water needs. The rest is propagated downstream. Indeed, since there is no feedback on the upstream discharge u_1, it does not adapt to the demand.

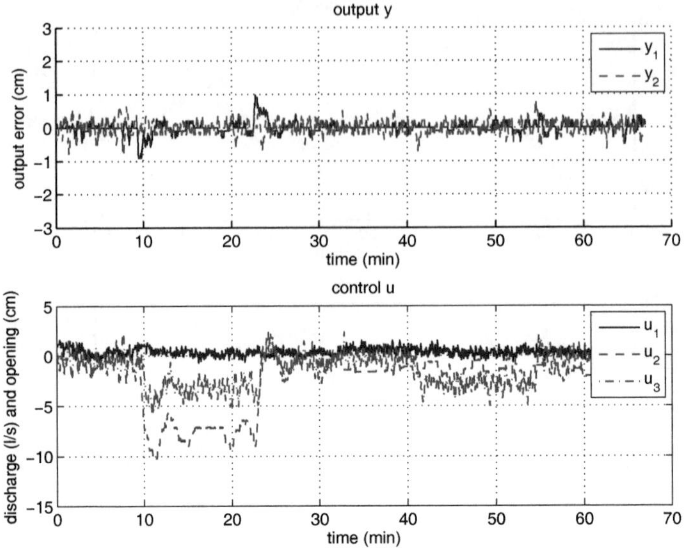

Fig. 11.28 Local upstream control of a canal with two pools. Response to an unpredicted withdrawal in pool 2

11.7 Mixed H_∞ Control of a Two-pool Canal

We now test a decentralized mixed H_∞ controller for two pools. The mixed controller of each pool has modulus input margins of ± 7 dB. The multivariable mixed controller has modulus input margins of ± 4.5 dB, and real structured input margins computed using the structured singular value of ± 8 dB. The frequency decoupling is effective, since the monovariable margins are recovered in the multivariable case. Structured robustness with respect to typical delay mismatch is also good.

The mixed controller is simulated on the same scenario (see Fig. 11.29). It is clear that the time domain performance is dramatically increased compared to the distant downstream controller. In this case, the performance is very similar to that obtained with a local upstream controller, but the water resources comes from upstream, as in the distant downstream control case. There is a substitution between u_2, u_3, and u_1.

To illustrate the ability of the mixed control scheme to achieve control objectives for irrigation canals, the following experiment was carried out: pool 1 is perturbed by high frequency but low amplitude water withdrawals, while pool 2 is subject to low frequency but high amplitude water withdrawals (see Fig. 11.30).

For comparison purposes, the experimental results of the decentralized distant downstream controller are first depicted in Fig. 11.31. The performance in this case is limited by the time-delay, and the controller cannot efficiently reject the high frequency perturbation p_1. However, the low frequency perturbation p_2 is slowly but efficiently rejected.

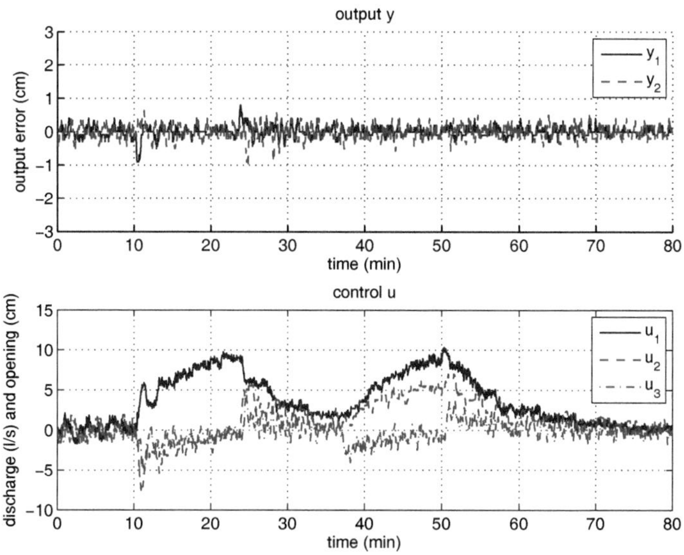

Fig. 11.29 Mixed local upstream/distant downstream control of a canal with two pools. Response to an unpredicted withdrawal in pool 2

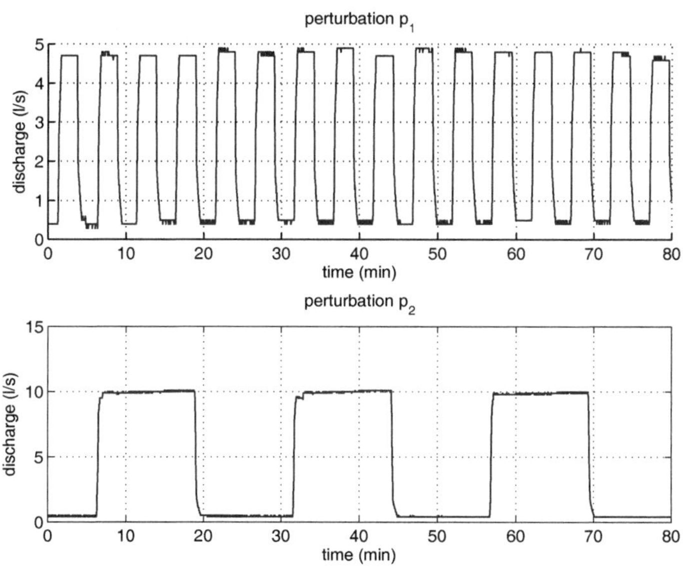

Fig. 11.30 Perturbations in pool 1 (high frequency low amplitude) and 2 (low frequency high amplitude)

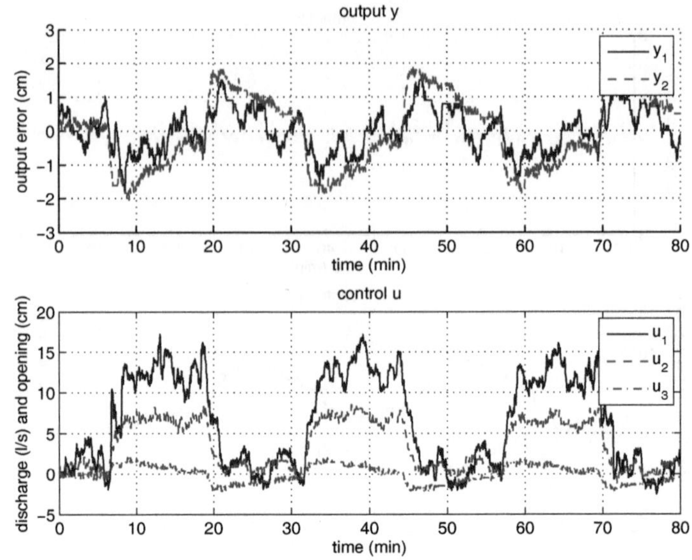

Fig. 11.31 Distant downstream control of a canal with two pools. Response to a high frequency perturbation in pool 1 and a low frequency perturbation in pool 2

The results of the mixed controller are depicted in Fig. 11.32. The local upstream controller of pool 1 efficiently rejects the high frequency perturbation, but at the same time is able to follow the low frequency water demand of pool 2. This is visible when the demand in pool 2 changes, corresponding to a change in the downstream discharge setpoint. In terms of water level control, the performance is more than ten times better with the mixed controller than with the distant downstream controller. One should note that this result was obtained while guaranteeing that the perturbations occurring in low frequencies are rejected with the upstream discharge, as in the distant downstream control case.

11.8 Conclusion

In this chapter, we have applied the methods developed in the book to an experimental canal. The proposed approach leads to remarkable results, in terms of modeling and control. The proposed method enables us to accurately predict the system behavior in terms of stability, instability, and performance. It shows the prediction ability of the Saint-Venant equations, which can be used together with hydraulic structures equations to model a canal for con-

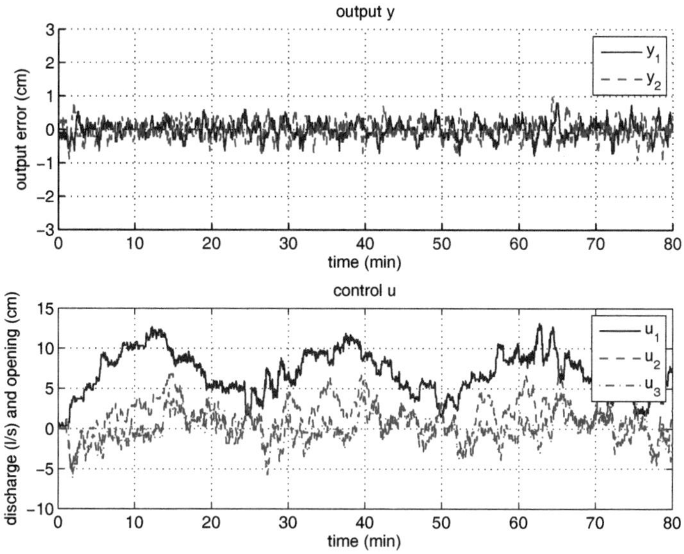

Fig. 11.32 Mixed local upstream/distant downstream control of a canal with two pools. Response to a high frequency perturbation in pool 1 and a low frequency perturbation in pool 2

troller design. The linearized Saint-Venant equations appear to be a very efficient model of open channel flow around steady flow regimes.

The experimental results are satisfactory for the design of classical controllers with a distant downstream control policy. This gives a clear validation of the proposed model-based methodology. We also validated the method for boundary control of oscillating modes and for the control of a canal with multiple pools.

One should also note that the experimental canal is rather complex to control, due to its oscillating nature. With respect to a real canal, this one is much more oscillating, which makes it more difficult to design efficient controllers.

References

[1] Franklin GF, Powell JD, Emami-Naeini A (2002) Feedback control of dynamic systems, 4th edn. Prentice Hall, Englewood Cliffs, NJ

[2] Fromion V, Scorletti G (2003) A theoretical framework for gain scheduling. Int J Robust Nonlin Contr 13:951–982

[3] Lencastre A (1996) Hydraulique générale. Eyrolles, SAFEGE, (in French)

[4] Litrico X, Fromion V (2002) Infinite dimensional modelling of open-channel hydraulic systems for control purposes. In: Proc. 41st IEEE Conf. decision and control, Las Vegas, pp 1681–1686

[5] Litrico X, Fromion V (2004) Analytical approximation of open-channel flow for controller design. Appl Math Mod 28(7):677–695

[6] Litrico X, Fromion V (2006) Boundary control of linearized Saint-Venant equations oscillating modes. Automatica 42(6):967–972
[7] Litrico X, Fromion V (2006) H_∞ control of an irrigation canal pool with a mixed control politics. IEEE Trans Control Syst Technol 14(1):99–111
[8] Litrico X, Fromion V, Baume JP, Arranja C, Rijo M (2005) Experimental validation of a methodology to control irrigation canals based on Saint-Venant equations. Contr Eng Pract 13(11):1425–1437
[9] Malaterre PO, Baume JP (1997) SIC 3.0, a simulation model for canal automation design. In: Int. Workshop on the Regulation of Irrigation Canals RIC'97, Marrakech (Morocco), pp 68–75
[10] Weyer E (2001) System identification of an open water channel. Contr Eng Pract 9:1289–1299

Chapter 12
Modeling and Control of Regulated Rivers

Abstract This chapter presents another application of modeling and control of a hydrosystem. We consider regulated rivers where one dam or multiple dams located upstream enable control of the upstream flow into the river. We first study the case of a river with one dam and multiple discharge measurement points. We then study the case of a regulated river controlled with multiple dams, each having a different capacity. The controlled output can be regulated with one dam acting rapidly, but with a low capacity, or with a large dam located further upstream, or both, if a suitable control strategy is applied.

12.1 Introduction

As has already been mentioned in the introductive chapter, regulated rivers are used in many countries to sustain summer low flows and supply water to different users (e.g., water for drinking, industries, irrigation, hydropower, and salubrity). In these systems, an upstream dam is used as a storage and the natural river is used as a channel to convey water to water users.

These dam–river systems are equipped with sensors, actuators, and communications means and can be managed in an automatic way. The main objective of the regulation is to satisfy, in spite of uncertainties, the water demand at each pumping station while guaranteeing a minimum discharge along the entire river and spending a minimum amount of water volume from the upstream reservoir.

In the following, the term "dam–river system" will be used for a regulated river with at least one dam at the upstream end and at least one measurement point at the downstream end of the river.

Using approximated linear models, [15] proposed design methods for a dam–river with one reach. [17] proposed a method by inverse propagation for the open-loop control of dam–river systems with intermediate measurements. As the system considered is dominated by long, varying time-delays, the robustness to time-delay variations is very important. [9] studied robust stability by looking at the poles of the closed-loop system, but did not end up with a reliable tuning method for robust control. [7] proposed a robust control method for a cascade of regulated river reaches.

In this chapter, we use an analytic modeling method based on physical equations of open channel hydraulics and design robust H_∞ controllers. The approach already developed for single input single output (SISO) systems [13] is here extended to the single input multiple outputs (SIMO) case, when there are multiple measuring points along the river. We also show how predictive H_∞ controllers can be designed, using the knowledge of future perturbations occurring on the system. This chapter provides another illustrative application of H_∞ controllers for hydrosystems.

12.2 Modeling of a Regulated River

12.2.1 Modeling of the System

In the case of rivers, the geometry is not known as accurately as in the case of canals. Furthermore, the inertia terms can be neglected in general, especially for relatively steep slopes. With these approximations, the Saint-Venant equations reduce to the diffusive wave equation (2.34). Linearizing this equation around uniform flow leads to the Hayami equation.

We focus on the analysis of a regulated river of total length X with a finite number of discharge measurement points at given locations x_i, $0 < x_1 < \ldots < x_n = X$ (see Fig. 12.1). The discharge is supposed to be measured at different locations along the river reach and there is a finite number of intermediate pumping stations distributed along the reach, which provide water to consumers (typically farmers who irrigate fields). The objective of the controller is to act on the upstream discharge $u(t) = q(0,t)$ in order to keep the measured downstream discharge $y_n(t) = q(X,t)$ close to a target value despite unmeasured users' withdrawals $w_i(t)$, using intermediate discharge measurements $y_i(t) = q(x_i,t)$.

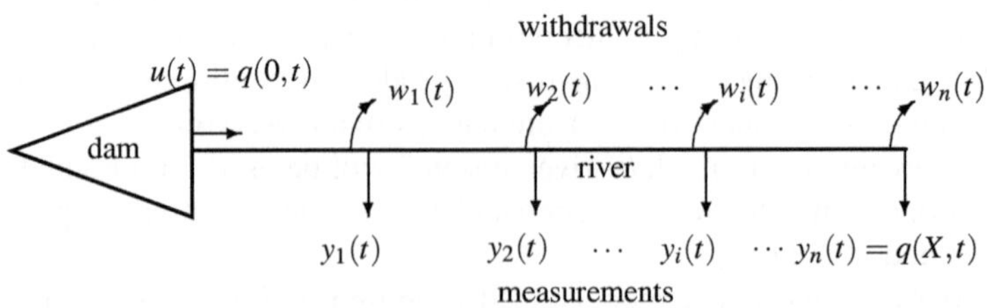

Fig. 12.1 Dam-river system

In other terms, the control objective is to use the upstream discharge u in order to keep the downstream discharge y_n as constant as possible, which means that the control should attenuate the perturbations w_i. This is a problem of regulation or desensitivity. This objective should be attained with a lower bound on gain and phase margins.

We recall here the Hayami equation, which can be obtained as the linearization of a simplified model of open channel flow:

$$\frac{\partial q}{\partial t} + C\frac{\partial q}{\partial x} - D\frac{\partial^2 q}{\partial x^2} = 0, \tag{12.1}$$

where

- q is the discharge (m^3/s)
- C the celerity coefficient (m/s)
- D the diffusion coefficient (m^2/s).

The relation between upstream and downstream discharge can also be expressed as a transfer function using Laplace transform. Thus, fixing a downstream limit condition of the type $\lim_{x\to\infty}\frac{\partial q}{\partial x} = 0$, one gets the Hayami transfer function:

$$G(s,x) = e^{(\frac{C-\sqrt{C^2+4Ds}}{2D})x}, \tag{12.2}$$

with x the length of the reach [m] and s the Laplace variable.

Remark 12.1. It can be shown that $G(s)$ is stable and belongs to the Callier–Desoer class of transfer functions [1].

A first-order with delay is a good approximation of the Hayami transfer function for low frequencies. This approximation is usually considered in hydrology [2].

This transfer function has the property $G(s,x_1+x_2) = G(s,x_1)G(s,x_2)$, which enables us to consider the system as a series of subsystems if there are some intermediate measurement points.

12.2.2 Control Objectives and Constraints

The control objectives are mainly to keep the flow rate at the downstream end of the reach close to a reference flow rate (a target denoted by z_c), defined for hygienic and ecological reasons. This has to be done while farmers and other users are withdrawing water from the river (pumping stations w_i). These not completely known withdrawals are considered as perturbations that have to be rejected by the feedback controller.

The water demand w_i can be predicted more or less accurately using weather forecast, soil-plant models and a data base of previous seasons. Water demand predictions are used in a feedforward controller. The variations in the demand are then considered as perturbations (due to unpredicted inflow or outflow), and have to be rejected by the feedback controller.

There are saturation limits on the actuator: the discharge delivered by the barrage has to be within the bounds $[u_{\min}, u_{\max}]$ with $u_{\min} = 0.1$ m^3/s and $u_{\max} = 5$ m^3/s. This is imposed on the system, but not taken into account explicitly in the control design.

12.3 Performance Versus Information

We have already shown that the performance of an open channel system is limited when the regulated variable is measured and perturbations are assumed unmeasured (see [11]). We investigate different ways to bypass this limitation:

- Additional measurement points (discharge measurement in the river)
- Measurement of disturbances (measurement of withdrawals)
- Predictions of disturbances (predictive control).

The performance is here considered in terms of attenuation of perturbations acting on the system. This requirement can be formalized by direct constraints on the transfer function between the perturbation w_i and the output y_n $T_{w_i \to y_n}$, which will be characterized in the sequel, for each perturbation w_i by the maximum frequency ω_{s_i} such that

$$\omega_{s_i} = \max\{\omega_1 : |T_{w_i \to y_n}(j\omega)| < 1, \forall \omega < \omega_1\}.$$

We first recall the result on the maximal achievable performance, illustrated in the case of a time-delay system.

12.3.1 Maximal Performance

Let us consider that the model of the river is a first-order system with time-delay (the same argument can be used for the general case), controlled by a SISO controller as in Fig. 12.2.

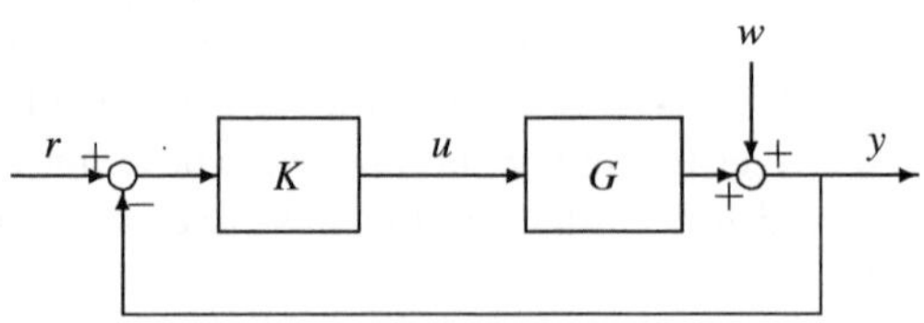

Fig. 12.2 SISO control and unmeasured perturbation

Following the result of [8], the complementary sensitivity function can be written as $T(s) = T_0(s)e^{-\tau s}$, where $T_0(s)$ is the rational part of the complementary sensitivity function. Since for our example and in the best case $T_0(s) = 1$ (perfect tracking), the sensitivity function becomes $T_{w_n \to y_n}(s) = 1 - e^{-\tau s}$. We then deduce a limit on the maximal bandwidth

$$|T_{w_n \to y_n}(j\omega)| = 1 \Rightarrow \left|\sin \frac{\omega\tau}{2}\right| = 0.5$$

which implies $\omega_{s_n} < \frac{\pi}{3\tau} \approx \frac{1}{\tau}$. This is consistent with the result obtained for distant downstream control of a canal pool in Chap. 7.

12.3.2 Incorporating Measurement Points

Incorporating additional measurement points in the river does not improve the global performance of the controlled system, but enables us to improve the perturbation rejection at some points and the robustness of the controller. To show this, we just consider the case of an additional measurement point at the middle of a river (corresponding to the river in Fig. 12.1 with $n = 2$). The general case can be treated in the same way. Let us consider a river represented by a rational system with a time-delay $G_r(s)e^{-\tau s}$. The intermediate measurement separates the system into two subsystems G_1 and G_2 with the same time-delay, i.e., $\tau/2$ (see Fig. 12.3).

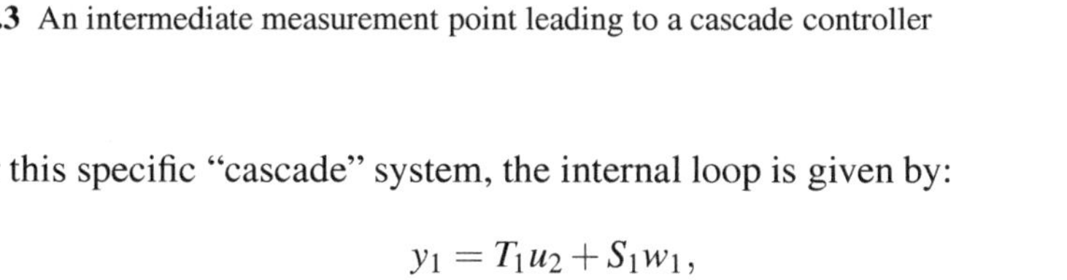

Fig. 12.3 An intermediate measurement point leading to a cascade controller

For this specific "cascade" system, the internal loop is given by:

$$y_1 = T_1 u_2 + S_1 w_1,$$

with $T_1 = G_1 K_1 (1 + G_1 K_1)^{-1}$ and $S_1 = 1 - T_1$.

For the external loop:

$$y_2 = T_{r \to y_2} r + T_{w_2 \to y_2}(w_2 + S_1 G_2 w_1),$$

with $T_{r \to y_2} = K_2 G_2 T_1 (1 + K_2 G_2 T_1)^{-1}$ and $T_{w_2 \to y_2} = 1 - T_{r \to y_2}$.

Since $T_1(s) = T_{0,1}(s)e^{-\tau/2s}$ and $T_1(s)G_2(s) \propto e^{-\tau s}$, we have:

$$T_{r\to y_2}(s) = T_{0,r\to y_2}(s)e^{-\tau s},$$

which yields the same constraint on the maximal achievable performance for rejecting perturbation w_2 as with the classical SISO controller (i.e., $\omega_s \leq 1/\tau$). However, the internal loop can reject higher frequency perturbations w_1. In fact, the sensitivity to $G_2 w_1$ is now $S_1 T_{w_2\to y_2}$ (to be compared to $T_{w_2\to y_2}$ in the SISO case). Assuming perfect tracking in both loops, one gets the following sensitivity:

$$\begin{aligned}|S_1(j\omega)T_{w_2\to y_2}(j\omega)| < 1 &\Rightarrow 4\left|\sin\left(\frac{\omega\tau}{4}\right)\sin\left(\frac{\omega\tau}{2}\right)\right| < 1\\ &\Rightarrow \sin^2\left(\frac{\omega\tau}{4}\right)\left|\cos\left(\frac{\omega\tau}{2}\right)\right| < \frac{1}{8},\end{aligned}$$

which implies the inequality

$$x - x^3 < \frac{1}{8},$$

with $x = \left|\cos\left(\frac{\omega\tau}{2}\right)\right|$.

Solving for x gives $x > 0.93$, which leads to $\omega_s\tau < 1.5$. This corresponds to a 50% increase of the maximum bandwidth (compare with the result obtained in Sect. 12.3.1). In this case, only the performance on intermediate perturbations is improved.

It should be noted that intermediate measurements are also interesting when the system is not perfectly known: the robustness of the controlled system is improved when another loop is introduced (see [6]).

12.4 H_∞ Control of a Dam–River System

In this section, we design two types of H_∞ controllers: a classical one, which rejects unmeasured perturbations with feedback control, and a predictive controller, which uses the knowledge of the perturbations in the future to include an open-loop part (see [14]).

12.4.1 Classical H_∞ Control

In the case when the perturbation is not known in the future, the considered control problem can be stated as a standard mixed sensitivity optimal control problem, using the H_∞ norm, as we have done in Chap. 7:

$$\gamma = \inf_K \left\|\begin{pmatrix} W_1 T_{w_n\to y_n} \\ W_2 K T_{w_n\to y_n}\end{pmatrix}\right\|_\infty, \qquad (12.3)$$

where W_1 and W_2 are weighting functions taking into account performance and robustness requirements. We select W_1 and W_2 in order to obtain the required robustness margins when $\gamma \leq 1$. Shaping $T_{w_n \to y_n}$ and $KT_{w_n \to y_n}$ allows us (i) to constrain the rejection time and to limit the control energy, and (ii) to ensure a desired modulus margin and a desired time-delay margin.

In our case, and due to the irrational nature of the system, criteria (12.3) cannot be solved in an easy way. To bypass this, we solve problem (12.3) using a rational approximation of (12.2) (see Chap. 4 and [11]).

This classical H_∞ controller design is very similar to that already developed in Chap. 7 and is not further detailed here. We rather focus on another interesting aspect, which uses the knowledge of future perturbations to design a predictive H_∞ controller.

12.4.2 Predictive Control

The vector of perturbations w is now assumed to be known in advance, with a prediction horizon equal to h. We propose to extend the previous H_∞ criterion in order to design a control law using the future behavior of the perturbation. In order to simplify the problem, we consider the design of a discrete-time controller. In this case, the future value of the perturbation is given by a finite dimensional vector $[w(k),\ldots,w(k+h)]$: in the continuous-time case, the corresponding vector is infinite dimensional. The control law has the corresponding structure:

$$u = K \begin{bmatrix} w(k) \\ \vdots \\ w(k+h) \\ y(k) \end{bmatrix} = \begin{bmatrix} K_{ol} | K_{cl} \end{bmatrix} \begin{bmatrix} w(k) \\ \vdots \\ w(k+h) \\ \hline y(k) \end{bmatrix}.$$

The corresponding closed-loop system is represented in Fig. 12.4. Let us denote by $T_{w \to y}$ the transfer function from w to y and let us denote by $T_{w \to u}$ the transfer function from w to u. The corresponding H_∞ problem can be written as:

$$\min_K \left\| \begin{pmatrix} W_1 T_{w \to y} \\ W_2 T_{w \to u} \end{pmatrix} \right\|_\infty . \tag{12.4}$$

The difficulty arises from the fact $T_{w \to y}$ and $T_{w \to y}$ are noncausal transfer functions since for instance:

$$T_{w \to y}(z) = (I - GK_{cl})^{-1} \left(I + GK_{ol} \begin{bmatrix} 1 \; z \cdots z^h \end{bmatrix} \right).$$

Nevertheless, we note that:

$$\left\| \begin{pmatrix} W_1 T_{w \to y} \\ W_2 T_{w \to u} \end{pmatrix} \right\|_\infty = \left\| \begin{pmatrix} W_1 T_{w \to y} z^{-h} \\ W_2 T_{w \to u} z^{-h} \end{pmatrix} \right\|_\infty .$$

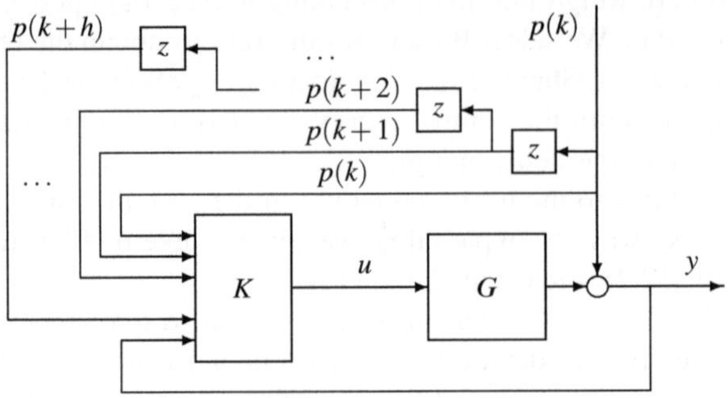

Fig. 12.4 Closed-loop system with predictive control law

In this case, $T_{w\to y}z^{-h}$ and $T_{w\to y}z^{-h}$ are causal transfer functions. As a consequence solving (12.4) is equivalent to the following problem:

$$\min_K \left\| \begin{pmatrix} W_1 T_{w\to y} z^{-h} \\ W_2 T_{w\to u} z^{-h} \end{pmatrix} \right\|_\infty, \tag{12.5}$$

which is a standard H_∞ problem [3].

In fact, the previous criterion does not take into account any direct constraint on the sensitivity function, i.e., $T_{w_n\to y_n}$, and therefore there is no constraint on the open-loop system bandwidth. In order to take into account desensitivity requirements, we add to the previous criteria a perturbation that models the unknown part of the perturbations acting on the system.

As specified in the above scheme, the controller can now use “measurements” of future perturbations (in fact, predictions) for real-time control. However, this problem cannot be solved with classical H_∞ algorithms using Riccati equations [3], since the criterion is singular (the perturbation measurement is perfectly known). This technical problem can be bypassed by using an LMI formulation (see [5] for details).

12.4.3 Application

The numerical applications are made for a river of 18 km, with two possible intermediate measurement points: y_1 at 6 km and y_2 at 12 km; the controlled point being the discharge y_3 at 18 km (*i.e.*, $n = 3$ in Fig. 12.1). The dynamic parameters $C = 0.7$ m/s and $D = 170$ m^2/s correspond to average values for sloping rivers at low flows, and for the whole river give a delay close to 7 hours. Figure 12.5 shows the Bode plot of the whole river (18 km), which is close to that of a first-order with delay.

Five different controllers were designed and compared on two different scenarios on the system described above. The two scenarios consist in a disturbance step of

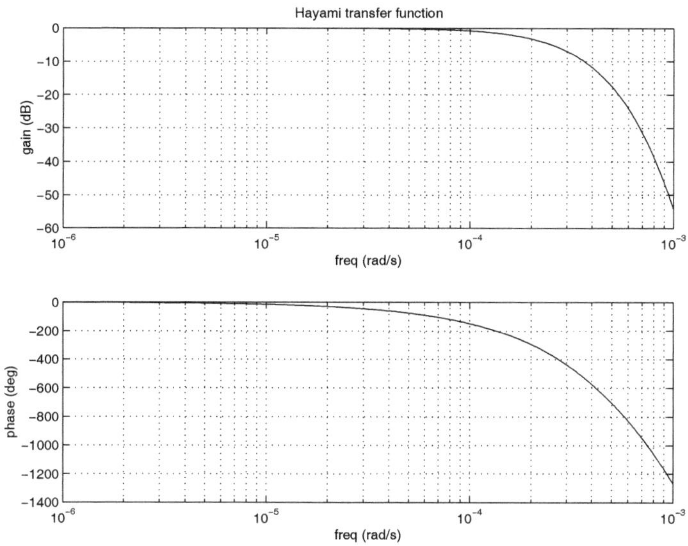

Fig. 12.5 Bode plot of $G(s)$

1 m^3/s applied at w_1 for the first scenario and w_3 for the second one. The results are summarized in Table 12.1, where the chosen indicator is the volume of water not delivered at the downstream end of the river. As was shown above, except for the predictive controller, the use of additional measurement points in the river or of perturbation measurement does not improve the performance for a perturbation acting at the downstream end (w_3) of the river. It is interesting to note that the three controllers SISO, SISO$_w$ (using perturbation measurement), MISO (using intermediate measurement in the river), and MISO$_w$ (using both) behave in the same way. However, the performance in the first scenario is clearly improved when additional information is used. Finally, the knowledge of future perturbations enables us to almost perfectly reject them (MISO predictive controller).

Table 12.1 Volume of water not delivered (in 10^3 m^3) for the different controllers and scenarios

	Scen. w_1	Scen. w_3
SISO	44.3	44.4
SISO$_w$	23.8	44.4
MISO	23.9	44.4
MISO$_w$	23.8	44.4
MISO pred.	-2.9	2.1

Figure 12.6 shows the results in simulation for a withdrawal of 1 m^3/s at the first measurement point for three different controllers: the MISO predictive controller using predicted perturbations, the MISO controller taking into account intermedi-

ate measurements (y_1 and y_2), and the SISO controller using only the downstream measurement point (y_3).

In this simulation, one can clearly see that the constraint on the actuator limits the achievable bandwidth in perturbation rejection.

Nevertheless, the effect of incorporating information in the controller is here clearly visible in terms of water savings.

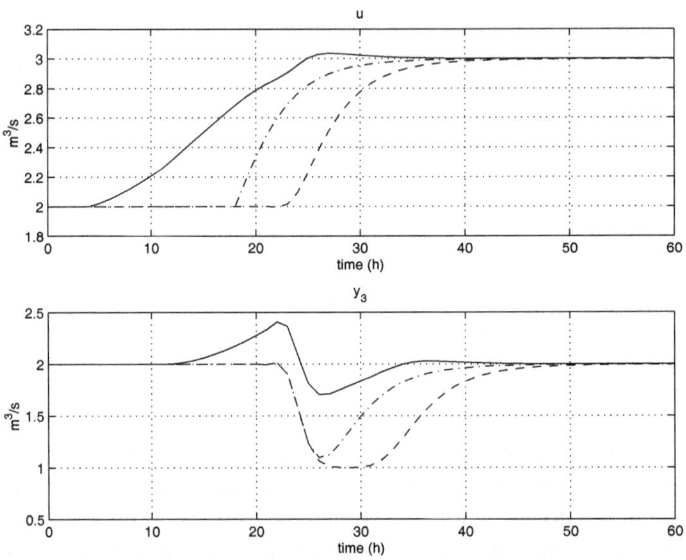

Fig. 12.6 Control u and controlled output y_3 for the three H_∞ controllers: MISO predictive controller (—), MISO controller (– – · – –) and SISO controller (– –), scenario w_1

Let us now consider a different system, where a river can be controlled using multiple reservoirs.

12.5 H_∞ Control of Multi-reservoir Regulated Rivers

The system studied here consists in a regulated river that can be controlled using two dams, each having a different available volume of water: the large dam is far upstream from the controlled point and the small one is close to the controlled point. Therefore, the best achievable performance (in terms of rejection of perturbations, which is directly linked to the sensitivity bandwidth) is obtained by using only the closest dam. But, since its available volume is limited, it cannot deliver water continuously to satisfy all water needs. The water demand can be satisfied by the large upstream dam, but with a much lower real-time performance. We present a way to recover the performance obtained with the small dam by using a combination of the two dams, allowing the small dam to compensate only for high frequency perturbations (see [12]).

We focus on the analysis of a controlled river of total length $X = X_1 + X_2 + X_3$ with two dams upstream, delivering discharges u_1 and u_2. We suppose that $X_1 \gg X_2$, and the controlled discharge is located downstream of the third reach (see Fig. 12.7).

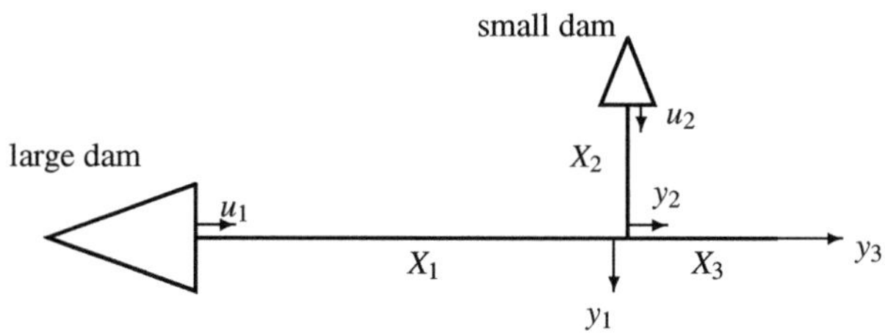

Fig. 12.7 System with two dams

The discharge is measured at different locations along the river reach and there is a finite number of intermediate pumping stations distributed along the reach that provide water to consumers. The objective of the controller is to act on the upstream discharges u_1 and u_2 in order to keep the measured downstream discharge y_3 close to a target despite unmeasured users' withdrawals w_i, using intermediate discharge measurements y_i.

We first design two H_∞ controllers for each SIMO subsystem. Then, we design a MIMO controller design, as an extension of the SIMO controllers: the proposed design naturally mixes the advantages of each controller. In order to use classical H_∞ control, we need a rational model. We will use a rational approximation of the transfer function (12.2) obtained by frequency least squares minimization.

The numerical applications of this paper are done for a model of the upstream part of the River Dropt. The river exhibits the physical parameters given in Table 12.2 (obtained from measurement and identification).

Table 12.2 Physical parameters of reaches

	Length X	C	D
Reach 1	23.1 km	0.32 m/s	98 m^2/s
Reach 2	1 km	0.43 m/s	108 m^2/s
Reach 3	3.3 km	0.51 m/s	128 m^2/s

12.5.1 Control Objectives

The control system should reach the following objectives:

- Use the measures on the system y_i in order to act on the control variables u_1 and u_2 to reject unknown perturbations w_i acting on the system.
- Follow a reference signal r.
- Achieve the above objectives without using too much water from the second dam.

In other terms, the control objective is to use the upstream discharges u_1 and u_2 in order to keep the downstream discharge y_3 as constant as possible, which means that the control should attenuate the perturbations w_i. This is a problem of regulation or desensitivity. As the "real" system may be different from the model used for controller design, the controller design should incorporate some robustness requirements.

12.5.2 H_∞ Controller Design in the SIMO Case

Following the approach used by [16], design specifications can be formulated using an H_∞ four-block type criterion. As a matter of fact, let us consider that the system is described by:

$$y = Gu + \tilde{G}w.$$

The closed-loop system that links the reference, r, and the perturbation, w, to the tracking error, e, and the controlled input, u, is given by

$$\begin{bmatrix} e \\ u \end{bmatrix} = \begin{pmatrix} S & S\tilde{G} \\ KS & KS\tilde{G} \end{pmatrix} \begin{bmatrix} r \\ w \end{bmatrix}, \tag{12.6}$$

where $S = (I + GK)^{-1}$ (sensitivity function).

The design specifications are then formulated using the following four-block criterion, where the goal is to find the smallest $\gamma > 0$ and the stabilizing controller K such that

$$\left\| \begin{matrix} W_1 S & W_1 \tilde{G} S W_3 \\ W_2 KS & W_2 \tilde{G} KS W_3 \end{matrix} \right\|_\infty \leq \gamma,$$

Following [16], $W_1, W_1^{-1} \in RH_\infty$ is used to specify tracking performances, perturbation rejection, and the modulus margin.

$W_2, W_2^{-1} \in RH_\infty$ is used to specify high frequency constraints on the controlled input. This allows us to constrain command effort and effects of sensor noise command. W_3 is a scaling factor acting on the perturbation.

The weighting functions W_1 and W_2 are chosen to be of the first-order, as in (7.27).

We first present two SIMO H_∞ designs corresponding, respectively, to the solution associated to the long and short river reaches.

12.5.2.1 Design for Water Savings: SIMO Controller ♯1

We first present the classical SIMO controller, taking water from the first (large) dam, which has the larger time-delay.

The chosen weighting function parameters are given in Table 12.3. The optimization resulted in $\gamma = 1$, with a gain margin of 7.5 dB. This controller takes water from the first dam, but reacts with a low performance to an unpredicted withdrawal w_3 (see Fig. 12.8).

Table 12.3 Parameters for weighting functions, controller SIMO ♯1

	W_1	W_2
G_0	1000	0.7
G_∞	0.2	100
ω_c	10^{-5} rad/s	10^{-4} rad/s

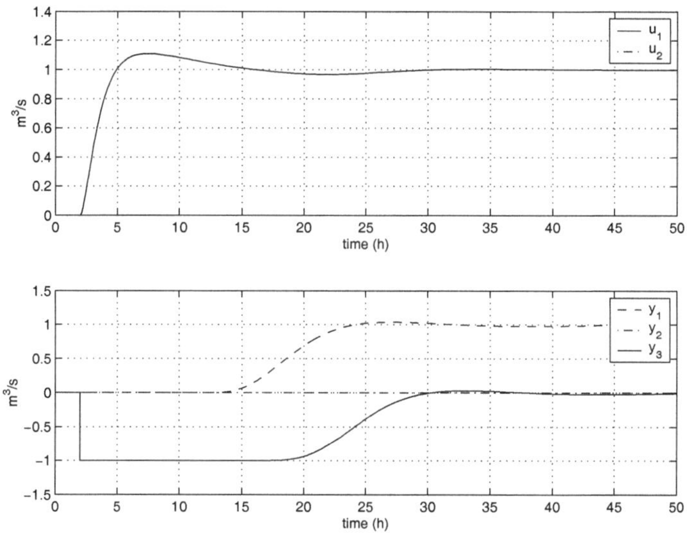

Fig. 12.8 SIMO controller ♯1 (large dam): reaction to an unmeasured withdrawal w_3

12.5.2.2 Design for High Performance: SIMO Controller ♯2

We now present the "best" controller with respect to perturbation rejection; this controller takes water from the second (small) dam, which is the one that can react quickly to a unmeasured perturbation. A withdrawal in w_3 is countered very quickly, as shown in Fig. 12.9. However, this solution is not satisfactory, as the second dam has a limited resource. The chosen weighting function parameters are given in Table 12.4. The optimization resulted in $\gamma = 1$, with a gain margin of 8 dB.

Table 12.4 Parameters for weighting functions, controller SIMO ♯2

	W_1	W_2
G_0	1000	0.1
G_∞	0.2	100
ω_c	9.10^{-5} rad/s	3.10^{-4} rad/s

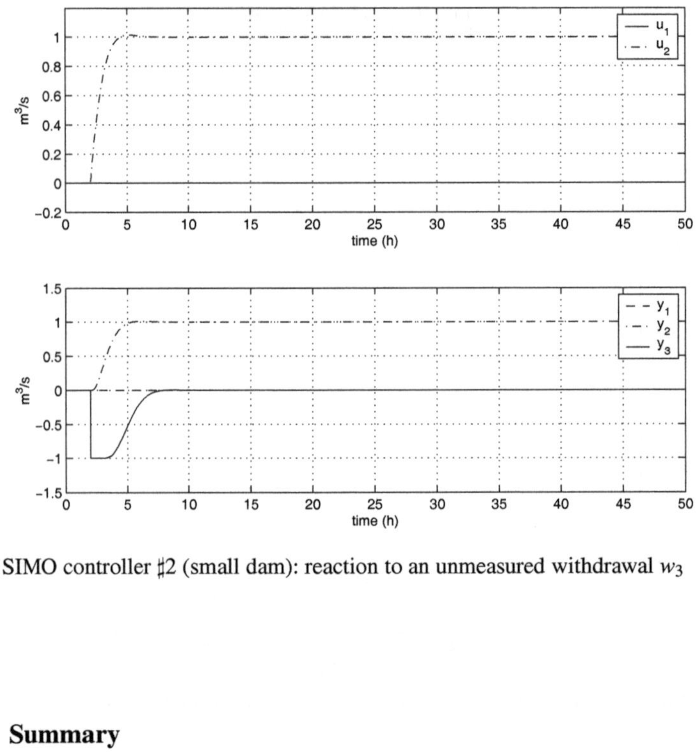

Fig. 12.9 SIMO controller ♯2 (small dam): reaction to an unmeasured withdrawal w_3

12.5.2.3 Summary

The preliminary results presented show that it is possible to design H_∞ SIMO controllers for each dam, ensuring robustness margins and good performance. However, our purpose is to take advantage of the two dams in order to recover the performance of the small one, without taking all the water from this one. This will be done with a MIMO controller.

12.5.3 MIMO H_∞ Design

We will first consider a simplified model of the system in order to clarify the MIMO design philosophy, then the design specifications will be written in terms of H_∞ constraints.

12.5.3.1 Simplified Representation

Below is a simplified representation of the control problem, supposing only the output discharge y is measured and there is only an output perturbation w. This simplified representation enables us to clearly explain the design trade-offs. With this hypothesis, the system can be written as the sum of two subsystems, as in Fig. 12.10. The controller is also the sum of two controllers, which can be tuned differently. The system $G_1(s)$ is the one with the larger time-delay, whereas the system G_2 exhibits a lower time-delay. G_3 is the common part, corresponding to the reach X_3.

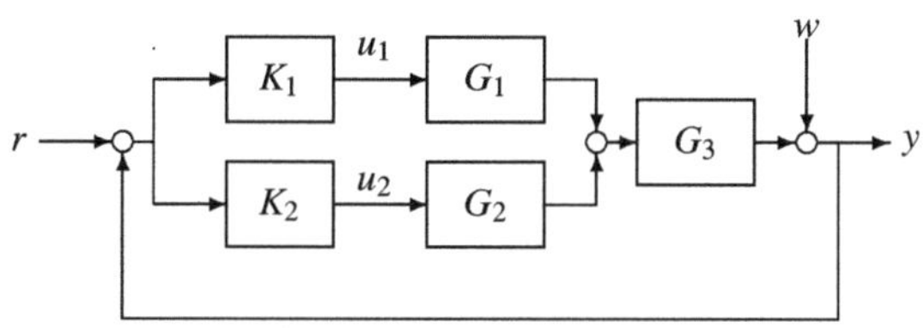

Fig. 12.10 Simplified representation of the control system

The objective is to recover the performance of the second dam (corresponding to the transfer G_2) with the sum of the two systems, one acting in low frequencies (K_1G_1) and the other one in high frequencies (K_2G_2). A naive solution would be to choose a perfect transfer function from a given desired closed-loop response (giving the best possible performance), and then find the corresponding open-loop transfers enabling achievement of this perfect transfer. Let $K_{\text{perfect}}G_2$ be the desired open-loop transfer for the system.

Then, we will try to identify K_1 and K_2 such that

$$K_{\text{perfect}}G_2G_3 = (K_1G_1 + K_2G_2)G_3.$$

The idea is to tune K_1 in order to fit to $K_{\text{perfect}}G_2G_3$ in low frequencies, and to tune K_2 in order to fit to $K_{\text{perfect}}G_2G_3$ in high frequencies. However, this is difficult to do "manually" (by trial and error) and to simultaneously ensure the required robustness margins for the controlled system. This is why the design is done using a systematic tool, by putting the design specifications in terms of H_∞ constraints.

12.5.3.2 Design Specifications

We now go back to the original system depicted in Fig. 12.7, with three measurement points (denoted by the vector y), three perturbations (w) and a reference for the output discharge (r). The augmented system is represented in Fig. 12.11.

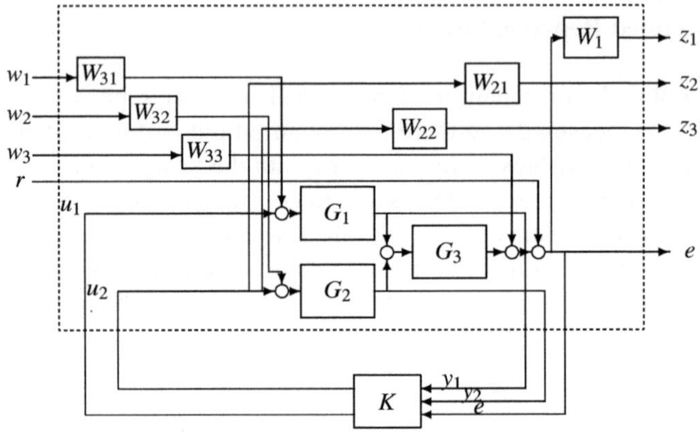

Fig. 12.11 Augmented system for H_∞ optimization

The transfer matrix between inputs $[r(s)\ w(s)]$ and outputs $[z(s)\ e(s)]$ is:

$$M = \begin{pmatrix} W_1 S & W_1 \tilde{G} S W_3 \\ W_2 K S & W_2 \tilde{G} K S W_3 \end{pmatrix}.$$

The proposed design is developed based on the chosen weighting functions for the SIMO controllers. This design consists in mixing the previous controllers in the frequency domain. The first controller (the rapid one) is used in high frequencies, while the slow controller is used to deliver water in low frequencies.

This is done by specifying two weighting functions for the transfer KS:

- The constraint on KS_1 (from r to u_1) is specified by a weighting function W_{21} identical to the one used for the SIMO controller ♯1 design.
- The constraint on KS_2 (from r to u_2) is specified by a weighting function W_{22}, keeping the high frequency constraint defined in the SIMO case ♯2 and adding a supplementary constraint in low frequencies, in order to impose a quasi zero mean value for the discharge u_2 (see Fig. 12.12).

We use a band filter, as given by [4], for the weighting function W_{22}:

$$W_{22}(s) = \frac{G_\infty s^2 + G_{\min} \xi \omega_{\min} s + G_\infty \omega_{\min}^2}{s^2 + \xi \omega_{\min} s + \omega_{\min}^2},$$

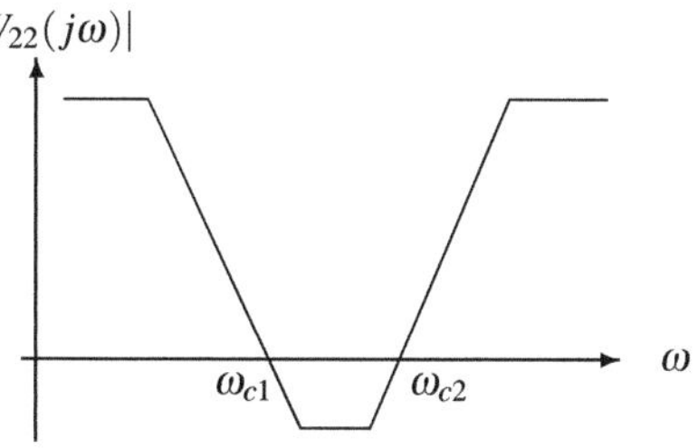

Fig. 12.12 Shape of frequency weighting function W_{22} for the MIMO controller

where ξ is used to tune the bandwidth, $G_{\min}$ is the minimum gain, and $\omega_{\min}$ the corresponding frequency. Let ω_{c1} and ω_{c2} be the cutting frequencies of this filter, one then has for any $k > 0$ (specifying the filter width): $\omega_{c1} = \frac{1}{k}\omega_{\min}$, $\omega_{c2} = k\omega_{\min}$ and $\xi = \frac{|k^2-1|}{k}\frac{\sqrt{1-G_\infty^2}}{\sqrt{G_{\min}^2-1}}$. The filter W_{22} is therefore completely determined if one chooses G_∞, $G_{\min}$, ω_{c1}, and ω_{c2}.

The weighting function parameters were chosen equal to $\omega_{c1} = 10^{-5}$ rad/s, $\omega_{c2} = 3.10^{-4}$ rad/s, $G_\infty = 100$, and $G_{\min} = 0.3$. The results of H_∞ optimization (transfer functions and corresponding constraints) are given in Fig. 12.13. The optimization resulted in $\gamma = 1.1$, with an input gain margin of 6.8 dB and an output gain margin of 7.6 dB.

The simulation in Fig. 12.14 shows that the desired behavior is obtained (compared to Fig. 12.8): the second dam reacts quickly to a perturbation in w_3, and while the water released by the first dam reaches the controlled point, the water delivered by the second dam diminishes and goes back to zero. Therefore, the MIMO controller enables us to recover the performance obtained by the SIMO controller ♯2 (see Fig. 12.9), but with a constraint on the available volume of each dam.

We considered the problem of designing controllers for MIMO systems (a river controlled by two dams) with a constraint on the available volume of each dam. This problem could not be solved easily with other classical approaches like LQG control [10]. We have shown that the best achievable performance in terms of output perturbation rejection can be obtained with the two dams while constraining the small dam to have a mean discharge equal to zero (this constraint is specified by frequency weighting functions in an H_∞ framework). The controller design can easily be parameterized in order to let the manager decide the required performance of each dam. The mixed sensitivity controller exhibits good robustness margins, which are needed in an application perspective. The proposed control design method is flexible and can easily incorporate additional information when available.

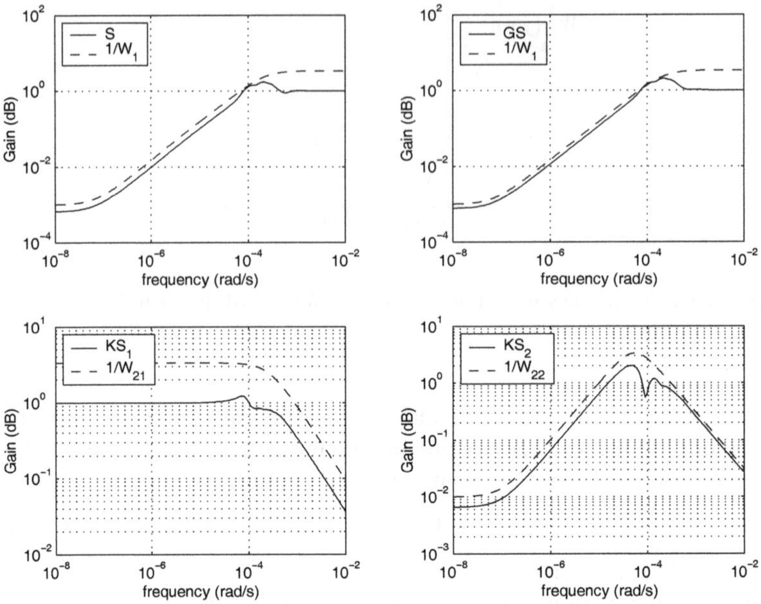

Fig. 12.13 Transfer functions and corresponding constraints for the MIMO controller

12.6 Conclusion

We have applied the frequency domain methods developed in this book to the control of a regulated river with multiple measurement point. We used the H_∞ framework to design controllers for open channel systems represented by a linear advection-diffusion equation. The proposed method is flexible and enables us to take into account more information when available. For this purpose, we extend the standard H_∞ criterion in order to design a "predictive" control law.

Our approach mixes the advantages of the H_∞ approach (a priori robustness and simplicity) with the advantages of the predictive approach (explicitly taking into account the future behavior or perturbations and/or references).

References

[1] Callier F, Desoer C (1978) An algebra of transfer functions for distributed linear time-invariant systems. IEEE Trans Circuits Syst CAS-25(9):651–662

[2] Dooge J, Napiórkowski J, Strupczewski W (1987) The linear downstream response of a generalized uniform channel. Acta Geophys Polon XXXV(3):279–293

[3] Doyle J, Glover K, Khargonekar P, Francis B (1989) State-space solutions to standard H_2 and H_∞ control problems. IEEE Trans Autom Contr 34(8):831 – 847

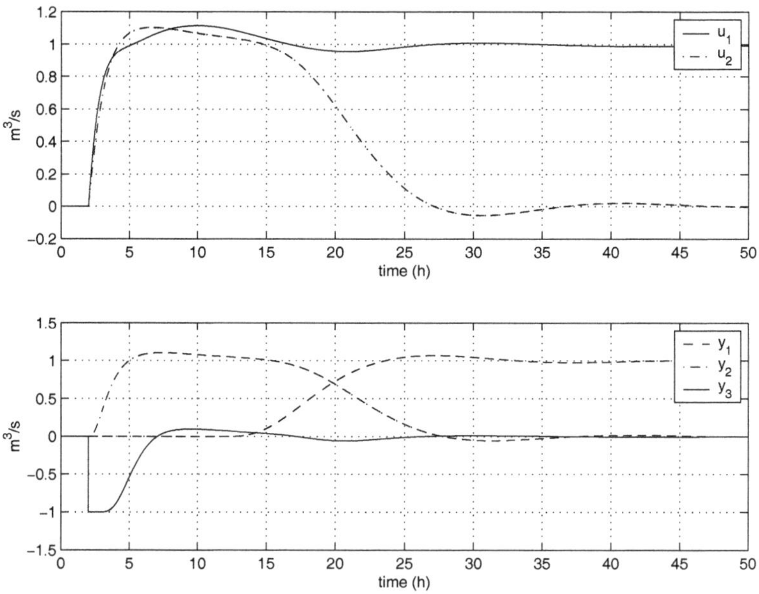

Fig. 12.14 Proposed MIMO controller: reaction to an unmeasured withdrawal w_3

[4] Font S (1995) Méthodologie pour prendre en compte la robustesse des systèmes asservis: optimisation H_∞ et approche symbolique de la forme standard. PhD thesis, Université de Paris-Sud Orsay, France

[5] Gahinet P, Apkarian P (1994) A linear matrix inequality approach to H_∞ control. Int J Robust Nonlin Control 4:421–448

[6] Horowitz I (1963) Synthesis of feedback systems. Academic Press, New York

[7] Jreij H (1997) Sur la régulation des cours d'eau aménagés. PhD thesis, Université Paris-XI Dauphine, (in French)

[8] Khargonekar P, Poolla K (1986) Robust stabilization of distributed systems. Automatica 22(1):77–84

[9] Kosuth P (1994) Techniques de régulation automatique des systèmes complexes: application aux systèmes hydrauliques à surface libre. PhD thesis, Institut National Polytechnique de Toulouse, (in French)

[10] Litrico X (1999) Modélisation, identification et commande robuste de systèmes hydrauliques à surface libre. PhD thesis, ENGREF–Cemagref, (in French)

[11] Litrico X, Fromion V (2001) About optimal performance and approximation of open-channel hydraulic systems. In: Proc. 40th IEEE Conf. on Decision and Control, Orlando, pp 4511–4516

[12] Litrico X, Fromion V (2002) Real-time management of multi-reservoir hydraulic systems using H_∞ optimization. In: IFAC World Congress, Barcelona

[13] Litrico X, Georges D (1999) Robust optimal control of a dam-river system with intermediate measurements. In: International European Control Conf., ECC'99, Karlsruhe, Germany, vol 2, pp 703–708

[14] Litrico X, Fromion V, Scorletti G (2001) Improved performance for open-channel hydraulic systems using intermediate measurements. In: IFAC Workshop on Time-Delay Systems, Santa Fe, pp 113–118

[15] Papageorgiou M, Messmer A (1989) Flow control of a long river stretch. Automatica 25(2):177–183
[16] Pognant-Gros P, Fromion V, Baume J (2001) Canal controller design: a multivariable approach using H_∞. In: European Control Conference, Porto, pp 3398–3403
[17] Sawadogo S (1992) Modélisation, commande prédictive et supervision d'un système d'irrigation. PhD thesis, LAAS-CNRS Toulouse

Appendix A
Derivation of the Saint-Venant Equations

Abstract In this appendix, we derive the Saint-Venant equations and their linearization around a nonuniform regime.

A.1 Derivation of the Saint-Venant Equations

In the sequel, $m(x,t)$ denotes the mass by unit length of the channel and it is defined as $m(x,t) = \rho A(x,t)$.

The variables are supposed to be continuous and differentiable, so that their partial derivatives with respect to x and t always exist and are continuous.

In order to derive the model, we first define a so-called control volume, denoted Ω in the sequel, which corresponds to an arbitrary volume of water in the flow field surrounded by an "imaginary" enclosing surface. This approach is very classical in fluid mechanics, where it is often more convenient to work with control volumes, as it is difficult to identify and follow a system of fluid particles. For the open channel, the control volume is defined as a stretch of channel of length Δx as described in in Fig. A.1.

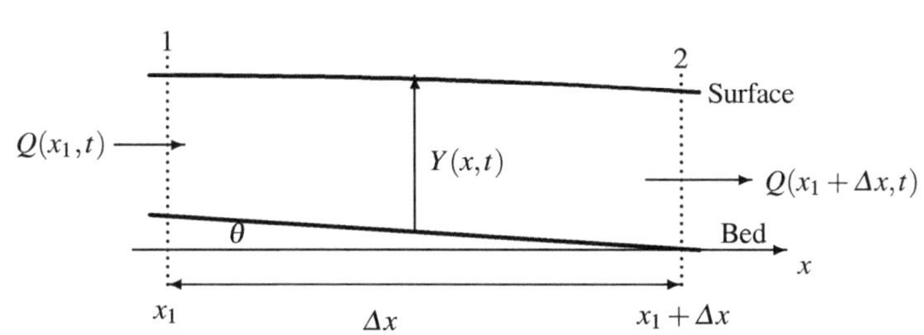

Fig. A.1 Longitudinal view of a small stretch of channel

A.1.1 Mass Conservation

The total mass of fluid within the control volume will increase or decrease depending on the net influx and outflux of water across the cross-sections 1 and 2. By definition of the control volume Ω and since the flow in the open channel is given by $Q(x,t)$ for each x, the incoming flow at the cross-section 1 ($x = x_1$) is equal to $Q(x_1,t)$ while the outgoing flow at the cross-section 2 ($x = x_1 + \Delta x$) is equal to $Q(x_1 + \Delta x, t)$. Consequently, the variation of the mass in the control volume is given by

$$M(t+\Delta t) - M(t) = \rho \int_t^{t+\Delta t} [Q(x_1,\tau) - Q(x_1+\Delta x,\tau)]\mathrm{d}\tau.$$

The use of the mean value theorem[1] implies that there exists $\xi \in (0,1)$ such that

$$M(t+\Delta t) - M(t) = \rho \Big[Q(x_1, t+\xi\Delta t) - Q(x_1+\Delta x, t+\xi\Delta t)\Big]\Delta t. \tag{A.1}$$

The total mass of water in the control volume can be evaluated as the integral of the mass per unit length. The mass at time t of the control volume Ω is thus given by

$$M(t) = \int_{x_1}^{x_1+\Delta x} m(x,t)\mathrm{d}x = \rho \int_{x_1}^{x_1+\Delta x} A(x,t)\mathrm{d}x.$$

Straightforward manipulations and the use of the mean value theorem allow us to deduce that there exists $\xi' \in (0,1)$ such that:

$$M(t+\Delta t) - M(t) = \Big[A(x_1+\xi'\Delta x, t+\Delta t) - A(x_1+\xi'\Delta x, t)\Big]\Delta x. \tag{A.2}$$

Let us now divide (A.1) and (A.2) by Δx and Δt to obtain:

$$\frac{A(x_1+\xi'\Delta x, t+\Delta t) - A(x_1+\xi'\Delta x, t)}{\Delta t} = \frac{Q(x_1, t+\xi\Delta t) - Q(x_1+\Delta x, t+\xi\Delta t)}{\Delta x}. \tag{A.3}$$

Taking the limit when Δx and Δt go to zero allows us to conclude. Indeed, for the left-hand side of relation (A.3) and since $A(x,t)$ is assumed to be continuously differentiable with respect to its arguments, we are able to deduce first that

$$\lim_{\Delta t \to 0} \frac{A(x_1+\xi'\Delta x, t+\Delta t) - A(x_1+\xi'\Delta x, t)}{\Delta t} = \frac{\partial A(x_1+\xi'\Delta x, t)}{\partial t},$$

[1] Let f a continuous function defined from $[a,b]$ into $\mathbb{R}$, then there exists $c \in (a,b)$ such that

$$\int_a^b f(x)\mathrm{d}x = f(c)(b-a).$$

and by continuity of its partial derivative with respect to the first argument[2]:

$$\lim_{\Delta x \to 0} \frac{\partial A(x_1 + \xi' \Delta x, t)}{\partial t} = \frac{\partial A(x_1, t)}{\partial t}.$$

Application of the same lines to the right-hand side of (A.3) leads to the final result:

$$\frac{\partial A(x_1,t)}{\partial t} + \frac{\partial Q(x_1,t)}{\partial x} = 0.$$

This equation is called the continuity equation.

A.1.2 Conservation of Momentum

The second law of Newton sets how the velocity of an object changes when it is subject to an external force: the law defines a force to be equal to change in momentum (mass times velocity) per change in time. This law can be rewritten as:

$$\frac{\mathrm{d}\mathscr{M}(t)}{\mathrm{d}t} = \sum_i F_{i,\text{external}}(t), \tag{A.4}$$

where the momentum is defined as $\mathscr{M}(t) = M(t)V(t)$ with $M(t)$ the mass of the object and $V(t)$ its velocity[3].

Equation (A.4) has been obtained with the so-called Lagrangian coordinates. These coordinates are widely used in the classical mechanics of particle dynamics, and they lead, in our context, to the description of all the quantities (velocity, acceleration) as those of a "moving fluid particle". In the context of open channel flow, one is more interested in the global behavior of the flow. In this case, the dynamics of the channel are described through quantities that are now viewed by an observer at a fixed position. This point of view corresponds to the the so-called Eulerian coordinates. In order to derive equations deduced from the second law of Newton, we have to rewrite (A.4) with the Eulerian coordinates. This problem is the main concern of the Reynolds transport theorem, which allows us to restate the momentum conservation law for a control volume. Since the goal of this section is to provide more insights about the conservation momentum equation derivation, we hereafter use a direct approach allowing us to avoid the use of the Reynolds transport theorem by providing its proof in a rather heuristic way (a rigorous derivation

---

[2] The mean value theorem ensures that $\xi' \Delta x \in (0, \Delta x)$ and then $\xi' \Delta x$ tends to zero when Δx tends to zero.

[3] A more familiar form of the second law is obtained when the object has a constant mass. Then, we have:

$$\frac{\mathrm{d}\mathscr{M}(t)}{\mathrm{d}t} = M \underbrace{\frac{\mathrm{d}V(t)}{\mathrm{d}t}}_{\text{acceleration}} = \sum_i F_{i,\text{external}}(t).$$

of the Reynolds transport theorem can be found in most classical textbooks on fluid mechanics).

As has already been pointed out, the second law of Newton in the form given by (A.4) can only be applied on a system described with the Lagrangian coordinates. So, in order to derive the equation on the basis of (A.4), we follow the evolution of fluid particles that belong at time t to the fixed control volume Ω.

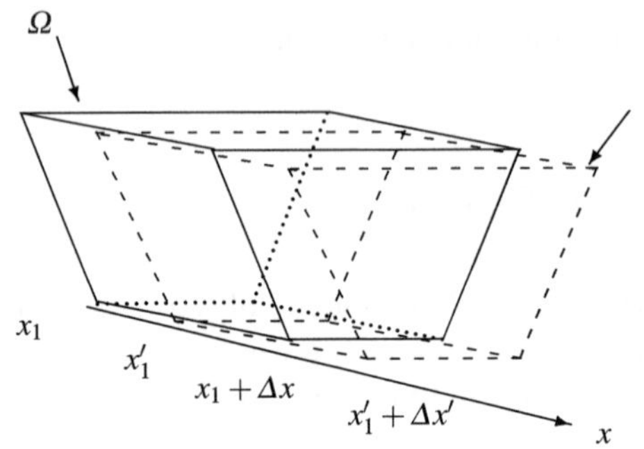

Fig. A.2 Definition of the control volumes Ω and Γ

The effect of external forces acting on this initial volume after a small period time Δt leads to the definition of a new stretch of channel denoted Γ, where x'_1 and $x'_1+\Delta x'$ denote the position of its two cross-sections, and $\mathscr{M}_\Gamma(t)$ the momentum of this moving volume (see Fig. A.2). We have by definition of the momentum:

$$\begin{aligned}\mathscr{M}_\Gamma(t+\Delta t)-\mathscr{M}_\Gamma(t) &= \int_{x'_1}^{x'_1+\Delta x'_1} m(x,t+\Delta t)V(x,t+\Delta t)\mathrm{d}x \\ &\qquad - \int_{x_1}^{x_1+\Delta x_1} m(x,t)V(x,t)\mathrm{d}x \\ &= \int_{x'_1}^{x'_1+\Delta x'_1} \rho Q(x,t+\Delta t)\mathrm{d}x - \int_{x_1}^{x_1+\Delta x_1} \rho Q(x,t)\mathrm{d}x\end{aligned}$$

and by the second law of Newton:

$$\mathscr{M}_\Gamma(t+\Delta t)-\mathscr{M}_\Gamma(t)=\sum_i \int_t^{t+\Delta t} F_{i,\text{external}}(\tau)\mathrm{d}\tau.$$

where $\mathscr{M}_\Gamma(t+\Delta t)$ corresponds to the momentum of Γ.

In order to link the momentum of the control volume, namely $\mathscr{M}_\Omega$, to that of Γ, we introduce tree elementary and specific volumes:

- $\Pi=\Omega\cap\Gamma$: fluid particles belonging to Ω and Γ
- $\Pi_i=\Omega\setminus\Gamma$: fluid particles only belonging to Ω

- $\Pi_o = \Gamma \setminus \Omega$: fluid particles only belonging to Γ

where by definition $\Pi_i \cap \Pi = \emptyset$, $\Pi_o \cap \Pi = \emptyset$, $\Pi_o \cap \Pi_i = \emptyset$, and $\Omega \cap \Gamma = \Pi_i \cup \Pi \cup \Pi_o$.

On this basis, the momentum of Γ and that of Ω are easily related, since by definition, we have $\Omega = \Pi_i \cup (\Gamma \setminus \Pi_o)$, which implies that[4]:

$$\mathscr{M}_\Omega(t+\Delta t) = \Delta \mathscr{M}_i + (\mathscr{M}_\Gamma(t+\Delta t) - \Delta \mathscr{M}_o)),$$

where $\Delta \mathscr{M}_i$ (or $\Delta \mathscr{M}_o$) corresponds to the momentum of fluid particles belonging to Π_i (or Π_o).

We have thus obtained the announced relation between the momentum of Ω and the external forces acting on it:

$$\mathscr{M}_\Omega(t+\Delta t) - \mathscr{M}_\Omega(t) + \Delta \mathscr{M}_o - \Delta \mathscr{M}_i = \sum_i \int_t^{t+\Delta t} F_{i,\text{external}}(\tau)\mathrm{d}\tau, \tag{A.5}$$

since by definition $\mathscr{M}_\Omega(t) = \mathscr{M}_\Gamma(t)$. It remains to compute the momentum $\Delta \mathscr{M}_i$ and $\Delta \mathscr{M}_o$,associated to Π_i and Π_o, respectively.

Actually, it can be shown that $\Delta \mathscr{M}_i$ is given by the following integral relation:

$$\Delta \mathscr{M}_i = \int_t^{t+\Delta t} \rho \frac{Q^2(x_1,\tau)}{A(x_1,\tau)} \mathrm{d}\tau.$$

In order to derive this expression, let assume that Δt is small enough, then $\Delta x' \approx V(x_1,t)\Delta t$ and the volume of Π_i is approximatively equal to $A(x_1,t)V(x_1,t)\Delta t$.

On this basis, since the momentum is the product of the mass by the velocity of the fluid particles, by assuming that all velocity of the particles in Π_i are close to $V(x_1,t)$, we deduce that the momentum of Π_i can be approximated by:

$$\Delta \mathscr{M}_i \approx \rho A(x_1,t)\, V(x_1,t)\, \Delta t\, V(x_1,t),$$

and thus $\Delta \mathscr{M}_i \approx \rho \frac{Q^2(x_1,t)}{A(x_1,t)} \Delta t$ since $A(x,t)V(x,t) = Q(x,t)$.

$\Delta \mathscr{M}_o$ can be computed following the same lines, and we have

$$\Delta \mathscr{M}_o = \int_t^{t+\Delta t} \frac{Q^2(x_1+\Delta x,\tau)}{A(x_1+\Delta x,\tau)} \mathrm{d}\tau.$$

The mean value theorem ensures that there exits $\xi \in (0,1)$ such that

$$\Delta \mathscr{M}_o - \Delta \mathscr{M}_i = \rho \left[\frac{Q^2(x_1+\Delta x, t+\xi\Delta t)}{A(x_1+\Delta x, t+\xi\Delta t)} - \frac{Q^2(x_1, t+\xi\Delta t)}{A(x_1, t+\xi\Delta t)} \right] \Delta t.$$

---

[4] For two domains D_1 and D_2 such that $D_1 \cap D_2 = \emptyset$, the momentum of the union is equal to the sum of their momentum:

$$\mathscr{M}_{D_1 \cup D_2} = \mathscr{M}_{D_1} + \mathscr{M}_{D_2}.$$

We finally conclude this section by noting that the variation of $\mathscr{M}_\Omega(t)$ can also be related to $Q(x,t)$. Indeed, by definition, we have

$$\begin{aligned}\mathscr{M}_\Omega(t+\Delta t)-\mathscr{M}_\Omega(t) &= \int_{x_1}^{x_1+\Delta x}\Big[m(x,t+\Delta t)V(x,t+\Delta t)-m(x,t)V(x,t)\Big]\mathrm{d}x\\ &= \int_{x_1}^{x_1+\Delta x}\rho\Big[Q(x,t+\Delta t)-Q(x,t)\Big]\mathrm{d}x.\end{aligned}$$

Finally, there exits $\xi'\in(0,1)$ such that

$$\mathscr{M}_\Omega(t+\Delta t)-\mathscr{M}_\Omega(t)=\rho\Big[Q(x_1+\xi'\Delta x,t+\Delta t)-Q(x_1+\xi'\Delta x,t)\Big]\Delta x. \quad \text{(A.6)}$$

A.1.2.1 Computation of the External Forces

The forces acting on the control volume in the x-direction are threefold:

- Gravity force
- Friction forces
- Pressure force

We first compute the values of the two first forces, which are easy to handle in our case.

Gravity Force Contribution

The force induced by the gravity in the x-direction corresponds to the projection of the gravity force on the bed slope of the canal. So if $\theta(x)$ defines the angle between the bed slope of the canal and the horizontal line, we deduce that

$$F_{gx}(t)=\int_{x_1}^{x_1+\Delta x} m(x,t)g\sin(\theta(x))\mathrm{d}x=\int_{x_1}^{x_1+\Delta x}\rho A(x,t)gS_b(x)\mathrm{d}x,$$

since by definition $S_b(x)=\sin(\theta(x))$. Then, there exists $\xi'\in(0,1)$ such that

$$F_{gx}(t)=\rho gA(x_1+\xi'\Delta x,t)S_b(x_1+\xi'\Delta x).$$

Friction Force Contribution

Friction force is applied to the control volume through shear along the channel bed and banks. We follow the classical viewpoint in expressing the shear force on a unit length of channel as $\rho gA(x,t)S_f(x,t)$, where S_f is so-called the friction slope. We then deduce that

$$F_{fx}(t)=-\int_{x_1}^{x_1+\Delta x}\rho gA(x,t)S_f(x,t)\mathrm{d}x,$$

and thus there exists $\xi' \in (0,1)$ such that

$$F_{fx}(t) = -\rho g A(x_1 + \xi'\Delta x, t) S_f(x_1 + \xi'\Delta x, t).$$

The friction slope is modeled using the Manning formula of (2.3).

Pressure Force Contribution

Since we have assumed that the pressure distribution is hydrostatic, the pressure at a given depth can be computed by using the rules attached to a static liquid. Consequently, the pressure at a given depth corresponds to the force induced by the water weight by unit area plus any pressure acting on the surface of the water and so the pressure at the depth h is then given by $\rho g h$ when the atmospheric pressure is neglected.

However, in order to derive the expression of forces induced by the pressure in the x-direction, we have to recall these two simple but fundamental facts concerning the pressure:

1. At a given depth, the pressure is independent of direction and it is actually the same in all directions.
2. The pressure on a submerged object is always perpendicular to the surface at each point on the surface.

For the control volume, there exist two main force contributions in the x-direction due to the pressure. The first one corresponds to the pressure forces associated to the surfaces defined at the cross-sections 1 and 2. A second term is due to the width variation in the x-direction.

Pressure Forces on the Cross-Sections

We first compute the force in the x-direction induced by the pressure acting on the surfaces associated to the cross-sections 1 and 2. The hydrostatic assumption allows us to compute the force induced by the pressure at a given depth. Indeed, this force by unit length at a given depth $Y(x_1,t) - \eta$ with $\eta \in [0, Y(x_1,t)]$ is given by

$$f_t(\eta, t) = \rho g (Y(x_1,t) - \eta) \sigma(x_1, \eta),$$

where $Y(x,t)$ is the water depth , $\sigma(x_1,\eta)$ is the width of the cross-section at the depth $Y(x_1,t) - \eta$ (see Fig. A.3).

We then deduce that the force associated to the surface of the cross-section 1 is given by

$$F_{p1x}(t) = \int_0^{Y(x_1,t)} f_t(\eta,t)\mathrm{d}\eta = \int_0^{Y(x_1,t)} \rho g (Y(x_1,t) - \eta)\sigma(x_1,\eta)\mathrm{d}\eta$$

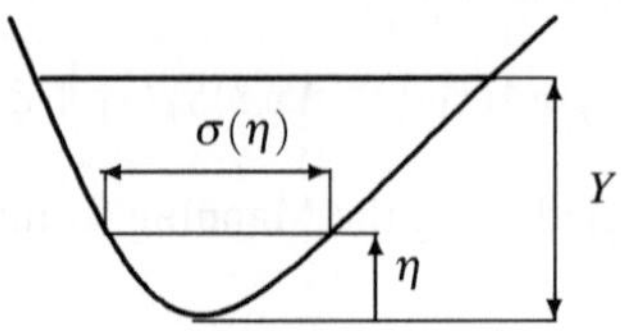

Fig. A.3 Section of an open channel. Definition of σ, η, and Y

and the force at the cross-section 2 by:

$$F_{p2x}(t) = \int_0^{Y(x_1+\Delta x,t)} \rho g(Y(x_1+\Delta x,t)-\eta)\sigma(x_1+\Delta x,\eta)\mathrm{d}\eta.$$

The resulting force is then given by $F_{px}(t) = F_{p1x}(t) - F_{p2x}(t)$. Since the expressions of the two forces are closely related, we develop F_{p2x} as a perturbation of F_{p1x}. For this purpose, we note that by the mean value theorem, there exits $\xi' \in (0,1)$ such that

$$Y(x_1+\Delta x,t) = Y(x_1,t) + \frac{\partial Y(x_1+\xi'\Delta x,t)}{\partial x}\Delta x,$$

and in the same way, there also exits $\theta' \in (0,1)$ such that

$$\sigma(x_1+\Delta x,\eta) = \sigma(x_1,\eta) + \frac{\partial \sigma(x_1+\theta'\Delta x,\eta)}{\partial x}\Delta x.$$

We then split the resulting force into four terms:

$$F_{px}(t) = F_{p1x}(t) - F_{p2x}(t) = I_1(t) + I_2(t) + I_3(t) + I_4(t),$$

where

$$\begin{aligned}
I_1(t) &= -\int_0^{Y(x_1,t)} \rho g(Y(x_1,t)-\eta)\frac{\partial \sigma(x_1+\theta'\Delta x,\eta)}{\partial x}\Delta x\mathrm{d}\eta, \\
I_2(t) &= -\int_0^{Y(x_1,t)} \rho g\frac{\partial Y(x_1+\xi'\Delta x,t)}{\partial x}\Delta x\sigma(x_1,\eta)\mathrm{d}\eta \\
&= -\rho g\Delta x\frac{\partial Y(x_1+\xi'\Delta x,t)}{\partial x}\int_0^{Y(x_1,t)} \sigma(x_1,\eta)\mathrm{d}\eta,
\end{aligned}$$

and then

$$I_2(t) = -\rho g\Delta x\frac{\partial Y(x_1+\xi'\Delta x,t)}{\partial x}A(x_1,t),$$

since by definition:

$$A(x_1,t) = \int_0^{Y(x_1,t)} \sigma(x_1,\eta)\mathrm{d}\eta.$$

In order to handle I_3, we first note that σ can be also rewritten as:

$$\sigma(x_1+\Delta x,\eta)-\sigma(x_1,\eta)=\int_0^1 \frac{\partial\sigma(x_1+\theta\Delta x,\eta)}{\partial x}\mathrm{d}\theta\Delta x.$$

From this previous expression, we then deduce that

$$I_3(t)=-\int_0^{Y(x_1,t)} \rho g\frac{\partial Y(x_1+\xi'\Delta x,t)}{\partial x}\Delta x\int_0^1 \frac{\partial\sigma(x_1+\theta\Delta x,\eta)}{\partial x}\Delta x\mathrm{d}\theta\mathrm{d}\eta,$$

and then

$$I_3(t)=-\rho g(\Delta x)^2\frac{\partial Y(x_1+\xi'\Delta x,t)}{\partial x}\int_0^{Y(x_1,t)}\int_0^1 \frac{\partial\sigma(x_1+\theta\Delta x,\eta)}{\partial x}\mathrm{d}\theta\mathrm{d}\eta.$$

The last term is given by:

$$I_4(t)=-\int_{Y(x_1,t)}^{Y(x_1+\Delta x,t)} \rho g(Y(x_1+\Delta x,t)-\eta)\sigma(x_1+\Delta x,\eta)\mathrm{d}\eta.$$

The mean value theorem guarantees that there exists $\xi\in(0,1)$ such that

$$I_4(t)=-\rho g(Y(x_1+\Delta x,t)-Y'(x_1,t))\sigma(x_1+\Delta x,Y'(x_1,t))\Delta Y,$$

with $\Delta Y=Y(x_1+\Delta x,t)-Y(x_1,t)$ and $Y'(x_1,t)=Y(x_1,t)+\xi'\Delta Y$.

Pressure Forces Due to Width Variation

Width variation of the channel leads to a contribution to the force in the x-direction. Actually, since the pressure is always perpendicular to the surface at each point on the surface, we then deduce that when the width of the channel increases (or decreases), there exits a contribution to the force in the x-direction. This contribution at a specific depth is then equal to the projection of the force pressure along the x-axis. It remains to define this angle. By definition, and with respect to the notation already introduced, we know that for a specific depth, the width is given by $\sigma(x,\eta)$. Since the pressure acts perpendicularly to this curve, we then deduce that the angle between force pressure and the x-axis is related to the gradient of $\sigma(x,\eta)$. Indeed, by definition, the tangent of the angle between the force and the x-axis is equal to the gradient of σ. So, if this gradient is small enough, then the tangent can be approximated by the angle value and then the contribution of the pressure force in the x-axis is then given by

$$p_w(x,t,\eta)=\underbrace{\rho g(Y(x,t)-\eta)}_{\text{pressure at depth }(Y(x,t)-\eta)}\frac{\partial\sigma(x,\eta)}{\partial x}.$$

The integration of $p_w(x,t,\eta)$ along the depth allows us to deduce the force by unit of length:

$$f_w(x,t) = \int_0^{Y(x,t)} \rho g(Y(x,t)-\eta)\frac{\partial \sigma(x,\eta)}{\partial x}\mathrm{d}\eta,$$

and finally, the total contribution of the width variation to the force in the x-direction corresponds to the integral of the previous quantity along x

$$\begin{aligned} F_{lx}(t) &= \int_{x_1}^{x_1+\Delta x} f_w(x,t)\mathrm{d}x \\ &= \int_{x_1}^{x_1+\Delta x}\int_0^{Y(x,t)} \rho g(Y(x,t)-\eta)\frac{\partial \sigma(x,\eta)}{\partial x}\mathrm{d}\eta\mathrm{d}x. \end{aligned}$$

Invoking again the mean value theorem allows us to deduce that there exists $\xi' \in (0,1)$ such that

$$F_{lx}(t) = \int_0^{Y(x,t)} \rho g(Y(x_1+\xi'\Delta x,t)-\eta)\frac{\partial \sigma(x_1+\xi'\Delta x,\eta)}{\partial x}\mathrm{d}\eta\Delta x.$$

A.1.2.2 Momentum Conservation

We have already shown that the second law of Newton implies that

$$\mathscr{M}_\Omega(t+\Delta t) - \mathscr{M}_\Omega(t) + \Delta\mathscr{M}_o - \Delta\mathscr{M}_i = \sum_i \int_t^{t+\Delta t} F_{i,\text{external}}(\tau)\mathrm{d}\tau,$$

which, by a simple application of the mean value theorem to the integral of forces leads to the existence of $\xi \in (0,1)$, such that

$$\underbrace{\mathscr{M}_\Omega(t+\Delta t) - \mathscr{M}_\Omega(t) + \Delta\mathscr{M}_o - \Delta\mathscr{M}_i}_{P_l} = \underbrace{\sum_i F_{i,\text{external}}(t+\xi\Delta t)\Delta t}_{P_r}.$$

Collecting all the terms in the previous sections, and dividing by Δx and Δt, we obtain for the left-hand side:

$$\frac{P_l}{\Delta x\Delta t} = \frac{\rho\left[Q(x_1',t+\Delta t) - Q(x_1',t)\right]}{\Delta t} + \rho\frac{\left[\frac{Q^2(x_1+\Delta x,t+\xi\Delta t)}{A(x_1+\Delta x,t+\xi\Delta t)} - \frac{Q^2(x_1,t+\xi\Delta t)}{A(x_1,t+\xi\Delta t)}\right]}{\Delta x}.$$

As in the case of the mass conservation equation, we then take the limit of the previous relation when Δx and Δt go to zero, and we obtain:

$$\lim_{\Delta t,\Delta x\to 0}\frac{P_l}{\Delta x\Delta t} = \rho\frac{\partial Q(x_1,t)}{\partial t} + \rho\frac{\partial}{\partial x}\left[\frac{Q^2(x_1,t)}{A(x_1,t)}\right].$$

The right hand side divided by Δx and Δt is equal to

$$\frac{P_r}{\Delta x\Delta t} = \frac{F_{gx}(t') + F_{fx}(t') + F_{px}(t') + F_{lx}(t')}{\Delta x},$$

where $t' = t + \xi \Delta t$.

Let us take the limit for each term of the previous expression. We then have

$$\lim_{\Delta t, \Delta x \to 0} \frac{F_{gx}(t')}{\Delta x} = \rho g A(x_1, t) S_b(x_1),$$

and also

$$\lim_{\Delta t, \Delta x \to 0} \frac{F_{fx}(t')}{\Delta x} = -\rho g A(x_1, t) S_f(x_1, t).$$

For $F_{px}(t')$, we have

$$\lim_{\Delta t, \Delta x \to 0} \frac{I_2(t')}{\Delta x} = -\rho g \frac{\partial Y(x_1, t)}{\partial x} A(x_1, t).$$

Furthermore, for I_3 and I_4, the limit is zero. It remains to consider I_1 and $F_{lx}(t')$. It is easy to show that

$$\lim_{\Delta t, \Delta x \to 0} \frac{I_1(t')}{\Delta x} = - \lim_{\Delta t, \Delta x \to 0} \frac{F_{lx}(t')}{\Delta x},$$

and then

$$\lim_{\Delta t, \Delta x \to 0} \frac{I_1(t') + F_{lx}(t')}{\Delta x} = 0.$$

We can now conclude and deduce the second equation which is the consequence of the second law of Newton:

$$\frac{\partial Q}{\partial t}(x_1, t) + \frac{\partial}{\partial x}\left[\frac{Q^2(x_1, t)}{A(x_1, t)}\right] + gA(x_1, t)\left(\frac{\partial Y(x_1, t)}{\partial x} + S_f(x_1, t) - S_b(x_1)\right) = 0.$$

We finally obtained the two Saint-Venant equations, expressing mass and momentum conservation for open channel flow.

One may also obtain these equations by integrating the Navier–Stokes equations for free surface open channel flow. We preferred this derivation because it enables us to show the underlying physics of the phenomenon. Moreover, this derivation can help the reader understand the importance of the assumptions used in order to derive the Saint-Venant equations (see the discussion in Sect. 2.1.1.1).

A.2 Derivation of the Linearized Equations

The four terms in (2.2) are denoted, respectively, (t_1), (t_2), (t_3), and (t_4), with a subscript zero for stationary values. The first term is directly linearized. For the second term, let us first expand $\frac{Q^2}{A}$ following (2.20):

$$\frac{Q^2}{A} - \frac{Q_0^2}{A_0} = 2\frac{Q_0}{A_0}q - \frac{Q_0^2}{A_0^2}\frac{\partial A_0}{\partial Y}y$$
$$= 2V_0 q - V_0^2 T_0 y \tag{A.7}$$

where $V_0 = \frac{Q_0}{A_0}$ and $\frac{\partial A_0}{\partial Y} = T_0$.

Taking the partial derivative of (A.7) with respect to x leads to:

$$(t_2) - (t_2)_0 = 2V_0\frac{\partial q}{\partial x} + 2\frac{\mathrm{d}V_0}{\mathrm{d}x}q - V_0^2 T_0\frac{\partial y}{\partial x} - \left(V_0^2\frac{\mathrm{d}T_0}{\mathrm{d}x} + 2V_0 T_0\frac{\mathrm{d}V_0}{\mathrm{d}x}\right)y. \tag{A.8}$$

Since $V_0 = Q_0/A_0$, and $\mathrm{d}Q_0/\mathrm{d}x = 0$, the derivative of V_0 with respect to x is given by:

$$\frac{\mathrm{d}V_0}{\mathrm{d}x} = -\frac{Q_0}{A_0^2}\frac{\mathrm{d}A_0}{\mathrm{d}x}.$$

The channel is supposed prismatic therefore $\frac{\mathrm{d}A_0}{\mathrm{d}x} = T_0\frac{\mathrm{d}Y_0}{\mathrm{d}x}$.

Replacing in (A.8) and collecting terms leads to:

$$(t_2) - (t_2)_0 = 2V_0\frac{\partial q}{\partial x} - 2\frac{V_0 T_0}{A_0}\frac{\mathrm{d}Y_0}{\mathrm{d}x}q - V_0^2 T_0\frac{\partial y}{\partial x} + \left(2\frac{V_0^2 T_0^2}{A_0}\frac{\mathrm{d}Y_0}{\mathrm{d}x} - V_0^2\frac{\mathrm{d}T_0}{\mathrm{d}x}\right)y.$$

The third term gives:

$$(t_3) - (t_3)_0 = gT_0\frac{\mathrm{d}Y_0}{\mathrm{d}x}y + gA_0\frac{\partial y}{\partial x}.$$

Following (2.20), the fourth term is expanded as:

$$(t_4) - (t_4)_0 = \left(gT_0(S_{f0} - S_b) + gA_0\frac{\partial S_f}{\partial Y}\right)y + gA_0\frac{\partial S_f}{\partial Q}q,$$

where the partial derivatives of S_f are evaluated for the equilibrium (Q_0, Y_0). These partial derivatives are obtained as:

$$\frac{\partial S_f}{\partial Q}(Q_0, Y_0) = 2\frac{S_{f0}}{Q_0},$$

and

$$\frac{\partial S_f}{\partial Y}(Q_0, Y_0) = -2\frac{S_{f0}}{A_0}\frac{\partial A_0}{\partial Y} - \frac{4}{3}\frac{S_{f0}}{R_0}\frac{\partial R_0}{\partial Y}.$$

Since $\frac{\partial A_0}{\partial Y} = T_0$ and $\frac{\partial R_0}{\partial Y} = \frac{T_0}{P_0} - \frac{A_0}{P_0^2}\frac{\partial P_0}{\partial Y}$, the overall expression for the fourth term is given by:

$$(t_4) - (t_4)_0 = gA_0 S_{f0}\left[2\frac{q}{Q_0} - \left(\left(\frac{7}{3} + \frac{S_b}{S_{f0}}\right)\frac{T_0}{A_0} - \frac{4}{3P_0}\frac{\partial P_0}{\partial Y}\right)y\right].$$

After collecting terms in q, y, $\frac{\partial q}{\partial x}$, and $\frac{\partial y}{\partial x}$, we obtain (2.21):

$$\frac{\partial q}{\partial t} + 2V_0 \frac{\partial q}{\partial x} + \delta q + (C_0^2 - V_0^2) T_0 \frac{\partial y}{\partial x} - \tilde{\gamma} y = 0,$$

with

$$\tilde{\gamma} = V_0^2 \frac{\mathrm{d}T_0}{\mathrm{d}x} + g T_0 \left(\kappa S_{f0} + S_b - (1 + 2F_0^2) \frac{\mathrm{d}Y_0}{\mathrm{d}x} \right),$$
$$\delta = \frac{2g}{V_0} \left(S_{f0} - F_0^2 \frac{\mathrm{d}Y_0}{\mathrm{d}x} \right),$$
$$\kappa = \frac{7}{3} - \frac{4A_0}{3T_0 P_0} \frac{\partial P_0}{\partial Y},$$

and $F_0^2 = \frac{V_0^2 T_0}{g A_0}$ is the Froude number for the equilibrium regime.

Appendix B
Proofs Related to Chap. 3

Abstract This appendix contains various proofs related to the frequency domain approach developed in Chap. 3.

B.1 Proof of the Modal Decomposition

To show that the transfer matrix $G(x,s)$ can be expressed as an infinite sum of simple elements, we apply the residue theorem to each element of the transfer matrix. The proof is closely related to the proof of the series decomposition of $\cot(z)$ in [2]. To simplify the exposition, we assume that the poles have simple multiplicity, but the proof can easily be adapted to the general case.

Let $\{\mathscr{C}_N; N \geq 0\}$ be a series of nested circular contours such that there are exactly two poles p_N and p_{-N} between $\mathscr{C}_{N-1}$ and $\mathscr{C}_N$. N is chosen sufficiently large such that the poles p_N and p_{-N} are complex conjugate.

Let us first define the function $s \mapsto f_{ij}(x,s) = g_{ij}(x,s) - \frac{a_{ij}^{(0)}(x)}{s}$, with $a_{ij}^{(0)}(x)$ the residue of the function $g_{ij}(x,s)$ in zero. This function is meromorphic and can be continuously extended in $s=0$ by $f_{ij}(x,0) = \frac{\mathrm{d}}{\mathrm{d}s}[s g_{ij}(x,s)]|_{s=0}$.

We apply the Cauchy residue theorem to the function $s \mapsto \frac{f_{ij}(x,s)}{z-s}$. For all $N > 1$, we have:

$$\frac{1}{2j\pi}\oint_{\mathscr{C}_N} \frac{f_{ij}(x,s)}{z-s}\mathrm{d}s = \sum_{n=-N, n\neq 0}^{N} \frac{a_{ij}^{(n)}(x)}{z-p_n} - f_{ij}(x,z), \tag{B.1}$$

with $a_{ij}^{(n)}(x) = \lim_{s\to p_n}(s-p_n) f_{ij}(x,s)$.

For $z=0$, (B.1) leads to:

$$\frac{1}{2j\pi}\oint_{\mathscr{C}_N} \frac{f_{ij}(x,s)}{s}\mathrm{d}s = f_{ij}(x,0) + \sum_{n=-N, n\neq 0}^{N} \frac{a_{ij}^{(n)}(x)}{p_n}. \tag{B.2}$$

Subtracting (B.2) from (B.1) gives:

$$f_{ij}(x,z) = f_{ij}(x,0) + \sum_{n=-N,n\neq 0}^{N} a_{ij}^{(n)}(x)\left(\frac{1}{z-p_n} + \frac{1}{p_n}\right) + \frac{1}{2j\pi}\oint_{\mathscr{C}_N} f_{ij}(x,s)\left(\frac{1}{z-s} + \frac{1}{s}\right)\mathrm{d}s,$$

or

$$f_{ij}(x,z) = f_{ij}(x,0) + \sum_{n=-N,n\neq 0}^{N} a_{ij}^{(n)}(x)\left(\frac{1}{z-p_n} + \frac{1}{p_n}\right) + \frac{z}{2j\pi}\oint_{\mathscr{C}_N} \frac{f_{ij}(x,s)}{s(z-s)}\mathrm{d}s.$$

Now, since $|f_{ij}(x,s)|$ is bounded, the integral on the right-hand side tends to zero as N tends to infinity.

Finally, going back to the original transfer functions, we obtain:

$$g_{ij}(x,s) = f_{ij}(x,0) + \frac{a_{ij}^{(0)}(x)}{s} + \sum_{n=-\infty,n\neq 0}^{\infty} \frac{s a_{ij}^{(n)}(x)}{p_n(s-p_n)},$$

which is the result we wanted to prove.

B.2 Proof of the Inner–Outer Decomposition

We first begin by a review of some useful results in complex analysis, before applying these to the case of the linearized Saint-Venant transfer matrix.

B.2.1 Review of Classical Results in Complex Analysis

Any function $G(s)$ analytic in the right half-plane can be factorized into two terms:

$$G(s) = G_o(s)G_i(s), \tag{B.3}$$

where $G_o(s)$ is the outer factor, and $G_i(s)$ the inner factor. An inner function belongs to H_∞ and is such that $|G_i(j\omega)| = 1$ for all $\omega \in \mathbb{R}$. An outer function has no zeros in the right half-plane.

The outer factor of $G(s)$ analytic in the right half-plane is obtained by [3]:

$$G_o(s_1) = \exp\left[\frac{1}{\pi}\int_{-\infty}^{+\infty} \frac{\omega s_1 + j}{\omega + j s_1} \log|G(j\omega)| \frac{\mathrm{d}\omega}{1+\omega^2}\right]. \tag{B.4}$$

B.2.1.1 Outer Factor Computation

The outer factor phase of $G(s)$ is obtained by [5]:

$$\arg(G_o(j\omega_1)) = \frac{2\omega_1}{\pi}\int_0^{\infty}\frac{\log|G(j\omega)| - \log|G(j\omega_1)|}{\omega^2-\omega_1^2}\mathrm{d}\omega. \tag{B.5}$$

B.2.1.2 Inner Factor Computation

Once obtained the outer factor phase, the inner factor can be computed using:

$$G_i(j\omega) = \frac{G(j\omega)}{|G_o(j\omega)|}\exp[-j\arg(G_o(j\omega))]. \tag{B.6}$$

B.2.2 Application to the Saint-Venant Transfer Matrix

Since the inner-outer decomposition leads to a product decomposition, we consider the factors of the transfer functions separately.

B.2.2.1 Case of Factor $G(s) = \exp((\lambda_1(s) - \lambda_2(s))L)$

We first consider the factor $G(s) = \exp(-\frac{(\alpha+\beta)L}{\alpha\beta}\sqrt{(s+b)^2-a^2})$. The second-order polynomial function $(s+b)^2-a^2$ has two roots given by $s=-b\pm a$. Since it can be easily checked that $a<b$, the roots are strictly negative real. Factor $G(s)$ therefore has no singularity in the right half-plane.

We now use the following lemma derived from the extended argument principle [1] in order to compute integral (B.4).

Lemma B.1. *Let $f(s)$ be a meromorphic function in $\mathbb{C}^+$ and $g(s)$ analytic in $\overline{\mathbb{C}}^+$, verifying the following hypotheses:*

- *g is odd on the imaginary axis: $g(j\omega) = -g(-j\omega)$*
- *f is conjugate symmetric : $\overline{f(s)} = f(\overline{s})$*
- $\forall\theta\in[-\frac{\pi}{2},\frac{\pi}{2}]$: $\lim_{R\to+\infty} R\left|\frac{f'(Re^{j\theta})}{f(Re^{j\theta})}g(Re^{j\theta})\right| = 0$

If $f(s)$ has no zeros or poles in $\mathbb{C}^+$ but N_z zeros z_i, $i=1,\dots,N_z$, and N_p poles p_i, $i=1,\dots,N_p$ on the imaginary axis, then

$$\int_{-\infty}^{+\infty}\frac{f'(j\omega)}{f(j\omega)}g(j\omega)d\omega = -\pi\Big(\sum_{i=1}^{N_z}g(z_i) - \sum_{i=1}^{N_p}g(p_i)\Big). \tag{B.7}$$

Let G be holomorphic on the right half-plane, lemma B.1 with $f(s) = \frac{s+j\omega_1}{s-j\omega_1}$ and $g(s) = \log \frac{G(s)}{|G(j\omega_1)|}$ leads to:

$$\int_{-\infty}^{+\infty} \frac{-2j\omega_1}{(j\omega - j\omega_1)(j\omega + j\omega_1)} \log \frac{G(j\omega)}{|G(j\omega_1)|} d\omega = \pi \log \frac{G(j\omega_1)}{|G(j\omega_1)|}.$$

Taking imaginary parts of both sides leads to (B.5).

We would like to apply this lemma to our case in order to compute the outer factor's phase. However, in our case, when $G(s) = \exp(-\frac{(\alpha+\beta)L}{\alpha\beta}\sqrt{(s+b)^2 - a^2})$, the third hypothesis of lemma B.1 on the integral convergence is not fulfilled, because:

$$\lim_{R\to+\infty} \frac{1}{R} \left|\log G(Re^{j\theta})\right| = \frac{(\alpha+\beta)}{\alpha\beta} L.$$

To have the hypothesis satisfied, we may multiply $G(s)$ by a function of unitary gain on the imaginary axis, which will not modify integral (B.4). The only possibility for this is $\exp(\frac{(\alpha+\beta)Ls}{\alpha\beta})$. Then, with $\tilde{G}(s) = \exp\left(\frac{(\alpha+\beta)s-\sqrt{d(s)}}{\alpha\beta} L\right)$, we have $\lim_{R\to\infty} \frac{1}{R} \left|\log \tilde{G}(Re^{j\theta})\right| = 0$.

Applying lemma B.1 to $\tilde{G}(s)$ shows that $\tilde{G}(s)$ is outer. The inner factor of $\exp(-\frac{\sqrt{d(s)}}{\alpha\beta}L)$ is, therefore, equal to the delay $e^{-\frac{(\alpha+\beta)}{\alpha\beta}Ls} = e^{-\frac{2C_0}{C_0^2-V_0^2}Ls}$.

B.2.2.2 Case of Other Factors

It can be readily checked that the other factors $\sqrt{d(s)}$, $\frac{1}{1-e^{(\lambda_1(s)-\lambda_2(s))L}}$ are outer: they have no zeros or poles in the right half-plane and they fulfill the hypothesis of lemma B.1.

B.2.2.3 Summary

Finally, we obtain the inner part of $p_{21}(s)$ as $\exp(\frac{\alpha-\beta-(\alpha+\beta)}{2\alpha\beta}sL) = \exp(-\frac{L}{\alpha}s)$, which is a pure delay $e^{-\tau_1 s}$. Similarly, the inner part of $p_{12}(s)$ is obtained as $\exp(\frac{-\alpha+\beta-(\alpha+\beta)}{2\alpha\beta}sL) = \exp(-\frac{L}{\beta}s)$, which is a pure delay $e^{-\tau_2 s}$.

Following the same line, one can show that the transfer functions $p_{11}(s)$ and $p_{22}(s)$ are outer.

B.3 Analysis of the Numerical Scheme

The numerical integration scheme is formally written as follows:

$$\begin{aligned}&\eta_0 := \hat{\xi}_0;\\ &\text{for } k = 0, 1, \ldots, n-1:\\ &\eta_{k+1} := \eta_k + h_k \Phi_d(x_k, \eta_k, h_k),\\ &x_{k+1} := x_k + h_k,\end{aligned} \tag{B.8}$$

with $h_k = x_{k+1} - x_k$ and

$$\Phi_d(x_k, \hat{\eta}_k, h_k) = \frac{(e^{\mathscr{A}(x_k,s)h_k} - I)\hat{\eta}_k}{h_k},$$

where I is the identity matrix.

We first prove that the proposed numerical scheme (B.8) is consistent, then that it is an integration method of order 1 and finally indicate its global convergence.

B.3.1 Consistency of the Numerical Scheme

Following [4], the numerical integration scheme (B.8) is consistent if for any $x \in [0, L]$ and any $\eta \in \mathbb{C}^2$ one has

$$\lim_{h \to 0} \Phi_d(x, \eta, h) = \mathscr{A}(x, s)\eta.$$

Since for any matrix A one has by definition of the exponential matrix function:

$$e^{Ah} = I + Ah + \frac{A^2h^2}{2!} + O(h^3),$$

where $O(h^3)$ is the remainder of the Taylor expansion of order 3, then Φ_d can be rewritten as

$$\Phi_d(x, \eta, h) = \mathscr{A}(x, s)\eta + \frac{\mathscr{A}(x, s)^2 h}{2!}\eta + O(h^2),$$

which allows us to conclude that the proposed numerical scheme is consistent.

B.3.2 Order of the Numerical Scheme

Let us now prove that the numerical scheme is of order 1. Let h be a discretization step and let us associate to a given $t \in [0, L-h]$ and a given $\eta \in \mathbb{C}^2$ two quantities: $\hat{\xi}(x)$, the exact solution of

$$\frac{\mathrm{d}\hat{\xi}(x)}{\mathrm{d}x} = \mathscr{A}(x,s)\hat{\xi} \;;\; \hat{\xi}(t) = \eta,$$

and $\tilde{\eta}$, the approximate solution, given by

$$\tilde{\eta} = \eta + h\Phi_d(t,\eta,h).$$

With these two variables as defined, the order of the method corresponds to the maximum value of $p \in \mathbb{N}$ such that

$$\lim_{h\to 0} \frac{\hat{\xi}(t+h) - \tilde{\eta}}{|h|^p} = 0$$

holds for any $t \in [0, L-h]$ and any initial condition $\eta \in \mathbb{C}^2$.

To evaluate this quantity in our case, let us write the Taylor expansion of $\hat{\xi}(x)$ about the point $x = t$:

$$\begin{aligned}\hat{\xi}(t+h) &= \hat{\xi}(t) + \tfrac{\mathrm{d}\hat{\xi}}{\mathrm{d}x}(t)h + \tfrac{\mathrm{d}^2\hat{\xi}}{\mathrm{d}x^2}(t)\tfrac{h^2}{2!} + O(h^3)\\ &= \eta + h\mathscr{A}(t,s)\eta + \tfrac{h^2}{2!}\left(\mathscr{A}(t,s)^2 + \tfrac{\partial\mathscr{A}(x,s)}{\partial x}(t)\right)\eta + O(h^3),\end{aligned}$$

and then

$$\hat{\xi}(t+h) - \tilde{\eta} = \frac{h^2}{2!}\frac{\partial\mathscr{A}(x,s)}{\partial x}\eta + O(h^3).$$

We finally deduce that the numerical scheme is an integration method of order 1.

B.3.3 Global Convergence of the Numerical Scheme

Following the result of the previous paragraph, and by Theorem 7.2.2.3 (page 440 in [4]), we can deduce that the global discretization error at $x = L$, is also of order 1, i.e., there exists a finite constant $K > 0$ such that

$$\|\eta_n - \hat{\xi}(L)\| \leq K|h|,$$

where n is the number of discretization space steps, and η_n is the last approximate value of the solution. The numerical scheme is such that the error goes to zero when $|h| \to 0$ (see [4] for details), which ensures its convergence.

References

[1] Chen J, Chen G, Ren Z, Qiu L (2001) Extended argument principle and integral design constraints part I: a unified formula for classical results. In: Proc. 40th IEEE Conf. on Decision

and Control, Orlando
[2] Fuchs B, Shabat B (1964) Functions of a complex variable and some of their applications, vol I. Pergamon, Oxford
[3] Hoffman K (1962) Banach spaces of analytic functions. Prentice Hall, London
[4] Stoer J, Burlisch R (1993) Introduction to numerical analysis, 2nd edn. Texts in applied mathematics, Springer, New York
[5] Zadeh L, Desoer C (1963) Linear system theory. McGraw-Hill, New York

Appendix C
Time Domain Expressions Using Bessel Functions

We develop here the expressions obtained in the uniform flow case with Bessel functions. Let us consider the dynamic characteristics change of variable. The relation between $\zeta_1(x,t)$ and $\zeta_2(x,t)$ and their values at the boundary are given by (3.73).

C.1 Explicit Expressions in the Time Domain

We use the expressions of the complex eigenvalues $\lambda_1(s)$ and $\lambda_2(s)$ to obtain the explicit expression of the generalized characteristics for the uniform flow case. The generalized characteristics variables are given by:

$$\hat{\zeta}_1(x,s) = \hat{q}(x,s) + \frac{T_0 s}{\lambda_2(s)}\hat{y}(x,s),$$
$$\hat{\zeta}_2(x,s) = \hat{q}(x,s) + \frac{T_0 s}{\lambda_1(s)}\hat{y}(x,s).$$

Using (3.72), expanding $s/\lambda_1(s)$ and $s/\lambda_2(s)$, and rearranging, we get:

$$\frac{s}{\lambda_1(s)} = -\alpha + \frac{\alpha^2 r_1}{s+\delta} + \frac{(\alpha+\beta)}{2(s+\delta)}\hat{F}(s), \tag{C.1a}$$
$$\frac{s}{\lambda_2(s)} = \beta - \frac{\beta^2 r_2}{s+\delta} - \frac{(\alpha+\beta)}{2(s+\delta)}\hat{F}(s). \tag{C.1b}$$

We now use the inverse Laplace transform to express the generalized characteristics $\zeta(x,t)$ as follows:

$$\zeta_1(x,t) = \chi_1(x,t) - \beta^2 r_2 T_0 \tilde{y}(x,t) - \frac{T_0(\alpha+\beta)}{2}\int_0^t F(t-\tau)\tilde{y}(x,\tau)\mathrm{d}\tau, \tag{C.2a}$$

$$\zeta_2(x,t) = \chi_2(x,t) + \alpha^2 r_1 T_0 \tilde{y}(x,t) + \frac{T_0(\alpha+\beta)}{2}\int_0^t F(t-\tau)\tilde{y}(x,\tau)\mathrm{d}\tau, \tag{C.2b}$$

where $\chi_1(x,t) = q(x,t) + \beta T_0 y(x,t)$ and $\chi_2(x,t) = q(x,t) - \alpha T_0 y(x,t)$ are the Riemann invariants for the solution in the horizontal frictionless case (see (2.29), $\tilde{y}(x,t)$ is defined by:

$$\tilde{y}(t) = \int_0^t e^{-\delta(t-\tau)} y(x,\tau)\mathrm{d}\tau, \tag{C.3}$$

and $F(t)$ is the convolution kernel given by (3.70).

We therefore observe that even if the characteristics approach fails to diagonalize the system with a static change of variable, it is still possible to diagonalize the system, with a *dynamic* change of variable. This dynamic change of variable can be considered as the extension of the change of variable that leads to the Riemann invariants for the horizontal frictionless channel case.

The change of variable includes a first-order filter of the water level, and a more complex infinite dimensional filter with convolution kernel $F(t)$. The convolution kernel $F(t)$ is displayed in Fig. C.1 for canals 1 and 2. This kernel is very similar to that of a first-order filter.

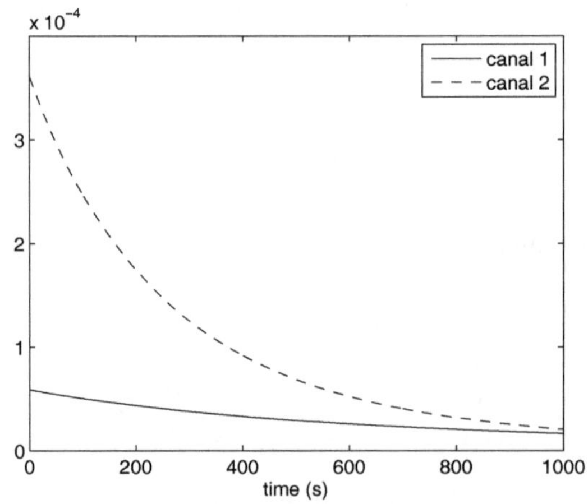

Fig. C.1 Convolution kernel $F(t)$ for canals 1 and 2

We have therefore obtained a dynamic change of variable, which is the extension of the characteristic form in the horizontal frictionless case. This change of variable includes a first-order filter, and an infinite dimensional filter, which are both linked to the nonzero values of parameters γ and δ. When $\gamma = \delta = 0$, we recover the horizontal frictionless case, and the variables $\zeta(x,t)$ coincide with the characteristics variables $\chi(x,t)$.

C.2 Inverse Transform

The inverse transform can also be expressed in the time domain as follows. Using the inverse transformation $\hat{\xi}(x,s) = \mathscr{X}(s)^{-1}\hat{\zeta}(x,s)$, we have:

$$\begin{aligned}
\hat{y}(x,s) &= \frac{\lambda_1(s)\lambda_2(s)}{T_0 s(\lambda_1(s)-\lambda_2(s))}(\hat{\zeta}_1(x,s)-\hat{\zeta}_2(x,s)),\\
\hat{q}(x,s) &= \frac{1}{\lambda_1(s)-\lambda_2(s)}(\lambda_1(s)\hat{\zeta}_2(x,s)-\lambda_2(s)\hat{\zeta}_1(x,s)).
\end{aligned}$$

Using (3.72), standard manipulations lead to

$$\begin{aligned}
\hat{y}(x,s) &= \frac{1}{T_0(\alpha+\beta)}(\hat{\zeta}_1(x,s)-\hat{\zeta}_2(x,s))\\
&\quad+\frac{1}{T_0(\alpha+\beta)}\left[\frac{\hat{F}(s)+c_y}{\sqrt{(s+b)^2-a^2}}\right](\hat{\zeta}_1(x,s)-\hat{\zeta}_2(x,s)),\\
\hat{q}(x,s) &= \frac{1}{\alpha+\beta}(\alpha\hat{\zeta}_1(x,s)+\beta\hat{\zeta}_2(x,s))\\
&\quad+\frac{\alpha-\beta}{2(\alpha+\beta)}\left[\frac{\hat{F}(s)+c_q}{\sqrt{(s+b)^2-a^2}}\right](\hat{\zeta}_1(x,s)-\hat{\zeta}_2(x,s)),
\end{aligned}$$

with

$$\begin{aligned}
c_y &= \frac{(\alpha^2+\beta^2)\delta-(\alpha-\beta)\gamma}{(\alpha+\beta)^2},\\
c_q &= \frac{2\alpha\beta}{(\alpha+\beta)^2}(2\gamma-(\alpha-\beta)\delta).
\end{aligned}$$

Using inverse Laplace transform tables, we find:

$$\mathscr{L}^{-1}\left\{\frac{\hat{F}(s)}{\sqrt{(s+b)^2-a^2}}\right\} = ae^{-bt}I_1(at),$$

and

$$\mathscr{L}^{-1}\left\{\frac{1}{\sqrt{(s+b)^2-a^2}}\right\} = e^{-bt}I_0(at),$$

with $I_0(t)$ and $I_1(t)$ the modified Bessel functions of orders 0 and 1, respectively.

Therefore, the inverse transformation from $(\zeta_1(x,t), \zeta_2(x,t))$ to $(y(x,t), q(x,t))$ can be expressed as:

$$\begin{aligned} y(x,t) = & \frac{1}{T_0(\alpha+\beta)}[\zeta_1(x,t) - \zeta_2(x,t)] \\ & + \frac{a}{T_0(\alpha+\beta)} \int_0^t e^{-b(t-\tau)} I_1(a(t-\tau))(\zeta_1(x,\tau) - \zeta_2(x,\tau)) \mathrm{d}\tau \\ & + \frac{c_y}{T_0(\alpha+\beta)} \int_0^t e^{-b(t-\tau)} I_0(a(t-\tau))(\zeta_1(x,\tau) - \zeta_2(x,\tau)) \mathrm{d}\tau, \end{aligned} \tag{C.4a}$$

$$\begin{aligned} q(x,t) = & \frac{1}{\alpha+\beta}[\alpha\zeta_1(x,t) + \beta\zeta_2(x,t)] \\ & + \frac{a(\alpha-\beta)}{2(\alpha+\beta)} \int_0^t e^{-b(t-\tau)} I_1(a(t-\tau))(\zeta_1(x,\tau) - \zeta_2(x,\tau)) \mathrm{d}\tau \\ & + \frac{c_q(\alpha-\beta)}{2(\alpha+\beta)} \int_0^t e^{-b(t-\tau)} I_0(a(t-\tau))(\zeta_1(x,\tau) - \zeta_2(x,\tau)) \mathrm{d}\tau. \end{aligned} \tag{C.4b}$$

This enables us to compute the physical variables $y(x,t)$ and $q(x,t)$ as functions of the characteristics variables $\zeta_1(x,t)$ and $\zeta_2(x,t)$.

Together, the formulas (3.73), (C.2), and (C.4) show that the linearized Saint-Venant equations around a uniform flow regime can be expressed in terms of generalized characteristics in a very similar way as the horizontal frictionless case. When expressed in the generalized characteristic variables $(\zeta_1(x,t), \zeta_2(x,t))$ with (C.2), the system becomes diagonal (see (3.73)). Therefore, the generalized characteristics can be viewed as an extension of the classical characteristics that diagonalize the system in the horizontal frictionless case.

C.3 Infinite Series Expansion

Using the expansion (3.31) with $z = e^{(\lambda_1(s)-\lambda_2(s))L}$, we can deduce an infinite series expansions for the $p_{ij}(s)$:

$$p_{11}(s) = \frac{1}{T_0}\sum_{k=0}^{\infty}\left[\frac{\lambda_2(s)}{s}e^{(\lambda_1(s)-\lambda_2(s))L} - \frac{\lambda_1(s)}{s}\right]e^{k(\lambda_1(s)-\lambda_2(s))L}, \tag{C.5a}$$

$$p_{12}(s) = \frac{(\lambda_1(s)-\lambda_2(s))}{T_0 s}e^{-\lambda_2(s)L}\sum_{k=0}^{\infty}e^{k(\lambda_1(s)-\lambda_2(s))L}, \tag{C.5b}$$

$$p_{21}(s) = \frac{(\lambda_2(s)-\lambda_1(s))}{T_0 s}e^{\lambda_1(s)L}\sum_{k=0}^{\infty}e^{k(\lambda_1(s)-\lambda_2(s))L}, \tag{C.5c}$$

$$p_{22}(s) = \frac{1}{T_0}\sum_{k=0}^{\infty}\left[\frac{\lambda_1(s)}{s}e^{(\lambda_1(s)-\lambda_2(s))L} - \frac{\lambda_2(s)}{s}\right]e^{k(\lambda_1(s)-\lambda_2(s))L}. \tag{C.5d}$$

Similarly to the horizontal frictionless case, the water levels at the boundaries result from the interaction of waves traveling upstream and downstream the channel. Indeed, one may use the inverse Laplace transforms given above to obtain an explicit expression for the time domain response of the system.

Let k_1 and k_2 be two positive integers, and let us consider the generic term $(k_1\lambda_1(s) - k_2\lambda_2(s))L$. Using (3.72) and (3.70), we have:

$$(k_1\lambda_1(s) - k_2\lambda_2(s))L = -(k_1r_1+k_2r_2)L - (k_1\tau_1+k_2\tau_2)s + (k_1+k_2)\frac{(\tau_1+\tau_2)}{2}\hat{F}(s).$$

Then, this yields:

$$e^{(k_1\lambda_1(s)-k_2\lambda_2(s))L} = e^{-(k_1r_1+k_2r_2)L-(k_1\tau_1+k_2\tau_2)s}\hat{K}\left(\frac{(k_1+k_2)}{2},s\right),$$

with $\hat{K}(k,s) = 1 + \hat{H}\Big(k(\tau_1+\tau_2),s\Big)$ and where $\hat{H}(l,s)$ is given by (3.69).

Finally, we obtain:

$$\frac{\lambda_1(s)}{s}e^{(k_1\lambda_1(s)-k_2\lambda_2(s))L} = e^{-(k_1r_1+k_2r_2)L-(k_1\tau_1+k_2\tau_2)s}\hat{\Lambda}_1\left(\frac{k_1+k_2}{2},s\right),$$
$$\frac{\lambda_2(s)}{s}e^{(k_1\lambda_1(s)-k_2\lambda_2(s))L} = e^{-(k_1r_1+k_2r_2)L-(k_1\tau_1+k_2\tau_2)s}\hat{\Lambda}_2\left(\frac{k_1+k_2}{2},s\right),$$

with $\hat{\Lambda}_1(k,s) = \frac{\lambda_1(s)}{s}\hat{K}(k,s)$ and $\hat{\Lambda}_2(k,s) = \frac{\lambda_2(s)}{s}\hat{K}(k,s)$.

Therefore, the time response of transfer functions $p_{ij}(s)$ can be computed in a very similar way as the horizontal frictionless case:

$$y(0,t) = y_{11}(t) + y_{12}(t), \tag{C.6a}$$
$$y(L,t) = y_{21}(t) + y_{22}(t), \tag{C.6b}$$

where the $y_{ij}(t)$ are given by:

- *The modal coefficients $a_{ij}^{(k)}(x)$ are given by:*

$$a_{ij}^{(k)}(x) = \frac{N_{ij}(x,p_k)}{\phi_{21}'(L,p_k)}, \tag{D.3}$$

where the $N_{ij}(x,p_k)$ are given by:

$$N_{11}(x,p_k) = -\frac{1}{T_0(x)}\phi_{11}(x,p_k)\phi_{22}(L,p_k) \tag{D.4a}$$

$$N_{12}(x,p_k) = \frac{1}{T_0(x)}\phi_{11}(x,p_k) \tag{D.4b}$$

$$N_{21}(x,p_k) = -\phi_{21}(x,p_k)\phi_{22}(L,p_k) \tag{D.4c}$$

$$N_{22}(x,p_k) = \phi_{21}(x,p_k), \tag{D.4d}$$

and $\phi_{21}'(L,p_k)$ is the term of the matrix $\Phi'(x,s)$ given by:

$$\Phi'(x,s) = \Phi(x,s)\int_0^x \Phi(v,s)^{-1}\frac{\partial\mathscr{A}(v,s)}{\partial s}\Phi(v,s)dv. \tag{D.5}$$

- *The coefficients $b_{ij}(x)$ are given by:*

$$b_{ij}(x) = \frac{1}{\tilde{\phi}_{21}(L,0)}\left(N_{ij}'(x,0) - \frac{\tilde{\phi}_{21}'(L,0)}{\tilde{\phi}_{21}(L,0)}N_{ij}(x,0)\right), \tag{D.6}$$

where the $N_{ij}'(x,0)$ are given by:

$$N_{11}'(x,0) = \frac{1}{T_0(x)}\Big(\phi_{12}(x,0)\phi_{21}'(L,0) - \phi_{11}'(x,0) - \phi_{11}(x,0)\phi_{22}'(L,0)\Big), \tag{D.7a}$$

$$N_{12}'(x,0) = \frac{1}{T_0(x)}\phi_{11}'(x,0), \tag{D.7b}$$

$$N_{21}'(x,0) = \phi_{21}'(L,0) - \phi_{21}'(x,0), \tag{D.7c}$$

$$N_{22}'(x,0) = \phi_{21}'(x,0), \tag{D.7d}$$

$\tilde{\phi}_{21}(L,0)$ is given by (D.2), and $\tilde{\phi}_{21}'(L,0)$ is given by:

$$\tilde{\phi}_{21}'(L,0) = -\int_0^L \phi_{11}'(x,0)dx. \tag{D.8}$$

Proof. We use in this proof the well-known fact that the state-transition matrix $\Phi(x,s)$ obtained from (3.77) verifies the differential equation (3.75).

Therefore, the $\phi_{ij}(x,s)$ verify the following linear differential system:

$$\frac{\partial \phi_{11}}{\partial x}(x,s) = \frac{(\alpha(x)-\beta(x))s+\gamma(x)}{\alpha(x)\beta(x)}\phi_{11}(x,s) - \frac{s+\delta(x)}{\alpha(x)\beta(x)}\phi_{21}(x,s), \tag{D.9a}$$

$$\frac{\partial \phi_{12}}{\partial x}(x,s) = \frac{(\alpha(x)-\beta(x))s+\gamma(x)}{\alpha(x)\beta(x)}\phi_{12}(x,s) - \frac{s+\delta(x)}{\alpha(x)\beta(x)}\phi_{22}(x,s), \tag{D.9b}$$

$$\frac{\partial \phi_{21}}{\partial x}(x,s) = -s\phi_{11}(x,s), \tag{D.9c}$$

$$\frac{\partial \phi_{22}}{\partial x}(x,s) = -s\phi_{12}(x,s), \tag{D.9d}$$

where by (3.77), the initial condition is given by the identity matrix $\Phi(0,s) = I$.

Let us first note that for $s = 0$ this system simplifies, and the solution can be expressed explicitly as follows:

$$\phi_{11}(x,0) = e^{\int_0^x \frac{\gamma(v)}{\alpha(v)\beta(v)}\mathrm{d}v}, \tag{D.10a}$$

$$\phi_{12}(x,0) = -\int_0^x \frac{\phi_{11}(x,0)}{\phi_{11}(v,0)}\frac{\delta(v)}{\alpha(v)\beta(v)}\mathrm{d}v, \tag{D.10b}$$

$$\phi_{21}(x,0) = 0, \tag{D.10c}$$

$$\phi_{22}(x,0) = 1. \tag{D.10d}$$

D.1.0.1 Computation of $a_{ij}^{(0)}(x)$

Let us introduce the modified transfer function $\tilde{\phi}_{21}(x,s)$ such that:

$$\tilde{\phi}_{21}(x,s) = \frac{1}{s}\phi_{21}(x,s).$$

This transfer function is well-defined in $s = 0$ since 0 is a root of $\phi_{21}(x,s)$.

Using (D.9c), $\tilde{\phi}_{21}(x,s)$ verifies the following differential equation:

$$\frac{\partial}{\partial x}\tilde{\phi}_{21}(x,s) = -\phi_{11}(x,s), \tag{D.11}$$

with an initial condition given by $\tilde{\phi}_{21}(0,s) = 0$.

Therefore $\tilde{\phi}_{21}(L,0)$ is obtained by integrating (D.10a):

$$\tilde{\phi}_{21}(L,0) = -\int_0^L \phi_{11}(v,0)\mathrm{d}v.$$

Finally (D.1) provide a way to compute the integrator gains $a_{ij}^{(0)}(x)$ for the general case of nonuniform flow.

D.1.0.2 Computation of $a_{ij}^{(k)}(x)$

Let us introduce the matrix $\Phi'(x,s) = \frac{\partial}{\partial s}\Phi(x,s)$, with $\Phi(x,s)$ the state-transition matrix given by (3.77).

Then, we have:

$$\frac{\partial \Phi'(x,s)}{\partial x} = \mathscr{A}(x,s)\Phi'(x,s) + \frac{\partial \mathscr{A}(x,s)}{\partial s}\Phi(x,s), \tag{D.12}$$

with

$$\frac{\partial \mathscr{A}(x,s)}{\partial s} = \begin{pmatrix} \frac{\alpha(x)-\beta(x)}{\alpha(x)\beta(x)} & -\frac{1}{\alpha(x)\beta(x)} \\ -1 & 0 \end{pmatrix}.$$

The solution of (D.12) is obtained as follows:

$$\Phi'(x,s) = \Phi(x,s)\Phi'(0,s) + \Phi(x,s)\int_0^x \Phi(v,s)^{-1}\frac{\partial \mathscr{A}(v,s)}{\partial s}\Phi(v,s)\mathrm{d}v.$$

By definition, we have

$$\begin{aligned} \Phi'(0,s) &\triangleq \lim_{|\delta s|\to 0} \frac{\Phi(0,s+\delta s) - \Phi(0,s)}{\delta s} \\ &= \frac{I-I}{\delta s} \\ &= 0. \end{aligned}$$

Therefore the value of $\Phi'(x,s)$ at $x=0$ is 0 and we finally get (D.5). This equation provides a way to compute $\Phi'(x,s)$ for any $s \in \mathbb{C}$.

Finally, in order to compute the coefficients $a_{ij}^{(k)}(x)$, we need to compute the $N_{ij}(x,p_k)$. For any $s \in \mathbb{C}$, the $N_{ij}(x,s)$ are given by:

$$N_{11}(x,s) = \frac{1}{T_0(x)}\Big(\phi_{12}(x,s)\phi_{21}(L,s) - \phi_{11}(x,s)\phi_{22}(L,s)\Big), \tag{D.13a}$$

$$N_{12}(x,s) = \frac{1}{T_0(x)}\phi_{11}(x,s), \tag{D.13b}$$

$$N_{21}(x,s) = \phi_{22}(x,s)\phi_{21}(L,s) - \phi_{21}(x,s)\phi_{22}(L,s), \tag{D.13c}$$

$$N_{22}(x,s) = \phi_{21}(x,s). \tag{D.13d}$$

The equations simplify since by definition of a pole, we have $\phi_{21}(L,p_k) = 0$, which yields (D.4).

D.1.0.3 Computation of $b_{ij}(x)$

The coefficients $b_{ij}(x)$ are defined by:

$$b_{ij}(x) = \frac{\partial}{\partial s}[sg_{ij}(x,s)]|_{s=0}. \tag{D.14}$$

Then, we have:

$$\begin{aligned} b_{ij}(x) &= \frac{\partial}{\partial s}[sg_{ij}(x,s)]|_{s=0} \\ &= \frac{\partial}{\partial s}\left[\frac{sN_{ij}(x,s)}{\phi_{21}(L,s)}\right]\Bigg|_{s=0} \\ &= \frac{\partial}{\partial s}\left[\frac{N_{ij}(x,s)}{\tilde{\phi}_{21}(L,s)}\right]\Bigg|_{s=0}, \end{aligned}$$

which yields (D.6).

The $N'_{ij}(x,0)$ can be computed using (D.13) and the solution $\Phi'(x,0)$ given by (D.5) for $s=0$. In that case, the expressions simplify, and lead to (D.7).

To compute $\tilde{\phi}'_{21}(L,0)$, we differentiate (D.11) with respect to s, which gives:

$$\frac{\partial}{\partial x}\tilde{\phi}'_{21}(x,s) = -\phi'_{11}(x,s),$$

with an initial condition given by $\tilde{\phi}'_{21}(0,s) = 0$.

Therefore, once $\phi'_{11}(x,0)$ is computed, $\tilde{\phi}'_{21}(L,0)$ is directly obtained by integration, as in (D.8).

Therefore, we have derived a numerical means to compute all the coefficients of the modal decomposition in the nonuniform case, which ends the proof. □

D.2 Modal Decomposition of the Delay-free Part

The modal decomposition of $\tilde{g}_{ij}(x,s)$ can also be obtained as

$$\tilde{g}_{ij}(x,s) = \tilde{b}_{ij}(x) + \frac{\tilde{a}_{ij}^{(0)}(x)}{s} + \sum_{k=-\infty,k\neq 0}^{\infty} \frac{\tilde{a}_{ij}^{(k)}(x)s}{p_k(s-p_k)},$$

where the coefficient $\tilde{a}_{ij}^{(k)}(x) = \lim_{s\to p_k}(s-p_k)\tilde{g}_{ij}(x,s)$ is the residue of transfer function $\tilde{g}_{ij}(x,s)$ at the pole p_k.

For the delay-free decomposition, we have: $\tilde{a}_{ij}^{(0)}(x) = a_{ij}^{(0)}(x)$, and for $k \neq 0$:

$$\tilde{a}_{11}^{(k)}(x) = a_{11}^{(k)}(x) e^{\int_0^x \frac{dv}{\alpha(v)} p_k}, \tag{D.15a}$$

$$\tilde{a}_{12}^{(k)}(x) = a_{12}^{(k)}(x) e^{\int_x^L \frac{dv}{\beta(v)} p_k}, \tag{D.15b}$$

$$\tilde{a}_{21}^{(k)}(x) = a_{21}^{(k)}(x) e^{\int_0^x \frac{dv}{\alpha(v)} p_k}, \tag{D.15c}$$

$$\tilde{a}_{22}^{(k)}(x) = a_{22}^{(k)}(x) e^{\int_x^L \frac{dv}{\beta(v)} p_k}, \tag{D.15d}$$

and the coefficients $\tilde{b}_{ij}(x)$ are given by:

$$\tilde{b}_{11}(x) = b_{11}(x) + a_{11}^{(0)}(x) \int_0^x \frac{dv}{\alpha(v)}, \tag{D.16a}$$

$$\tilde{b}_{12}(x) = b_{12}(x) + a_{12}^{(0)}(x) \int_x^L \frac{dv}{\beta(v)}, \tag{D.16b}$$

$$\tilde{b}_{21}(x) = b_{21}(x), \tag{D.16c}$$

$$\tilde{b}_{22}(x) = b_{22}(x). \tag{D.16d}$$

These equations enable us to directly compute the coefficients $\tilde{a}_{ij}^{(k)}(x)$ and $\tilde{b}_{ij}(x)$ once $a_{ij}^{(k)}(x)$ and $b_{ij}(x)$ are computed.

Appendix E
Boundary Control of a Hyperbolic System

E.1 A Frequency Domain Approach for Infinite Dimensional Systems

We introduce and summarize the properties of a class of linear time invariant systems allowing to handle the infinite dimensional systems considered in this book: the Callier–Desoer algebra $\hat{\mathscr{B}}$, introduced in [5] as a convenient algebra of irrational transfer functions.

E.1.1 A Set of Irrational Transfer Functions

Let us first consider $\mathscr{F}$, a LTI system defined by the following convolution:

$$y(t) = (f * u)(t) \triangleq \int_0^t f(t-\tau)u(\tau)\mathrm{d}\tau, \tag{E.1}$$

where $f(t)$ is by definition the impulse response of $\mathscr{F}$ and is such that $f(t) = 0$ when $t < 0$ and has the form:

$$f(t) = \sum_{i=0}^{\infty} f_i \delta(t - t_i) + f_a(t) \text{ if } t \geq 0, \tag{E.2}$$

where $\delta(.)$ denotes the unit delta distribution (a Dirac delta), f_i denotes a real coefficient belonging to $\mathbb{R}$, $0 \leq t_0 < t_1 < \cdots$ are constants, and $f_a(.)$ is a (Lebesgue) integrable function on any finite support. This class, which is enough for our purpose, does not coincide with the class of all the possible linear operators since there exists linear systems where f_a is not a (Lebesgue) integrable function but a distribution [7, Example 5.2.6].

The presence of a unit delta distribution in the impulse response is classical. Indeed, that is already the case for this non strictly proper LTI system:

$$y = \mathscr{F}(u) : \begin{cases} \dot{x}(t) = Ax(t) + bu(t) \\ y(t) = cx(t) + du(t) \end{cases},$$

which has the following impulse response:

$$f(t) = ce^{At}b + d\delta(t).$$

More generally, let us recall that the linear operator associated to a delayed impulse response, $f_i(t) = \delta(t - t_i)$ corresponds to a delay t_i. Indeed, the properties of the convolution product allows us to easily show that

$$y(t) = \mathscr{F}_i(u) = (f_i * u)(t) = u(t - t_i),$$

and thus the announced result. We then conclude that the first term of the right-hand side of (E.2) corresponds to the sum of a countable number of linear delayed systems.

Let us now introduce $\mathscr{A}$, the set of impulse responses $f(t)$ defined by (E.2) such that

$$\int_0^{\infty} |f_a(t)| \mathrm{d}t + \sum_{i=0}^{\infty} |f_i| < \infty.$$

This class is of interest since it defines the set of stable impulse responses. Indeed, we have the following important result:

Proposition E.1. *Consider the LTI operator $\mathscr{F}$ defined by (E.1). Then the two following statements are equivalent:*

(i) $f \in \mathscr{A}$,
(ii) *$\mathscr{F}$ is $\mathscr{L}_p$ gain stable for all $p \in [1, \infty]$, i.e., there exists $\gamma_p > 0$ such that*

$$\|y\|_p \leq \gamma_p \|u\|_p \textit{ for any } u \in \mathscr{L}_p$$

and $\gamma_p \leq \int_0^{\infty} |f_a(t)| dt + \sum_{i=0}^{\infty} |f_i|$.

The upper bound provided in the previous proposition for the $\mathscr{L}_2$ gain is known to be conservative. When $p = 2$, the $\mathscr{L}_2$ gain of $\mathscr{F}$ can be computed exactly:

$$\gamma_2 = \sup_{\omega \in \mathbb{R}} |\hat{F}(j\omega)|,$$

where $\hat{F}(s)$ is the Laplace transform of $f(t)$ (see the definition below). Let us recall that the $\mathscr{L}_2$ gain and the H_∞ norm of $\mathscr{F}$ are two equivalent notions.

In most cases, and especially in the control context, we need to consider possibly unstable open-loop systems, which means that we have to extend the set $\mathscr{A}$. For this purpose, we define the set of impulse responses $f(t)$ defined by (E.2) such that

$$\int_0^{\infty} |f_a(t)e^{-\sigma t}| \mathrm{d}t + \sum_{i=0}^{\infty} |f_i| e^{-\sigma t_i} < \infty,$$

with $\sigma \in \mathbb{R}$. This set denoted $\mathscr{A}(\sigma)$ can include unstable impulse responses when σ is positive. That leads us to define the set of LTI systems of interest:

$$\mathscr{A}(\infty) \triangleq \bigcup_{\sigma<\infty} \mathscr{A}(\sigma).$$

Let us now introduce the Laplace transform of elements of $\mathscr{A}(\infty)$. By definition, each impulse response in $\mathscr{A}(\sigma)$ has the following Laplace transform:

$$\hat{F}(s) \triangleq \int_0^{\infty} f(t)e^{-st}\mathrm{d}t = \sum_{i=0}^{\infty} f_i e^{st_i} + \hat{F}_a(s),$$

which exists and is well-defined on the closed half-plane $\overline{\mathbb{C}_\sigma^+} \triangleq \{s \in \mathbb{C} | \Re(s) \geq \sigma\}$. Let us recall that by contrast to a rational transfer function, an irrational transfer function does not necessarily possess a meromorphic extension on the whole complex plane and the transfer function is then only defined on $\overline{\mathbb{C}_\sigma^+}$. In the following, we then denote by $\hat{\mathscr{A}}(\sigma)$ the set of Laplace transforms associated to the impulse responses belonging to $\mathscr{A}(\sigma)$. By construction, transfer functions belonging to $\hat{\mathscr{A}}(\sigma)$ are bounded for $\Re(s) \geq \sigma$, analytic on the open half-plane $\mathbb{C}_\sigma^+ \triangleq \{s \in \mathbb{C} | \Re(s) > \sigma\}$, and continuous on $s = \sigma + j\omega$ for $\omega \in \mathbb{R}$ (see [8], Theorem A.7.47, p. 663). Let us recall that a complex function is analytic if it is locally given by a convergent power series. Cauchy's theorem allows us to prove that the class of analytic functions coincides with the class of holomorphic functions. Indeed, any holomorphic function that is a differential complex function has derivatives of every order and it coincides with its own Taylor series.

Obviously, the interest of such a set is strengthened by this result:

Lemma E.1. *If $f(t)$ and $g(t)$ belong to $\mathscr{A}(\sigma)$ then $(f * g)(t)$ belongs to $\mathscr{A}(\sigma)$.*

Actually $\mathscr{A}(\sigma)$ is a commutative Banach algebra, i.e., $(f * g)(t) = (g * f)(t)$, with unitary element δ, i.e., $(f * \delta)(t) = f(t)$, when it is equipped with the norm:

$$\|f\|_\sigma \triangleq \int_0^{\infty} |f_a(t)e^{-\sigma t}|\mathrm{d}t + \sum_{i=0}^{\infty} |f_i| e^{-\sigma t_i}.$$

The fact that $\mathscr{L}_2$ gain stable irrational transfers are only analytic on the open right complex-plane has important implications about the behavior of its poles in the neighborhood of the imaginary axis. Actually, the $\mathscr{L}_2$ gain stability does not generally guarantee that the poles of the irrational transfer function are uniformly stable, i.e., there exists $\varepsilon > 0$ such that the real part of any pole is less than $-\varepsilon$.

In consequence, even if Proposition E.1 has shown that LTI systems with an impulse responses in $\mathscr{A}(0)$ are also $\mathscr{L}_2$ gain stable, many researchers prefer to define stability through the use of a slightly stronger notion. In order to introduce this alternative definition, we first have to introduce the following set of LTI systems:

$$\mathscr{A}_-(0) \triangleq \bigcup_{\sigma_1<0} \mathscr{A}(\sigma_1) \text{ and } \hat{\mathscr{A}}_-(0) \triangleq \bigcup_{\sigma_1<0} \hat{\mathscr{A}}(\sigma_1).$$

Clearly, $f(t)$ belongs to $\mathscr{A}_-(0)$ if there exists $\varepsilon > 0$ such that

$$\int_0^\infty |f_a(t)|e^{\varepsilon t}\mathrm{d}t + \sum_{i=0}^\infty |f_i|e^{\varepsilon t_i} < \infty.$$

In accordance with the stability definition considered in [8] and [13] to cite a few, in the following, we use this definition of stability:

Definition E.1. Consider the LTI operator $\mathscr{F}$ defined by (E.1). Then $\mathscr{F}$ is said to be (strongly) stable if $f(t) \in \mathscr{A}_-(0)$.

Remark E.1. From a practical point of view, this definition is *almost* equivalent to require the $\mathscr{L}_2$ gain stability and the impact of this modification may be viewed as minor. Actually, this is not true from a theoretical point of view where this slight modification has important consequences of the system properties.

Let us first note that any rational transfer function in $\hat{\mathscr{A}}(0)$ also belongs to $\hat{\mathscr{A}}_-(0)$. Indeed, it is stable and necessarily possesses a finite number of stable poles which ensures that there necessarily exists $\varepsilon > 0$ such that all its poles have a real part less than $-\varepsilon$. It is then straightforward to conclude that this rational transfer function is analytic and bounded on any open domain contained in $\mathbb{C}^+_{-\varepsilon}$ and thus belongs to $\hat{\mathscr{A}}_-(0)$.

We will now show that any irrational transfer function in $\hat{\mathscr{A}}_-(0)$ has uniformly stable poles. It is sufficient to recall that it is analytic and bounded on $\mathbb{C}^+_{-\varepsilon}$ for a given $\varepsilon > 0$, and thus all its poles necessarily have a real part less than $-\varepsilon$.

We conclude this section by a brief discussion on MIMO LTI systems. Let $\mathscr{F}$ be an m inputs, l outputs system defined by the convolution:

$$y(t) = (F * u)(t) \triangleq \int_0^t F(t-\tau)u(\tau)\mathrm{d}\tau, \tag{E.3}$$

where $F(t)$ is by definition the impulse response matrix of $\mathscr{F}$ belonging to $\mathscr{M}\mathscr{A}(\sigma)$, which denotes the set of matrices with elements in $\mathscr{A}(\sigma)$. In the same vein, $\mathscr{M}\hat{\mathscr{A}}(\sigma)$ denotes the set of matrices with elements in $\hat{\mathscr{A}}(\sigma)$.

Finally $\mathscr{F}$ is $\mathscr{L}_p$ gain stable if and only if $F(t) \in \mathscr{M}\mathscr{A}$ (with $\mathscr{M}\mathscr{A} \triangleq \mathscr{M}\mathscr{A}(0)$) and $\mathscr{F}$ is said to be (strongly) stable if $F(t)$ belongs to $\mathscr{M}\mathscr{A}_-$ (with $\mathscr{M}\mathscr{A}_- \triangleq \mathscr{M}\mathscr{A}_-(0)$). Furthermore, if $F(t) \in \mathscr{M}\mathscr{A}_-$, then its $\mathscr{L}_2$ gain or equivalently its H_∞ norm is given by:

$$\gamma_2 = \sup_{\omega\in\mathbb{R}} \bar{\sigma}(\hat{F}(j\omega)),$$

where $\bar{\sigma}(F(j\omega)) \triangleq \sqrt{\lambda_{\max}(\hat{F}(j\omega)^*\hat{F}(j\omega))}$ is the maximal singular value of $\hat{F}(j\omega)$ and $\hat{F}(j\omega)^*$ corresponds to the complex conjugate of the transpose of the matrix $\hat{F}(j\omega)$.

E.1.2 From Nyquist Criteria to a Graphical Test

In order to derive a stability test for the closed-loop systems, we first restrict our attention to the SISO case where the feedback loop is unitary (see Fig. E.1). The closed-loop system is then defined as the interconnection between a linear time invariant systems, $\mathscr{G}$ belonging to $\mathscr{A}(\infty)$ and a unitary feedback.

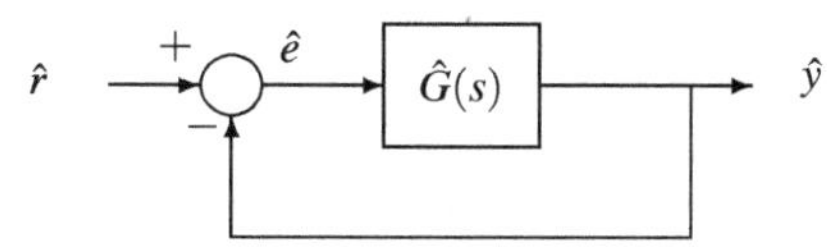

Fig. E.1 Closed-loop system with unitary feedback

The input r, the error e and the output y are functions from $\mathbb{R}^+$ into $\mathbb{R}$. On $\overline{\mathbb{C}_\sigma^+}$ where the Laplace transforms of $\mathscr{G}$ is defined, the transfer matrix of the feedback-loop is formally defined by the equations:

$$\begin{cases} \hat{y}(s) = \hat{G}(s)\hat{e}(s) \\ \hat{r}(s) = \hat{e}(s) + \hat{G}(s)\hat{e}(s) \end{cases}, \tag{E.4}$$

which leads us to deduce that the closed-loop system, $\mathscr{H}$, linking r to y, possesses the following Laplace transform:

$$\hat{H}(s) = \hat{G}(s)(1+\hat{G}(s))^{-1}.$$

An important issue in the context of control design is to ensure that the closed-loop system is stable. By itself the stability of the closed-loop is not necessarily a real issue since in many cases the controlled plant is already stable. Actually, feedback control is mainly used to reduce the effect of uncertainties acting on the system. From this point of view stability is thus a minimal requirement that has to be achieved in order for the feedback loop to ensure stronger requirements concerning the system performance.

When the open-loop system is assumed to be stable, we have this fundamental result (see [14, 21, 8]):

Theorem E.1. *Let* $\hat{G}(s)$ *be in* $\hat{\mathscr{A}}_-(0)$. *The closed-loop system defined by (E.4) is stable, i.e.,* $\hat{H}(s) \in \hat{\mathscr{A}}_-(0)$ *if and only if* $\inf_{\Re(s)\geq 0} |1+\hat{G}(s)| > 0$.

As noted in [21], the necessity of the previous condition is quite intuitive. Indeed, if $(1+\hat{G}(s))^{-1}$ belongs to $\hat{\mathscr{A}}_-(0)$, then $(1+\hat{G}(s))^{-1}$ has to be bounded over the closed-right half plane and that is equivalent to requiring that there exist $\varepsilon > 0$ such that $|1+\hat{G}(s)| \geq \varepsilon$ for any $s \in \overline{\mathbb{C}_0^+}$ that is $\inf_{\Re(s)\geq 0}|1+\hat{G}(s)| > 0$.

In the previous results, we assumed that the open-loop transfer function is stable, which could be restrictive. Actually, if the open-loop system is unstable but can be decomposed as the sum of a stable distributed part and a strictly proper and unstable rational transfer function, then the closed-loop stability property can be deduced from the open-loop one through this result:

Theorem E.2. *Let* $\hat{G}(s)$ *be in* $\mathscr{A}(\infty)$ *and such that* $\hat{G}(s) \triangleq \hat{G}_s(s) + \hat{R}_u(s)$ *where* $\hat{G}_s(s) \in \hat{\mathscr{A}}_-(0)$ *and* $\hat{R}_u(s)$ *is a strictly proper rational transfer function with only unstable poles. The closed-loop system defined by (E.4) is stable, i.e.,* $\hat{H}(s) \in \hat{\mathscr{A}}_-(0)$ *if and only if* $\inf\limits_{\Re(s)\geq 0} |1+\hat{G}(s)| > 0$.

With the restriction to the class of LTI systems satisfying the decomposition of the proposition, we now have a necessary and sufficient condition for proving the closed-loop stability. The main issue is now to find a way to test the proposition conditions easily. When the open-loop is a rational transfer function, it is possible to reduce the condition on the complex half-plane, i.e., $\inf_{\Re(s)\geq 0} |1+\hat{G}(s)| > 0$ to a simpler condition on the imaginary axis. This test, the so-called Nyquist criteria, is a consequence of the argument principle theorem and forms the basis of a well-known graphical test in the classical control approach. We now recall that such a test applied on rational transfer functions can be extended to our class of interest (Theorem A.1.14 in [8] provides a version of the Nyquist criteria that can be applied to a more general class of linear systems possessing a meromorphic transfer functions on $\overline{\mathbb{C}_0^+}$).

For the sake of simplicity, we assume in the sequel that the open-loop system has no pole on the imaginary axis (see Remark E.2 when this condition is not satisfied). We then introduce the phase of the return difference, denoted in the sequel by $\theta(j\omega)$ and defined through the implicit relation:

$$e^{j\theta(j\omega)}|1+\hat{G}(j\omega)| \triangleq 1+\hat{G}(j\omega),$$

and such that $\theta(j0) = 0$ if $1+\hat{G}(j0) > 0$ or π otherwise. The properties of $\hat{G}(s)$ on the imaginary axis allow to show that $\theta(j\omega)$ is a well-defined and continuous function of $\omega \in \mathbb{R}$. Furthermore, this phase function corresponds to the argument of $1+\hat{G}(s)$ on the imaginary axis:

$$\theta(j\omega) = \arg(1+\hat{G}(j\omega)) \triangleq \Im \log(1+\hat{G}(j\omega)),$$

where some care has to be taken when the logarithm of a complex function is defined (see [9], p. 93 for details). Following this definition, we are then able to recall this extended version of the Nyquist theorem extracted from [4], which paves the way of the definition of a graphical test for irrational transfer functions.

Theorem E.3 (Extended Nyquist Criteria). *Let* $\hat{G}(s)$ *be in* $\hat{A}(\infty)$ *such that* $\hat{G}(s) \triangleq \hat{G}_s(s) + \hat{R}_u(s)$ *where* $\hat{G}_s(s) \in \hat{\mathscr{A}}_-(0)$ *and* $\hat{R}_u(s)$ *is a strictly proper rational transfer function with unstable poles (not on the imaginary axis). Then* $\inf\limits_{\Re(s)\geq 0} |1+\hat{G}(s)| > 0$ *if and only if* *(i)* $\lim\limits_{\Re(s)\to\infty} [1+\hat{G}(s)] \neq 0,$

(ii) $\inf\limits_{\omega\in\mathbb{R}} |1+\hat{G}(j\omega)| > 0,$

(iii) $\omega \mapsto \theta(j\omega)$ *is bounded on* $\mathbb{R}^+$,

(iv) $\lim\limits_{\Omega\to\infty} \frac{1}{\Omega}\int_{\omega_0}^{\omega_0+\Omega} \theta(j\omega)d\omega - \lim\limits_{\Omega\to\infty} \frac{1}{\Omega}\int_{-\omega_0-\Omega}^{-\omega_0} \theta(j\omega)d\omega = 2m_p\pi,$

where ω_0 *is any nonnegative number and* m_p *denotes the numbers of poles of* $\hat{G}(s)$ *in the open right-half plane, counting multiplicity.*

Remark E.2. As in the case of rational transfer functions, when $\hat{G}(s)$ has pole(s) on the imaginary axis, the $j\omega$-axis has to be indented in order to derive the test (see [21], p. 315).

In the case of rational transfer functions, the previous conditions can easily be handled graphically. To illustrate this point, let us assume that the open-loop transfer function is given by:

$$\hat{G}(s) = \frac{1000}{(s-0.5)(s+10)(s+100)}.$$

The open-loop is unstable since it has a pole at $s = 0.5$. Let us prove the closed-loop stability by checking the conditions of the extended Nyquist criteria.

Since $\lim_{\Re(s)\to\infty} \hat{G}(s) = 0$, $\lim_{\Re(s)\to\infty}[1+\hat{G}(s)] = 1$ and (i) holds true. The second condition can be tested graphically by plotting the Nyquist plot of $G(s)$, i.e., the graph in the complex plane of $\hat{G}(j\omega)$ for $\omega \in \mathbb{R}$. For the chosen rational transfer function, the Nyquist plot is depicted in Fig. E.2. Since clearly the Nyquist plot does not pass through -1, condition (ii) is then satisfied. Condition (iii) is always satisfied for rational transfer functions. It remains to test condition (iv). Due to the properties of the phase in this case (see [4] for details), this condition can be checked through the computation of the number of encirclement of -1 by the Nyquist plot. In our case, a simple examination of the Nyquist plot in Fig. E.2 allows us to conclude that condition (iv) also holds true and thus that the closed-loop system is stable.

One of the main difficulty with using graphical test in the case of an irrational open-loop is related to conditions (ii) and (iv) because of the possible complex behavior of the Nyquist plot at infinity. In order to understand this point, we consider a simple example concerning condition (ii). Indeed, in the previous example, and more generally for the rational transfer function case, this condition is verified if the Nyquist plot does not pass through -1. By contrast, for the irrational case, such a simplified test does not necessarily ensure that condition (ii) is fulfilled. To illustrate this simple claim, we consider that the open-loop transfer function is given by:

$$\hat{G}(s) = \frac{e^{-s}(s+1)}{s+10^{-4}}.$$

The Nyquist plot of $\hat{G}(s)$ is depicted in Fig. E.3. By Lemma E.1, it is routine to show that $\hat{G}(s)$ belongs to $\hat{\mathscr{A}}_-(0)$. Indeed, the product of two Laplace transforms

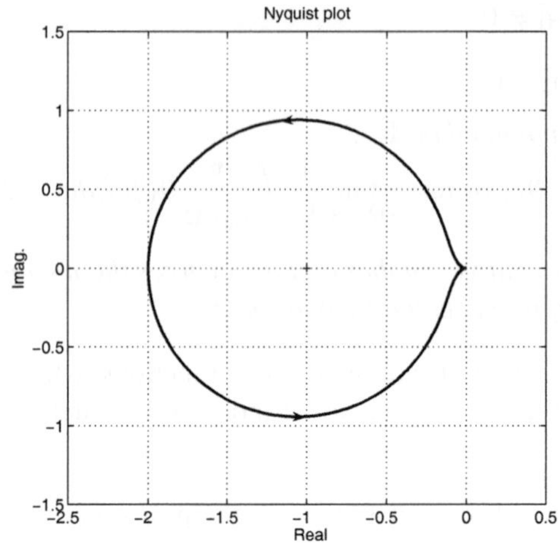

Fig. E.2 Nyquist plot of $\hat{G}(s) = \dfrac{1000}{(s-0.5)(s+10)(s+100)}$

is equivalent to the convolution of the two associated impulse responses, and the result is obtained since a delay and a rational and stable transfer function both have an impulse response belonging to $\mathscr{A}_-(0)$.

By definition, the module of $\hat{G}(j\omega)$ is given by:

$$|\hat{G}(j\omega)| = \sqrt{\frac{1+\frac{1}{\omega^2}}{1+\frac{10^{-8}}{\omega^2}}},$$

and it is always greater than 1 for any $\omega \in \mathbb{R}$. Consequently, since

$$|\hat{G}(j\omega)+1| \geq |\hat{G}(j\omega)| - 1,$$

we then deduce that $|\hat{G}(j\omega)+1| > 0$ for any $\omega \in \mathbb{R}$, and thus that the Nyquist plot of $\hat{G}(j\omega)$ does not pass through -1. Nevertheless, condition (ii) is not satisfied. To prove this claim, we have to show that for any $\varepsilon > 0$, there always exists $\omega \in \mathbb{R}$ such that $|\hat{G}(j\omega)+1| \leq \varepsilon$. Let us then first rewrite the rational part of $\hat{G}(s)$ as:

$$\frac{j\omega+1}{j\omega+10^{-4}} = m_R(j\omega)e^{j\theta_R(j\omega)}.$$

Let us now define an infinite and increasing series $\{\omega_i\}$ such that $\theta_R(j\omega_i)+\omega_i \equiv -\pi$ (mod 2π) and with $\lim_{i\to\infty}\omega_i = \infty$. In this way, we have

$$\hat{G}(j\omega_i) = m_R(j\omega_i)e^{j(\theta_R(j\omega_i)+\omega_i)} = -m_R(j\omega_i) = -\sqrt{\frac{1+\frac{1}{\omega_i^2}}{1+\frac{10^{-8}}{\omega_i^2}}},$$

and then the announced result is obtained since for any $\varepsilon > 0$, there always exists ω_i such that $|\hat{G}(j\omega_i)+1| \leq \varepsilon$.

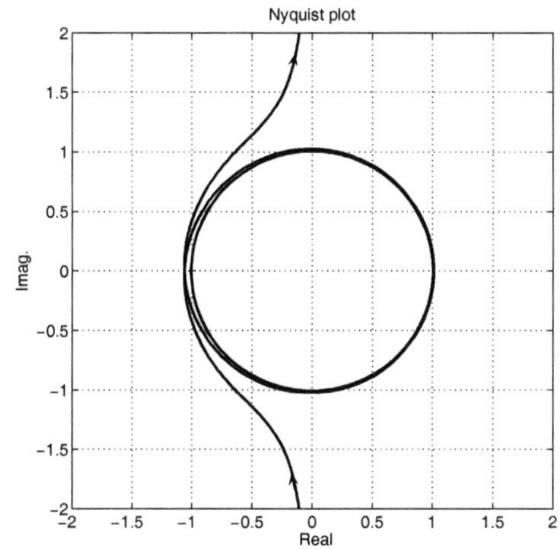

Fig. E.3 Nyquist plot of $\hat{G}(s) = \frac{e^{-s}(s+1)}{s+10^{-4}}$

Such an example perfectly explains why a graphical test cannot be considered in such a case, since it could lead to a wrong conclusion. It moreover explains why a small gain type condition on $\hat{G}(s)$ or a stronger condition ensuring that the module of $\hat{G}(j\omega)$ goes to zero at infinity is very useful, since it allows us the use of the graphical test. Indeed, in these both cases, there exist $\omega_0 \geq 0$ and $\eta < 1$ such that $|\hat{G}(j\omega)| \leq \eta$ for any $\omega \geq \omega_0$ and condition (ii) holds true if the plot of $\hat{G}(j\omega)$ does not pass through -1 on $[0, \omega_0)$.

In the same spirit, the phase behavior of the return difference at infinity can also be complicated and it may be impossible to check graphically if the number of encirclements of -1 is in accordance with the number of unstable poles of the open-loop transfer (counting multiplicity). In this context, condition (iv) provides an alternative to this difficulty since it generally possesses a very interesting feature at infinity (see [4] for details and also [23, 9]).

To summarize, in contrast with the study of the stability of the closed-loop system associated to a rational open-loop transfer function, the use of a simple graphical test for an irrational open-loop transfer function may be delicate, even worse, impossible. Nevertheless and generally, the strict properness of the controller enables us to avoid such a complication and allows us to easily verify the Nyquist conditions through a graphical test.

After the derivation of stability conditions for the SISO case, we now consider the MIMO case depicted in Fig. E.4, corresponding to a closed-loop system defined as the interconnection between two linear time invariant systems, $\mathscr{G}$ and $\mathscr{F}$ belonging to $\mathscr{M}\mathscr{A}(\infty)$ and $\mathscr{M}\mathscr{A}_-(0)$, respectively. The input r, the error e, and the output y are functions from $\mathbb{R}^+$ into $\mathbb{R}^n$. On the common half-plane where the Laplace transforms of $\mathscr{F}$ and $\mathscr{G}$ are defined, the transfer matrix of the feedback-loop is formally defined by these equations:

$$\begin{cases} \hat{y}(s) = \hat{G}(s)\hat{e}(s) \\ \hat{r}(s) = \hat{e}(s) + \hat{F}(s)\hat{G}(s)\hat{e}(s) \end{cases}, \tag{E.5}$$

which leads to defined the closed-loop operator, $\mathscr{H}$, linking r to y, and which possesses the following Laplace transform:

$$\hat{H}(s) \triangleq \hat{G}(s)[I + \hat{F}(s)\hat{G}(s)]^{-1},$$

where $I \in \mathbb{R}^{n\times n}$ is the identity matrix.

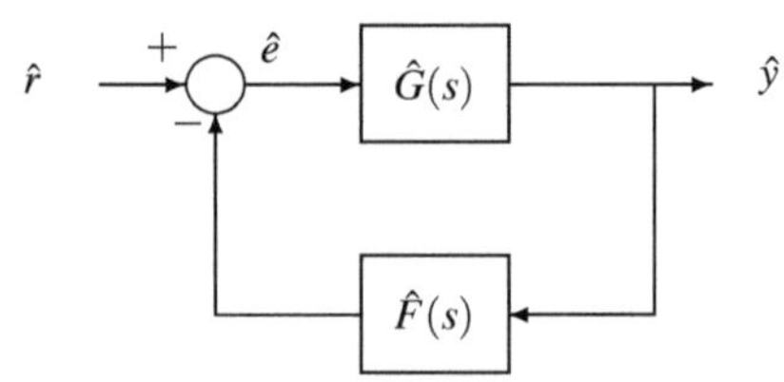

Fig. E.4 Closed-loop system as the feedback interconnection of two MIMO systems

There also exists a simple stability condition when the feedback loop is assumed to be unitary and $\mathscr{G}$ to be stable. Indeed, we have this result:

Theorem E.4. *Let $\hat{F}(s)$ be the identity matrix and $\hat{G}(s)$ be in $\mathscr{M}\hat{\mathscr{A}}_-(0)$. The closed-loop system defined by (E.5) is stable, i.e., $\hat{H}(s) \in \mathscr{M}\hat{\mathscr{A}}_-(0)$ if and only if* $\inf_{\Re(s)\geq 0} |\det[I + \hat{G}(s)]| > 0$.

When the open-loop system is unstable, and by contrast to the SISO case, the previous condition has to be modified in order to derive necessary and sufficient conditions for the closed-loop stability. The main difficulty is not related to the irrational feature of the open-loop system, but only to the multivariable one. Let us assume that $\hat{F}(s) = I$ and $\hat{G}(s)$ is a given matrix of rational transfer functions. Then it is possible to prove (see [9] p. 60 for details) that if p_c is a zero of $\det[I + \hat{G}(s)]$ then it is a pole of $\hat{H}(s)$, but by contrast to the SISO case, the converse is not necessary true. Actually, when p_c is a pole of $\hat{H}(s)$ it is either a zero of $\det[I+\hat{G}(s)]$ or a pole of $\hat{G}(s)$. This problem has been resolved through the introduction of suitable conditions depending on the coprime factorization of open-loop systems. Actually, we have (from [3], see also [9]):

Theorem E.5. *Let $\hat{F}(s) \in \mathscr{M}\hat{\mathscr{A}}_-(0)$ and let $G(s) \in \mathscr{M}\hat{\mathscr{A}}(\infty)$ be of the form:*

$$\hat{G}(s) = \hat{G}_s(s) + \hat{R}_u(s),$$

where $\hat{G}_s(s) \in \mathscr{M}\hat{\mathscr{A}}_-(0)$ and $\hat{R}_u(s)$ is a strictly proper rational transfer matrix with m_p poles p_i such that $\Re(p_i) \geq 0$ for $i \in \{1, \ldots, m_p\}$. Let $N(s)$ and $D(s)$ two polynomial matrices such that $\hat{R}_u(s) \stackrel{\Delta}{=} N(s)D(s)^{-1}$ for all $s \in \mathbb{C}$ and such that $N(s)$ and $D(s)$ are right coprime, i.e., there exist two polynomial matrices $P(s)$ and $Q(s)$ such that $P(s)N(s) + Q(s)D(s) = I$ for all $s \in \mathbb{C}$. Then the closed-loop system is stable, i.e., $\hat{H}(s) \in \mathscr{M}\hat{\mathscr{A}}_-(0)$, if and only if

(i) $\inf_{\Re(s)\geq 0} |\det[I + \hat{F}(s)\hat{G}(s)]| > 0$

(ii) $\det[D(p_i) + \hat{F}(p_i)N(p_i) + \hat{F}(p_i)\hat{G}_s(p_i)D(p_i)] \neq 0$ *for* $i \in \{1, \cdots, m_p\}$

Since the right coprime factor always exists for any strictly proper matrix of rational transfer functions, the previous theorem provides necessary and sufficient conditions for the closed-loop stability of any open-loop system that can be written as the sum of a stable part belonging to $\mathscr{M}\hat{\mathscr{A}}_-(0)$ and a strictly proper unstable matrix of rational transfers.

Remarks.

1. Condition (ii) can be simplified by only considering the condition locally for each pole (see [9], p. 91 for details).
2. The graphical test provided in the SISO case can be also used to test condition (i) (see [9], p. 92 for details).

Finally, it is interesting to apply the previous result to the SISO case, since it allows us to obtain stability criteria for feedback loops which are not unitary. Actually, when $n = 1$, the condition (ii) of the previous theorem reduces to:

(ii') $\hat{F}(p_i) \neq 0$ for $i \in \{1, \cdots, m_p\}$,

and corresponds to a condition preventing an instability due to the cancelation of an unstable pole of $\hat{G}(s)$ by an unstable zero of $\hat{F}(s)$.

E.1.3 The Callier–Desoer Algebra

The initial motivation for introducing this specific class of irrational transfer functions can be found in the so-called algebraic control theory (see [22, 20, 5, 6, 8]). We prefer hereafter to follow the line of [13], which allows us to explain why the

Callier–Desoer class corresponds to the largest subset of $\mathscr{A}(\infty)$ that is reasonable to consider for practical applications.

Indeed, in most practical applications, the open-loop system is strictly proper. Let us recall that when there exists a strictly proper element in a feedback loop, then the open-loop system is also necessarily strictly proper. The strict properness of the open-loop system can then be implied by the strict properness of the initial plant, of its associated actuators or sensors, or by the use of strict proper controllers. Furthermore, the use of a digital implementation of the control law generally also leads to the definition of a strict proper open-loop. Finally, the most important argument to justify our interest in this restricted class of open-loop systems can be found in the robustness issue which will be discussed later.

The main purpose of this section is to show that the Callier–Desoer set of irrational transfer functions corresponds to the largest set of controlled plants that can be stabilized in spite of a strictly proper open-loop.

Before stating the main result, let us first recall that for a rational transfer function, its strict properness is ensured if the number of its zeros is strictly lower than the number of its poles. For an irrational transfer function, the definition is more subtle. The set $\mathscr{S}\mathscr{P}$ of strictly proper LTI systems belonging to $\mathscr{A}(\infty)$ is defined by:

$$\mathscr{S}\mathscr{P} = \{F(s) \in \hat{\mathscr{A}}(\infty) | \lim_{\substack{|s| \to \infty \\ \text{with } \Re(s) > \sigma}} |f(s)| = 0\}.$$

The results extracted from [13] summarize the main features of the strictly proper class:

Proposition E.2. *Let $f(t), g(t)$ be in $\mathscr{A}(\infty)$. Then*

(i) $f(t)(\triangleq \sum_{i=0}^{\infty} f_i \delta(t - t_i) + f_a(t))$ *belongs to* $\mathscr{S}\mathscr{P}$ *if and only if* $f(t) = f_a(t)$
(ii) *if* $f(t)$ *belongs to* $\mathscr{S}\mathscr{P}$ *then* $(f * g)(t)$ *belongs to* $\mathscr{S}\mathscr{P}$
(iii) *if* $f(t) \in \mathscr{A}_-(0) \cap \mathscr{S}\mathscr{P}$ *then* $\lim_{\omega \to \infty} |f(j\omega)| = 0$

Property (i) states that the output of the system does not depend directly on the input value or of any of its delayed value. Property (iii) implies that for any stable system, strict properness implies a finite bandwidth property, i.e., there exists $w_0 \in \mathbb{R}^+$ such that $|\hat{F}(j\omega)| < 1$ for any $\omega \in [\omega_0, \infty)$.

The fact that an open-loop system is strictly proper and that the associated closed-loop system is stable has a very interesting consequence on the form of the possible controlled plant. Indeed, we have:

Proposition E.3. *Let us consider the closed-loop system depicted in Fig. E.5 and that $\hat{G}, \hat{K} \in \hat{\mathscr{A}}(\infty)$. Then if the open-loop transfer function, i.e., $\hat{L}(s) \triangleq \hat{G}(s)\hat{K}(s)$ belongs to $\mathscr{S}\mathscr{P}$ and the closed-loop system is stable i.e. $\mathscr{H} \in \mathscr{A}_-(0)$, then*

(i) $G(s) \triangleq \hat{N}(s)\hat{M}(s)^{-1}$ *where $\hat{N}(s)$ and $\hat{M}(s)$ belong to $\hat{\mathscr{A}}_-(0)$ and $\hat{M}(s)$ is bounded away from zero at infinity, i.e., there exists $r > 0$ such that*

$$\inf_{\Re(s) \geq 0, |s| \geq r} |\hat{M}(s)| > 0$$

(ii) $\hat{G}(s)$ *has a finite number of unstable poles*

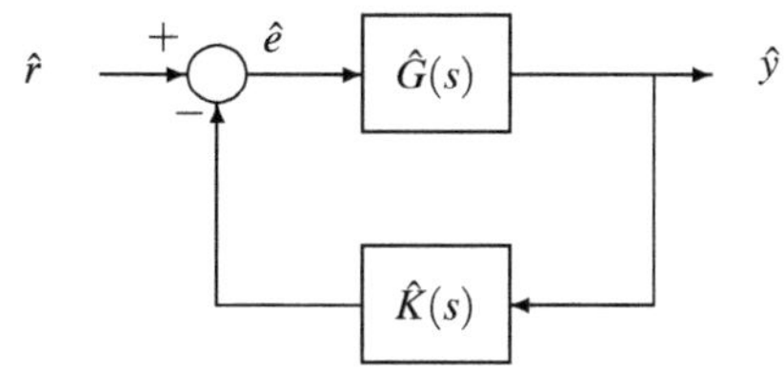

Fig. E.5 Closed-loop system with controller K

Proof. **Proof of** (i). Let us recall that the closed-loop transfer function is given by $\hat{H}(s) = \hat{G}(s)(I+\hat{G}(s)\hat{K}(s))^{-1}$ and then $\hat{G}(s) = \hat{H}(s)(I+\hat{G}(s)\hat{K}(s))$. We can then set $\hat{N}(s) \triangleq \hat{H}(s)$, which belongs to $\hat{\mathscr{A}}_{-}(0)$ since the closed-loop system is assumed to be stable; and then take $\hat{M}(s) \triangleq (I+\hat{G}(s)\hat{K}(s))^{-1}$. $\hat{M}(s)$ also belongs to $\hat{\mathscr{A}}_{-}(0)$ since the stability of the closed-loop system implies that $(I+\hat{G}(s)\hat{K}(s))^{-1} \in \hat{\mathscr{A}}_{-}(0)$. It remains to show that $\hat{M}(s)$ is bounded away from zero at infinity. For this purpose, we recall that[1]

$$\begin{aligned}(I+\hat{G}(s)\hat{K}(s))^{-1} + \hat{G}(s)\hat{K}(s)(I+\hat{G}(s)\hat{K}(s))^{-1} &= (I+\hat{G}(s)\hat{K}(s))^{-1}(I+\hat{G}(s)\hat{K}(s)) \\ &= 1\end{aligned}$$

and then

$$\hat{M}(s) = 1 - \hat{G}(s)\hat{K}(s)(I+\hat{G}(s)\hat{K}(s))^{-1}.$$

$\hat{G}(s)\hat{K}(s)$ belongs to $\mathscr{S}\mathscr{P}$ and we then have the announced result since

$$\lim_{|s|\to\infty} |1 - \hat{G}(s)\hat{K}(s)(I+\hat{G}(s)\hat{K}(s))^{-1}| = 1.$$

Proof of (ii). It is a direct consequence of point (i) and the properties of zeros of analytic transfer functions. Indeed, any analytic transfer function possesses isolated zeros of finite multiplicity on its domain of analyticity, and the number of these zeros in any compact domain is finite (counting their multiplicity) [8, Theorem A.1.14 p. 566]. Clearly, the boundedness away from zero at infinity ensures that if $\hat{M}(s)$ has zeros, they necessary belong to a bounded domain, which is compact since any bounded set is compact in a finite dimensional space. We thus have proved that $\hat{M}(s)$ has a finite number of zeros on $\overline{\mathbb{C}_0^+}$ since it is analytic on $\mathbb{C}_{-\varepsilon}^+$ with $\varepsilon > 0$. Consequently, $\hat{G}(s)$ can only have a finite number of unstable poles on $\overline{\mathbb{C}_0^+}$. Indeed,

[1] $\hat{S}(s) + \hat{T}(s) = I$ where $\hat{S}(s)$ is the so-called sensitivity transfer function and $\hat{T}(s)$ is the complementary sensitivity transfer function of the closed-loop system.

$\hat{N}(s)$ has no unstable pole, and the set of unstable poles of $\hat{G}(s)$ is a subset of the unstable zeros of $\hat{M}(s)$. □

The class of controlled plant identified in the previous proposition forms the so-called Callier–Desoer algebra. Indeed, we have

Definition E.2. The Callier–Desoer algebra $\hat{\mathscr{B}}(0)$ consists of all transfer functions $\hat{F}(s) \triangleq \hat{N}(s)\hat{M}(s)^{-1}$, where $\hat{N}(s), \hat{M}(s) \in \hat{\mathscr{A}}_{-}(0)$ and in addition $\hat{M}(s)$ is bounded away from zero at infinity.

By construction, any transfer function belonging to $\hat{\mathscr{B}}(0)$ possesses a finite number of isolated poles of finite multiplicity on the closed half-plane $\overline{\mathbb{C}_0^+}$. Actually, this way of parameterizing the irrational transfer functions of interest is very efficient, since we have this alternative definition of the Callier–Desoer class (see Theorem 7.1.16 p. 347 in [8] for details and proof):

Proposition E.4. *Consider the LTI operator $\mathscr{F}$ defined by (E.1). These two statements are equivalent:*

(i) $\hat{F}(s)$ belongs to $\hat{\mathscr{B}}(0)$
(ii) There exist $\hat{F}_s(s) \in \hat{\mathscr{A}}_{-}(0)$ and $\hat{R}_u(s)$, a strictly proper rational transfer function with all its poles in $\overline{\mathbb{C}_0^+}$ such that $\hat{F}(s) = \hat{F}_s(s) + \hat{R}_u(s)$

This proposition shows that the Callier–Desoer class of transfer functions, $\hat{\mathscr{B}}(0)$, corresponds exactly to the class necessary in Theorem E.2 to derive necessary and sufficient conditions for the closed-loop stability.

Another strong interest of the Callier–Desoer class is related to the fact that it generalizes the algebra of proper rational transfer functions, which forms the basis of the so-called algebraic control theory of control ([20]). In order to derive a very important result in this context, it is first necessary to introduce a way to obtain a "minimal" realization of the transfer functions belonging to $\hat{\mathscr{B}}(0)$ as a quotient of two stable transfer functions that have common unstable zeros. Indeed, in Proposition E.4, there may exist cancelations between the unstable zeros of $\hat{N}(s)$ and that of $\hat{M}(s)$, which leads to have representation of $\hat{G}(s)$ as a nonminimal quotient of two stable transfer functions, where all the unstable zeros of $\hat{D}(s)$ do not necessarily correspond to unstable poles of the controlled plant.

Lemma E.2. *For any $\hat{F}(s)$ in $\hat{\mathscr{B}}(0)$, there always exist $\hat{N}(s), \hat{M}(s), \hat{X}(s)$ and $\hat{Y}(s)$ belonging to $\hat{\mathscr{A}}_{-}(0)$ such that*

$$\hat{F}(s) \triangleq \hat{N}(s)\hat{M}(s)^{-1},$$

where the Bezout equation holds:

$$\hat{X}(s)\hat{N}(s) - \hat{Y}(s)\hat{M}(s) = 1 \textit{ for } s \in \overline{\mathbb{C}_0^+}, \tag{E.6}$$

and such that:

$$\inf_{s \in \overline{\mathbb{C}_0^+}} (\det(|\hat{N}(s)| + |\hat{M}(s)|) > 0. \tag{E.7}$$

The pair $(\hat{N}(s), \hat{M}(s))$ corresponds to the so-called coprime factorization of $\hat{F}(s)$ over $\hat{\mathscr{A}}_-(0)$. Beyond the existence of a solution to the Bezout equation, the condition (E.7) guarantees that $\hat{N}(s)$ $\hat{M}(s)$ have no common zero on the closed right plane and thus ensure the "minimality" of the quotient. We then have certainty that the unstable zeros of $\hat{M}(s)$ define the unstable poles of $\hat{F}(s)$.

After this short introduction to coprime factorization, the next result shows that it allows us to parameterize all the controllers $K(s) \in \hat{\mathscr{B}}(0)$ ensuring that the closed-loop system depicted in Fig. E.5 is stable. Actually, we have:

Theorem E.6 (Youla parameterization). *Let $\hat{G}(s)$ be in $\hat{\mathscr{B}}(0)$, $(\hat{N}(s), \hat{M}(s))$ its coprime factorization, and $\hat{X}(s)$ and $\hat{Y}(s)$ in $\hat{\mathscr{A}}_-(0)$ such that the Bezout equation (E.6) holds. Then the closed-loop system depicted in Fig. E.5 belongs to $\hat{\mathscr{A}}_-(0)$ if and only if the controller $\hat{K}(s) \in \hat{\mathscr{B}}(0)$ is such that:*

$$\hat{K}(s) = (\hat{Y}(s) + \hat{M}(s)\hat{Q}(s))(\hat{X}(s) + \hat{N}(s)\hat{Q}(s))^{-1},$$

where $\hat{Q}(s)$ belongs to $\hat{\mathscr{A}}_-(0)$ and it is such that $\hat{X}(s) + \hat{N}(s)\hat{Q}(s)$ is bounded away from zero at the infinity.

E.1.3.1 The Linearized Saint-Venant Transfer Matrix Belonging to Callier–Desoer Algebra

We now consider the Saint-Venant transfer matrix in the uniform flow case, and show that it belongs to the Callier–Desoer algebra. Let us recall that the transfer matrix is given by $P(s)$, with:

$$\begin{pmatrix} \hat{y}(0,s) \\ \hat{y}(L,s) \end{pmatrix} = P(s) \begin{pmatrix} \hat{q}(0,s) \\ \hat{q}(L,s) \end{pmatrix} \tag{E.8}$$

and the elements $p_{ij}(s)$ of $P(s)$ are given by (3.50).

Using the above definitions we state the following proposition.

Proposition E.5. *Each element $p_{ij}(s)$ of the transfer matrix $P(s)$ belongs to the Callier–Desoer algebra $\hat{\mathscr{B}}(0)$ if and only if $\gamma > 0$ or $\delta > 0$.*

Proof. We have shown in Chap. 3, Sect. 3.3.2.3 that each element of $P(s)$ can be decomposed as an infinite sum (3.56). Then, $p_{ij}(s)$ is the sum of an unstable finite dimensional part and a stable infinite dimensional part belonging to $\hat{\mathscr{A}}_-(\sigma)$ with $\sigma > -\frac{2\alpha\beta}{(\alpha+\beta)^2}\left[\delta + \left(\frac{1}{\beta} - \frac{1}{\alpha}\right)\frac{\gamma}{2}\right]$.

Then by PropositionE.4 $p_{ij}(s)$ belongs to $\hat{\mathscr{B}}(0)$. Finally, $P(s) \in \mathscr{M}\hat{\mathscr{B}}(0)$, which is the multivariable extension of $\hat{\mathscr{B}}(\sigma)$. □

Remark E.3 (Horizontal frictionless channels). If $\gamma = \delta = 0$, we have seen that the open-loop poles of the system are located on the imaginary axis. Therefore the system has an infinite number of marginally stable poles, and does not belong to $\hat{\mathscr{B}}(0)$.

E.1.4 Robustness Issues for Infinite Dimensional Systems

The issue of robustness is central in any design method. Indeed, the model used to design the controller is always an approximation of the true system. The task of control engineers is to ensure that the closed-loop system is stable when the controller is connected with the real system, and even more so that the closed-loop system achieves a suitable level of performance. Clearly the fulfillment of this task can only be tested experimentally. Nevertheless, this is a real issue for the control engineer, who has to develop methods allowing the design of controllers possessing a priori well-defined properties, so called robustness.

The first task is to represent the discrepancy between the true system and the model used for controller design. There are two main kinds of discrepancies:

- The first kind is intrinsic to any modeling process: there always exists some discrepancy between the true system and its mathematical representation. As an example, parametric uncertainties affect some parameters of a system, whose values are not perfectly known. Other uncertainties called "dynamical uncertainties" are often introduced during the modeling process. Indeed, in most applications, the control designer has chosen a priori a time scale of interest to control the considered system. This typically leads to neglect some rapid dynamics of the system (with respect to the chosen time scale). Let us note that even if these two situations are the most common, they do not cover all situations leading to model uncertainties.
- A second kind of uncertainties is classically introduced and it is due to the simplifications/reductions of the initial mathematical model to provide the model used for designing the controller.

In order to handle the problem in the mathematical field, it is then necessary to describe in some way the discrepancy between the model used for the design and the true system. This question has been deeply investigated in the literature and it is clearly beyond the scope of this section to present all the aspects related to robustness analysis. We will only provide in the sequel some elements explaining how the classical approach used in the context of rational transfer functions can be used for infinite dimensional systems.

For the presentation of the classical tools for robustness analysis, we refer to existing textbooks such as the classical book [19], providing a continuum view between the classical concepts attached the frequency approach to the more advanced ones attached to the so-called H_∞ control [16], including advanced aspects associated to the control of infinite dimensional/delayed systems, and [11] presenting a good overview and practical applications of the μ analysis, which is probably the most achieved tool developed for the performance and robustness analysis of LTI systems. Finally, most of the theoretical results and associated proofs can be found in [27] and [10], among others.

As we have already pointed out, a central issue in the robustness analysis is to link the nominal model to the true plant. In this context, a key concept is the notion of the set of uncertain models. Let us now explain this central idea by considering the so-called *additive model uncertainty*. We assume in the sequel that the nominal model is $\hat{G}(s)$ and we associate to the nominal model the following set of uncertain LTI systems:

$$\hat{\mathscr{G}}(\beta) \stackrel{\Delta}{=} \{\tilde{G}(s) \stackrel{\Delta}{=} \hat{G}(s) + \hat{\Delta}(s) | \hat{\Delta}(s) \in \hat{\mathscr{A}}_{-}(0) \text{ s.t. } \|\hat{\Delta}\|_{\infty} < \beta\},$$

where $\|\hat{\Delta}\|_{\infty}$ corresponds to the H_{∞} norm of $\hat{\Delta}(s)$. $\hat{\mathscr{G}}(\beta)$ then defines an "open ball of LTI systems" in $\mathscr{B}(0)$ with a radius $\beta > 0$ and where the center is the nominal model $\hat{G}(s)$.

In this context, the main feature of the robustness approach is not to prove that the closed-loop system remains stable for a given uncertain system belonging to $\hat{\mathscr{G}}(\beta)$ but to compute the (maximal) size of the open ball of uncertain systems in the neighborhood of the nominal model for which all the perturbed closed-loop systems remain stable.

In order to consider this problem, we first note that after straightforward block manipulations, it is possible to transform the perturbed closed-loop system associated to any uncertain plant belonging to $\hat{\mathscr{G}}(\beta)$ and the controller $\hat{K}(s)$ as the interconnection between $\hat{\Delta}(s)$ and $\hat{M}(s)$ where $\hat{M}(s) \stackrel{\Delta}{=} \hat{K}(s)(1 + \hat{G}(s)\hat{K}(s))^{-1}$.

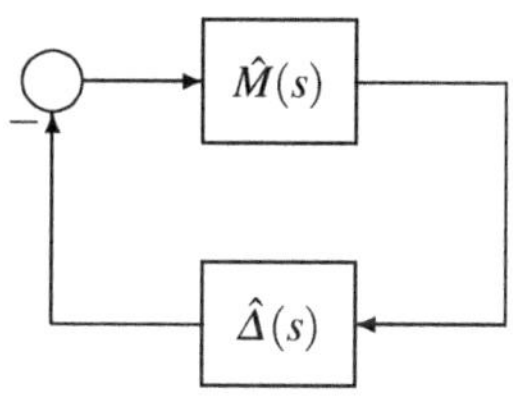

Fig. E.6 Interconnection with model uncertainty

By construction, the nominal closed-loop system is stable and $\hat{M}(s)$ belongs to $\hat{\mathscr{A}}_{-}(0)$, and then we look for the (maximal) value of β such that the closed-loop system depicted in Fig. E.6 remains stable for any $\hat{\Delta}(s) \in \hat{\mathscr{A}}_{-}(0)$ such that $\|\hat{\Delta}\|_{\infty} < \beta$. The following result provides a first estimate of β:

Proposition E.6. *Let $\hat{G}, \hat{K}$ be in $\hat{\mathscr{B}}(0)$ and let us assume that $\hat{M}(s) \stackrel{\Delta}{=} \hat{K}(s)(1 + \hat{G}(s)\hat{K}(s))^{-1}$ is stable and $\gamma \stackrel{\Delta}{=} \|\hat{M}\|_{\infty}$. Then any perturbed closed-loop systems associated to an uncertain plant belonging to $\hat{\mathscr{G}}(1/\gamma)$ is stable.*

Actually, the previous condition corresponds to a small gain condition (see [26, 9] for details), since by definition of the set $\hat{\mathscr{G}}(1/\gamma)$, the following condition holds:

$$\|\hat{M}\|_{\infty} \|\hat{\Delta}\|_{\infty} < \gamma \frac{1}{\gamma} = 1. \tag{E.9}$$

Proof. Let us now show that the small gain condition (E.9) implies that the conditions of the extended Nyquist criteria (Theorem E.3) are satisfied. For this purpose, we then first introduce the perturbed open-loop system defined by $\hat{L}(s) \triangleq \hat{\Delta}(s)\hat{M}(s)$ and recall this essential property of the H_∞ norm:

$$\|\hat{L}\|_\infty \leq \|\hat{\Delta}\|_\infty \|\hat{M}\|_\infty.$$

Consequently, if the conditions of the proposition hold then for any $\hat{\Delta} \in \mathscr{A}_-(0)$ such that $\|\hat{\Delta}\|_\infty < 1/\gamma$, we have $\|\hat{L}\|_\infty < 1$, which means that there exists $\varepsilon > 0$ such that $\|\hat{L}\|_\infty \leq 1-\varepsilon$.

We can now check the conditions of the extended Nyquist criteria. To prove condition (i), let us recall that by Lemma A.6.15 in [8], $\|\hat{L}\|_\infty = \sup_{\Re(s)>0}|L(s)|$ and then $|\hat{L}(s)| \leq 1-\varepsilon$ for any $s \in \mathbb{C}_0^+$. Then, since $|1+\hat{L}(s)| \geq 1-|\hat{L}(s)| \geq \varepsilon$, we deduce that condition (i) of Theorem E.3 is satisfied. Let us now consider condition (ii). Since $\hat{L}(s)$ belongs to $\hat{\mathscr{A}}_-(0)$, then $\hat{L}(j\omega)$ is a continuous function of its argument and we have

$$|\hat{L}(j\omega)| \leq \|\hat{L}\|_\infty \text{ for any } \omega \in \mathbb{R},$$

and then condition (ii) is also satisfied, since $|1+\hat{L}(j\omega)| \geq 1-|\hat{L}(\omega)| \geq \varepsilon$. Conditions (iii) and (iv) are easily verified. Indeed, the Nyquist plot of $\hat{L}(j\omega)$ belongs to a circle centered at 0 with a radius strictly less than 1. We first deduce that the phase of the return difference transfer function is necessarily a bounded function of ω and condition (iv) holds true. We moreover deduce that the number of encirclements of point -1 by the Nyquist plot is equal to 0, which is in accordance with the fact that the perturbed open-loop system has no unstable pole. We have thus achieved the proof of the proposition. □

Remarks.

1. It is easy to show that the small gain condition given in the previous proposition is necessary when $\hat{M}(s)$ is strictly proper. To prove this claim, let us assume that the small gain condition is not satisfied, i.e., that $\|\hat{M}\|_\infty = \tilde{\gamma} > \gamma$ and let us show that there exists $\|\hat{\Delta}_u(s)\|_\infty < 1/\gamma$, which leads to an unstable perturbed closed-loop system. For this purpose, let us define

$$\hat{\Delta}_u(s) \triangleq \frac{e^{-\tau s}}{\tilde{\gamma}},$$

with $\tau > 0$. Clearly $\|\hat{\Delta}_u\|_\infty = 1/\tilde{\gamma} < 1/\gamma$ since by definition $\tilde{\gamma} > \gamma$. Since $\hat{M}(s) \in \hat{\mathscr{A}}_-(0)$ then $|\hat{M}(j\omega)|$ is a continuous function of its argument, and since $\hat{M}(s)$ is assumed to be strictly proper then by point (iii) of Proposition E.2, we have $\lim_{\omega\to\infty}|\hat{M}(j\omega)| = 0$. We thus deduce that the supremum of $|\hat{M}(j\omega)|$ is a maximum and it is obtained at a finite frequency point $\omega_0 \in \mathbb{R}$ where $|\hat{M}(j\omega_0)| = \tilde{\gamma}$. Consequently, the perturbed open-loop system at this frequency point is given by

$$\hat{L}(j\omega_0) = \hat{M}(j\omega_0)\hat{\Delta}_u(j\omega_0) = \tilde{\gamma}e^{j\theta_{\hat{M}}(j\omega_0)}\frac{e^{-j\tau\omega_0}}{\tilde{\gamma}} = e^{j\theta_{\hat{M}}(j\omega_0)}e^{-j\tau\omega_0},$$

where $\theta_{\hat{M}}(j\omega_0)$ is the phase (argument) of $\hat{M}(j\omega_0)$. Since there always exists $\tau > 0$ such that

$$\theta_M(j\omega_0) - \tau\omega_0 \equiv -\pi \pmod{2\pi},$$

we have identified an uncertainty $\hat{\Delta}_u(s)$ belonging to $\mathscr{A}_-(0)$ such that $\|\hat{\Delta}_u\|_\infty < 1/\gamma$ and ensuring that the Nyquist plot of $\hat{M}(j\omega)\hat{\Delta}_u(j\omega)$ passes through -1. Consequently, we have proven that the small gain condition (E.9) corresponds to a necessary and sufficient condition to ensure the closed-loop stability for any uncertain plant belonging to the set $\hat{\mathscr{G}}(1/\gamma)$.
2. Even if the condition given in the previous proposition is associated to a set of uncertain LTI plants, it is possible to show that the small gain type condition guarantees the existence of an open ball of systems which could be, for example, nonlinear (see, e.g., [12]).

Since the the stability of the nominal closed-loop system always implies the existence of an open ball of LTI uncertain systems for which the perturbed closed-loop system remains stable, we have handled the robust stability problem if the "true system" belongs to this open set of systems. Obviously, by definition, this last claim is only an expectation since the true plant is not exactly known. But this sentence summarizes the main way to handle the uncertainty problem in the robust control approach: since we are not able to exactly describe the true plant, we choose to guarantee the existence of a suitable neighborhood of the nominal plant to which it is expected that the true plant belongs.

In order to increase the interest of small gain type condition, two classical improvements have been introduced:

1. The use of suitable weighting transfer functions in order to refine the description of the set of possible uncertain plants
2. The use of other types of parameterization for describing the set of possible uncertain plants

The use of weighting transfer functions allows us to module the level of plant uncertainty as a function of the frequency. For example, we define this following set of possible uncertain plants:

$$\hat{\mathscr{G}}_{\hat{W}}(\beta) \triangleq \{\tilde{G}(s) \triangleq \hat{G}(s) + \hat{W}(s)\hat{\Delta}(s) | \hat{\Delta}(s) \in \hat{\mathscr{A}}_-(0) \text{ s.t. } \|\hat{\Delta}\|_\infty < \beta\},$$

where $\hat{W}(s)$ is assumed to be stable, i.e., $\hat{W}(s) \in \hat{\mathscr{A}}_-(0)$. A robust stability condition is given by the following result:

Proposition E.7. *Let $\hat{G}(s), \hat{K}(s)$ be in $\hat{\mathscr{B}}(0)$ and $\hat{W}(s)$ be in $\hat{\mathscr{A}}_-(0)$ and let us define $\gamma_{\hat{W}} = \|\hat{W}\hat{K}(1+\hat{G}\hat{K})^{-1}\|_\infty$. Then any perturbed closed-loop systems associated to an uncertain plant belonging to $\hat{\mathscr{G}}_{\hat{W}}(1/\gamma_{\hat{W}})$ is stable.*

The previous condition can be rewritten in a small gain form since it is equivalent to $\|\hat{W}\hat{M}\|_\infty \|\hat{\Delta}\|_\infty < 1$, where $\hat{M}(s) \triangleq \hat{K}(s)(1+\hat{G}(s)\hat{K}(s))^{-1}$.

The second improvement is obtained by considering others uncertainty structures like the *additive inverse uncertainty*:

$$\mathscr{G}_{\hat{W}}(\beta) \triangleq \{\tilde{G}(s) \triangleq (\hat{G}(s)+\hat{W}(s)\hat{\Delta}(s))^{-1} | \hat{\Delta}(s) \in \mathscr{A}_-(0) \text{ s.t. } \|\hat{\Delta}\|_\infty < \beta\},$$

the *multiplicative (direct) uncertainty*:

$$\hat{\mathscr{G}}_{\hat{W}}(\beta) \triangleq \{\tilde{G}(s) \triangleq \hat{G}(s)(1+\hat{W}(s)\hat{\Delta}(s)) | \hat{\Delta}(s) \in \hat{\mathscr{A}}_-(0) \text{ s.t. } \|\hat{\Delta}\|_\infty < \beta\},$$

or the *multiplicative inverse uncertainty*:

$$\hat{\mathscr{G}}_{\hat{W}}(\beta) \triangleq \{\tilde{G}(s) \triangleq \hat{G}(s)(1+\hat{W}(s)\hat{\Delta}(s))^{-1} | \hat{\Delta}(s) \in \hat{\mathscr{A}}_-(0) \text{ s.t. } \|\hat{\Delta}\|_\infty < \beta\}.$$

In all these cases, straightforward manipulations on blocks allow us to rewrite the perturbed closed-loop system as an interconnection between the uncertainty $\hat{\Delta}(s)$ and a suitable transfer function, only depending on the nominal model, the controller, and the weighting function. In Table E.1, we have summarized the associated robustness test. Let us note that it is possible to follow the proof made in Remark 1 to show that each small gain condition given in Table E.1 also provides a necessary condition when the nominal model or the controller are assumed to be strictly proper.

Table E.1 Definition of robustness conditions for the various structures of uncertainty where $\hat{S}(s) \triangleq (I+\hat{G}(s)\hat{K}(s))^{-1}$ and $\hat{T}(s) \triangleq \hat{G}(s)\hat{K}(s)(I+\hat{G}(s)\hat{K}(s))^{-1}$.

Uncertainty structure	Uncertain plants	Robustness test
Additive (direct) uncertainty	$\tilde{G}(s) = \hat{G}(s)+\hat{W}(s)\hat{\Delta}(s)$	$\|\hat{W}\hat{K}\hat{S}\|_\infty \|\hat{\Delta}\|_\infty < 1$
Additive inverse uncertainty	$\tilde{G}(s) = (\hat{G}(s)+\hat{W}(s)\hat{\Delta}(s))^{-1}$	$\|\hat{W}\hat{G}\hat{S}\|_\infty \|\hat{\Delta}\|_\infty < 1$
Multiplicative (direct) uncertainty	$\tilde{G}(s) = \hat{G}(s)(1+\hat{W}(s)\hat{\Delta}(s))$	$\|\hat{W}\hat{T}\|_\infty \|\hat{\Delta}\|_\infty < 1$
Inverse Multiplicative uncertainty	$\tilde{G}(s) = \hat{G}(s)(1+\hat{W}(s)\hat{\Delta}(s))^{-1}$	$\|\hat{W}\hat{S}\|_\infty \|\hat{\Delta}\|_\infty < 1$

To summarize, and as it has already pointed out, the use of the Nyquist criteria in the infinite dimensional case allows us to recover most of the results developed in the finite dimensional case for systems described by rational transfers. This explains why we do not detail results further.

E.2 Input–Output Stability Versus Lyapunov Stability

The main goal of this section is to prove that if the closed-loop system associated to the linearized Saint-Venant transfer matrix is input–output stable, then it implies the

$$\tilde{u}_2(t) = \chi_2(L,t) = \begin{cases} 0 & \text{if } t < T_r - \tau_2 \\ \chi_{2T_r}\Big(L - \beta(T_r - t)\Big) & \text{if } t \geq T_r - \tau_2 \end{cases} \tag{E.15}$$

ensures that $\chi_2(x,T_r) = \chi_{2T_r}(x)$ for all $x \in [0,L]$. This is depicted in Fig. E.8.

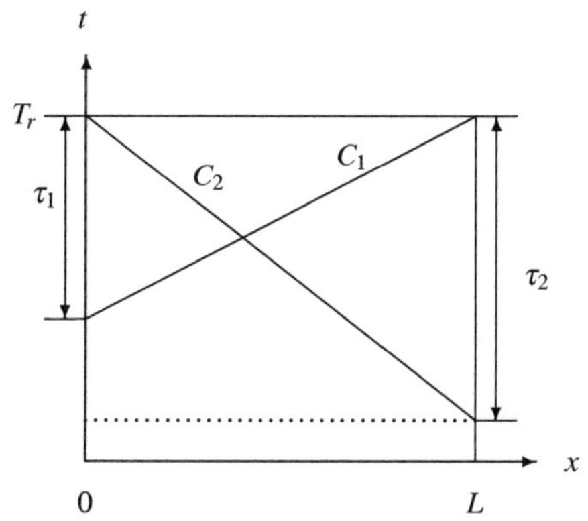

Fig. E.8 Characteristic curves for reachability with boundary conditions in the case of a horizontal frictionless rectangular channel

This solution is defined when $T_r \geq \tau_2 \geq \tau_1$. Therefore, we have obtained the controls $\tilde{u}_1(t) = \chi_1(0,t)$ and $\tilde{u}_2(t) = \chi_2(L,t)$ steering the state from 0 to $\chi_{T_r}(x)$.

Moreover, we know that:

$$\begin{aligned} \int_0^{T_r} \|\tilde{u}(t)\|^2 \mathrm{d}t &= \int_0^{T_r} (\tilde{u}_1^2(t) + \tilde{u}_2^2(t)) \mathrm{d}t \\ &= \int_{T_r-\tau_1}^{T_r} \chi_{1T_r}^2(\alpha(T_r - t)) \mathrm{d}t + \int_{T_r-\tau_2}^{T_r} \chi_{2T_r}^2(L - \beta(T_r - t)) \mathrm{d}t. \end{aligned}$$

With the change of variables $x_1 = \alpha(T_r - t)$ and $x_2 = L - \beta(T_r - t)$ for the first and second integrals, respectively, the right-hand side becomes:

$$\begin{aligned} \int_0^{T_r} \|\tilde{u}(t)\|^2 \mathrm{d}t &= \int_0^L \chi_{1T_r}^2(x_1) \mathrm{d}x_1 + \int_0^L \chi_{2T_r}^2(x_2) \mathrm{d}x_2 \\ &= \int_0^{L-\alpha T_r} \chi_{1T_r}^2(x_1) \mathrm{d}x_1 + \int_0^{\beta T_r} \chi_{2T_r}^2(x_2) \mathrm{d}x_2 \\ &= \int_0^L (\chi_1^2(x,T_r) + \chi_2^2(x,T_r)) \mathrm{d}x \\ &= \|z_{T_r}\|_Z^2. \end{aligned}$$

The open channel is then reachable according to Definition E.3 with two boundary controls in terms of the characteristics variables for any $T_r \geq \tau_2$. As stated above, one may also show that it is reachable with only one boundary control, for any $T_r \geq \tau_1 + \tau_2$.

Now, it is easy to see that $\chi_1(L,t) = 0$ and $\chi_2(0,t) = 0$. Therefore, to show the reachability with boundary discharges, we notice that once the controls have been obtained in terms of $\chi_1(0,t)$ and $\chi_2(L,t)$, the boundary discharges can be computed using (2.29), leading to:

$$q(0,t) = \frac{\alpha}{\alpha+\beta}\chi_1(0,t), \tag{E.16a}$$

$$q(L,t) = \frac{\beta}{\alpha+\beta}\chi_2(L,t). \tag{E.16b}$$

The system is well-defined and the initial condition is assumed to be zero. Then by (E.12), it is easy to prove that for any $T_r \geq 0$, there exists a finite constant M, such that

$$\int_0^{T_r} \|y_\Sigma(t)\|^2 \mathrm{d}t \leq M \int_0^{T_r} \|\tilde{u}(t)\|^2 \mathrm{d}t.$$

Thus by (E.16), there exists $M' > 0$ such that

$$\begin{aligned}\int_0^{T_r} (|q(t,0)|^2 + |q(t,L)|^2)\mathrm{d}t &\leq M' \int_0^{T_r} (|\tilde{u}_1(t)|^2 + |\tilde{u}_2(t)|^2)\mathrm{d}t \\ &\leq M' \int_0^{T_r} \|\tilde{u}(t)\|^2 \mathrm{d}t,\end{aligned}$$

which allows us to deduce that the open channel is reachable with controls in terms of boundary discharges.

Observability

We assume a nonzero initial condition $\chi_0(x)$, and we want to show that this initial state is observable with boundary measurements in terms of characteristic variables $\chi_2(0,t)$ and $\chi_1(L,t)$. A sketch of characteristic curves is depicted in Fig. E.9.

To show that the state is observable, we need to prove that the inequality of definition E.4 is verified when $u = 0$. Using (E.13), and assuming $T_o \geq \tau_2$, we have:

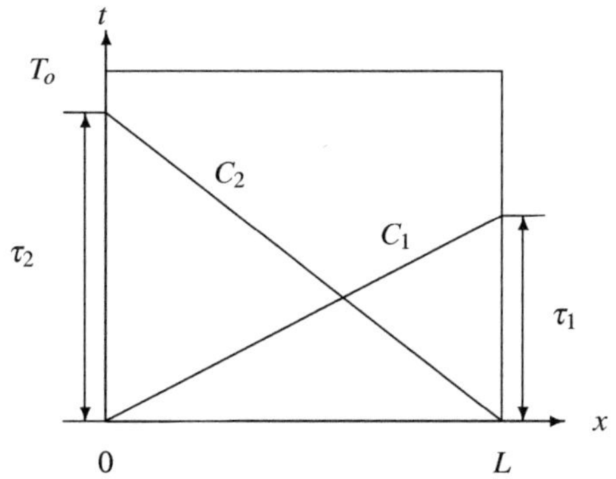

Fig. E.9 Sketch of the characteristic curves for observability with boundary conditions in the case of a horizontal frictionless rectangular channel

$$\begin{aligned}
\int_0^{T_o} \|y(t)\|^2 \mathrm{d}t &= \int_0^{T_o} (\chi_1^2(L,t) + \chi_2^2(0,t))\mathrm{d}t \\
&= \int_0^{\tau_1} \chi_{10}^2(L-\alpha t)\mathrm{d}t + \int_0^{\tau_2} \chi_{20}^2(\beta t)\mathrm{d}t \\
&+ \int_{\tau_1}^{T_o} \chi_1^2(L,t)\mathrm{d}t + \int_{\tau_2}^{T_o} \chi_2^2(0,t)\mathrm{d}t \\
&\geq \int_0^{\tau_1} \chi_{10}^2(L-\alpha t)\mathrm{d}t + \int_0^{\tau_2} \chi_{20}^2(\beta t)\mathrm{d}t \\
&\geq \int_0^{L} \chi_{10}^2(x_1)\mathrm{d}x_1 + \int_0^{L} \chi_{20}^2(x_2)\mathrm{d}x_2 = \|z_0\|_Z^2.,
\end{aligned}$$

where we have used the change of variables $x_1 = L - \alpha t$ and $x_2 = \beta t$ to compute the last equality.

The open channel is then observable according to Definition E.4 with two boundary measurements in terms of the characteristic variables, for any $T_o \geq \tau_2$. Using similar arguments, one may show that it is also observable with only one boundary measurement, for any $T_o \geq \tau_1 + \tau_2$.

Following the same line as above, we can show that the open channel is observable with boundary measurements in terms of water level and discharge.

E.2.3 Link Between Input–Output Stability and Exponential Stability

We now recall a result allowing us to strongly relate input–output stability and Lyapunov stability for systems possessing a minimal state-space realization. This result finds its roots in the dissipativity framework introduced by Willems in his seminal paper [25]. It should be noted that Willems mentioned in [24] that his work extended a previous result of Baker and Bergen [1] to dynamical systems in the context of linear infinite dimensional systems.

In the sequel, Σ is a causal linear time-invariant system such that for any input u in $\mathscr{L}_2([0,t],\mathbb{R}^p)$, its output given by $y=\Sigma(u)$ belongs to $\mathscr{L}_2([0,t],\mathbb{R}^m)$ (Σ is thus assumed well-defined). Z is a normed vectorial space equipped with the norm $\|\cdot\|_Z$ and corresponds to the the state-space of Σ. Finally, the state of Σ at time $t\in[0,\infty)$ belonging to Z is denoted by $z(t)$ and it is formally related to the input u and the initial condition by the following causal relation: $z(t)=\phi(t,0,z(0),u^{(t)})$.

The following definition corresponds to the uniform reachability and the uniform observability defined by Willems in [24] for causal linear invariant systems.

Definition E.5. Σ is said to be minimal if it is uniformly reachable from $z(0)=0$ and uniformly observable.

Proposition E.8. *Let Σ be a causal linear time invariant system defined from $\mathscr{L}_2([0,t],\mathbb{R}^p)$ into $\mathscr{L}_2([0,t],\mathbb{R}^m)$. If Σ is finite gain stable on $\mathscr{L}_2$, i.e if there exists $\eta\geq 0$ such that $\|y\|_2\leq\eta\|u\|_2$ for any $u\in\mathscr{L}_2([0,\infty),\mathbb{R}^p)$ and if its state-space realization is minimal then Σ is uniformly exponentially stable, i.e., there exist a and b positive such that for any $z(0)\in Z$, we have $\|z(t)\|_Z\leq ae^{-bt}\|z(0)\|_Z$ for any $t\geq 0$.*

The proof of the proposition is a consequence of results given in [24, 25]. We explain how exponential stability can be deduced based on a dissipativity argument without any assumption on the regularity of the storage function.

Let us first recall that the available storage, S_a, of a time-invariant dynamical system, Σ defined from $\mathscr{L}_2([0,t],\mathbb{R}^p)$ into $\mathscr{L}_2([0,t],\mathbb{R}^m)$, with supply rate $w(t)$, is the function from Z into $\overline{\mathbb{R}}^+$ defined by [25]:

$$S_a(z)=\sup_{\substack{z\to\\ t\geq 0}} -\int_0^t w(\tau)\mathrm{d}\tau, \tag{E.17}$$

where the supremum is taken on any interval of time $[0,t]$ with $t\in[0,\infty)$ over all motions starting in state z at time $t=0$ under any input u belonging to $\mathscr{L}_2([0,t],\mathbb{R}^p)$.

For systems with an $\mathscr{L}_2$ gain lower than η, the supply rate is defined by

$$w(t)=\eta^2\|u(t)\|^2-\|y(t)\|^2.$$

The main interest of the dissipativity framework is to link the behavior of the state and its input–output properties and especially characterize Lyapunov-like properties. We now state the proof of Proposition E.8.

Proof. Since the state-space of Σ is minimal, it is routine to deduce the following properties: if Z is uniformly reachable from $z = 0$, then $S_a(z) \leq \eta^2 \alpha_r^2 \|z\|_Z^2$ for all $z \in Z$. Furthermore, if Σ is uniformly observable, then $S_a(z) \geq \beta_o^2 \|z\|_Z^2$ and $S_a(z(T)) - S_a(z) \leq -\beta_o^2 \|z\|_Z^2$ for any $T \geq T_o$ and any $z \in Z$, where $z(T)$ is the state of the system associated to the null input and the initial condition z.

Following these preliminary results, we deduce that S_a has the following upper and lower bounds:

$$\beta_o^2 \|z\|_Z^2 \leq S_a(z) \leq \eta^2 \alpha_r^2 \|z\|_Z^2,$$

and moreover that $S_a(z(t+T)) - S_a(z(t)) \leq -\beta_o^2 \|z(t)\|_Z^2$ where $T \geq T_o$. On this basis, after straightforward manipulations we obtain the following inequality:

$$S_a(z(t+T)) \leq \left(1 - \frac{\beta_o^2}{\eta^2 \alpha_r^2}\right) S_a(z(t)).$$

Finally, by the dissipativity inequality, one may show that $S_a(z(t))$ is a nonincreasing function of time. One therefore has $S_a(z(\tau)) \leq S_a(z(0))$ for any $\tau \in [0, T]$ and thus for any $\tau \in [0, T]$ and any $k \in \mathbb{N}$:

$$\|z(\tau + kT)\|_Z^2 \leq \left(1 - \frac{\beta_o^2}{\eta^2 \alpha_r^2}\right)^k \left(\frac{\eta \alpha_r}{\beta_o}\right)^2 \|z(0)\|_Z^2.$$

Let us now introduce $\rho \triangleq 1 - \frac{\beta_o^2}{\eta^2 \alpha_r^2}$ ($\rho < 1$ since by construction $\beta_o^2 \|z\|_Z^2 \leq S_a(z) \leq \eta^2 \alpha_r^2 \|z\|_Z^2$) and $d \triangleq \left(\frac{\eta \alpha_r}{\beta_o}\right)^2$ (≥ 0) in order to rewrite the last inequality as

$$\|z(\tau + kT)\|_Z^2 \leq d \rho^k \|z(0)\|_Z^2,$$

which implies that for any $t \geq 0$, we have

$$\|z(t)\|_Z \leq a e^{-bt} \|z(0)\|_Z,$$

with $b = -\log(\rho)/(2T)$ and $a = d^{1/2}$, which corresponds to the announced exponential stability result. □

E.2.3.1 Application to the Linearized Saint-Venant Transfer Matrix Stabilized by a Finite Dimensional Controller

We now consider the system of Fig. E.10, where a system represented by the linearized Saint-Venant transfer is controlled by a finite dimensional controller $K(s)$.

We have the following proposition:

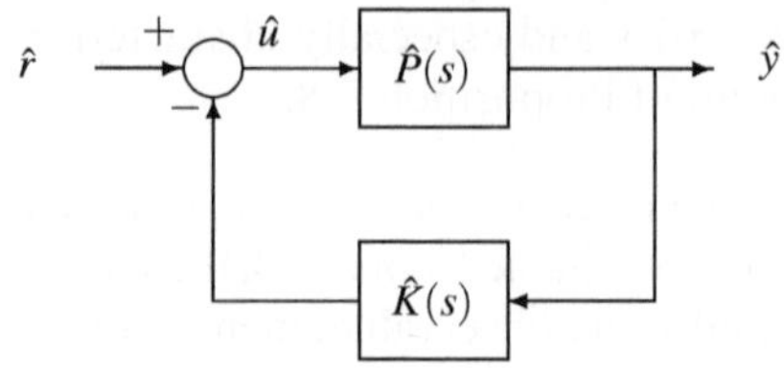

Fig. E.10 Linearized Saint-Venant equation connected with a finite dimensional controller

Proposition E.9. *Let $K(s)$ be a finite dimensional controller with a minimal realization $K = \left[\begin{array}{c|c} A & B \\ \hline C & D \end{array}\right]$. If the closed-loop is stable (it belongs to $\mathscr{A}(0)$) then the closed-loop system is uniformly exponentially stable.*

Proof. In our context, the state-space of the closed-loop system is given by the concatenation of the state-space of the hyperbolic system given by (E.10) and the one of the controller K. We then deduce that $Z = \mathscr{L}_2([0,L],\mathbb{R}^2)$ and $z = \xi$ when a constant feedback is considered. When K is a finite dimensional time-invariant linear controller of order n, then $Z = \mathscr{L}_2([0,L],\mathbb{R}^2) \times \mathbb{R}^n$ with $z = (\xi, x_K)$, where x_K is the state of K. In this last case, Z is equipped with the following norm:

$$\|z\|_Z = \left(\|\xi\|^2_{\mathscr{L}_2([0,L],\mathbb{R}^2)} + \|x_K\|^2\right)^{1/2}.$$

Since the state-space realization of the controller is such that (A,B) is reachable and (A,C) is observable, it is straightforward to prove that the state-space realization of K is minimal following Definition E.5 [2]. Based on the results presented in the previous sections, the minimality of state-space realization of the closed-loop system holds, since the hyperbolic system given by (E.10) is also minimal, and the result is a direct consequence of Proposition E.8. □

References

[1] Baker R, Bergen AR (1969) Lyapunov stability and Lyapunov functions of infinite dimensional systems. IEEE Trans Autom Contr 14:325–334
[2] Brockett RW (1970) Finite dimensional linear systems. Wiley, New York
[3] Callier F, Desoer C (1972) A graphical test for checking the stability of a linear time-invariant feedback system. IEEE Trans Autom Contr 17:773–780
[4] Callier F, Desoer C (1976) On simplifying a graphical stability criterion for linear distributed feedback system. IEEE Trans Autom Contr 21:128–129
[5] Callier F, Desoer C (1978) An algebra of transfer functions for distributed linear time-invariant systems. IEEE Trans Circuits Syst CAS-25(9):651–662

[6] Callier F, Desoer C (1980) Simplifications and clarifications on the paper "an algebra of transfer functions for distributed linear time-invariant systems". IEEE Trans Circuits Syst CAS-27(5):320–323
[7] Curtain R (1992) A synthesis of time and frequency domain methods for the control of infinite dimensional systems: a system theoretic approach, in Banks HT (ed.), *Control and Estimation in Distributed Parameter Systems*, Frontiers in Applied Mathematics, vol 11, SIAM
[8] Curtain R, Zwart H (1995) An Introduction to Infinite-Dimensional Linear Systems Theory. Springer, New York
[9] Desoer CA, Vidyasagar M (1975) Feedback systems: input ouput properties. Academic Press, New York
[10] Dullerud GE, Paganini F (2000) A course in robust control theory: a convex approach, Texts in applied mathematics, vol 36. Springer, New York
[11] Ferreres G (1999) A practical approach to robustness analysis. Kluwer Academic Publishers, Norwell, MA
[12] Fromion V, Monaco S, Normand-Cyrot D (2001) The weighting incremental norm approach: from linear to nonlinear H_∞ control. Automatica 37:1585–1592
[13] Helmicki AJ, Jacobson CA, Nett CN (1991) Ill-posed distribued parameter systems: a control view point. IEEE Trans Autom Contr 36(9):1053–1057
[14] Hille E, Phillips RS (1957) Functional analysis and semi-groups, Colloquium Publications, vol 31. American Mathematical Society
[15] Kailath T (1980) Linear systems. Information and system sciences series, Prentice Hall, London
[16] Ozbay H (1999) Introduction to feedback control theory. CRC Press, Boca Raton, FL
[17] Russell DL (1973) Quadratic performance criteria in boundary control of linear symmetric hyperbolic systems. SIAM Journal on Control 11(3):475–509
[18] Russell DL (1978) Controllability and stabilizability theory for linear partial differential equations: recent progress and open questions. SIAM Rev 20(4):639–739
[19] Skogestad S, Postlethwaite I (1996) Multivariable feedback control. Analysis and design. Wiley, New York
[20] Vidyasagar M (1985) Control system synthesis: A factorization approach. MIT Press
[21] Vidyasagar M (1992) Nonlinear systems analysis, 2nd edn. Prentice Hall, Englewood Cliffs, NJ
[22] Vidyasagar M, Schneider H, Francis BA (1982) Algebraic and topological aspects of feedback stabilization. IEEE Trans Autom Contr 27(4):880–894
[23] Willems JC (1971) The analysis of feedback systems, Research monographs, vol 62. MIT Press, Cambridge, MA
[24] Willems JC (1971) The generation of Lyapunov functions for input-output stable systems. SIAM Journal on Control 9(1):105–134
[25] Willems JC (1972) Dissipative dynamical systems Part I/Part II. Archive for Rational Mechanics and Analysis 45:321–341/352–393
[26] Zames G (1966) On the input-output stability of time-varying nonlinear feedback systems–parts I, II. IEEE Trans Autom Contr 11:228–238, 465–476
[27] Zhou K, Doyle J, Glover K (1995) Robust and Optimal Control. Prentice Hall, Englewood Cliffs, NJ

Appendix F
Computation of Structured Singular Values for the Mixed Controller

F.1 Input Sensitivity Functions of the Mixed Controller

We consider the mixed controller for one canal pool, and compute the structured singular value of the input sensitivity functions, given by:

$$S_u = S_y \begin{pmatrix} 1+\tilde{G}_1 K_{21} & -\tilde{G}_1 K_{11} \\ -G_1 K_{21} & 1+G_1 K_{11} \end{pmatrix},$$

and

$$T_u = S_y \begin{pmatrix} G_1 K_{11} & \tilde{G}_1 K_{11} \\ G_1 K_{21} & \tilde{G}_1 K_{21} \end{pmatrix},$$

with $S_y = (1+G_1K_{11}+\tilde{G}_1K_{21})^{-1}$ and $T_y = (1+G_1K_{11}+\tilde{G}_1K_{21})^{-1}(G_1K_{11}+\tilde{G}_1K_{21})$.

To evaluate the multivariable robustness margins, we compute the structured singular value of the input sensitivity function for complex diagonal perturbations. In our case, since we are dealing with 2×2 matrices, the structured singular value μ_Δ is equal to the upper bound:

$$\mu_\Delta(A) = \inf_{D\in\mathscr{D}} \overline{\sigma}(DAD^{-1}),$$

with $D = \begin{pmatrix} 1 & 0 \\ 0 & d \end{pmatrix}$, and $d \in \mathbb{C}$.

For any complex matrix $A = (a_{ij}) \in \mathbb{C}^{2\times 2}$:

$$M = DAD^{-1} = \begin{pmatrix} a_{11} & d^{-1}a_{12} \\ da_{21} & a_{22} \end{pmatrix}$$

and

$$MM^* = \begin{pmatrix} |a_{11}|^2 + |d|^{-2}|a_{12}|^2 & a_{11}d^*a_{21}^* + d^{-1}a_{12}a_{22}^* \\ da_{21}a_{11}^* + a_{22}d^{-*}a_{12}^* & |a_{22}|^2 + |d|^2|a_{21}|^2 \end{pmatrix}.$$

F.1.1 Structured Singular Value of T_u

We want to compute the singular values of $M = DT_eD^{-1}$, or the square root of the eigenvalues of MM^*.

Given the particular form of the problem, we notice that

$$\det(T_u) = 0,$$

which implies that, for $M = DT_uD^{-1}$, we have:

$$\det(MM^*) = 0.$$

Therefore 0 is an eigenvalue of T_u and a singular value of DT_uD^{-1}. The nonzero singular value is equal to the square root of the trace of MM^*, i.e.:

$$\overline{\sigma}(DT_uD^{-1}) = |S_y|\sqrt{|G_1K_{11}|^2 + |\tilde{G}_1K_{21}|^2 + |dG_1K_{21}|^2 + |d|^{-2}|\tilde{G}_1K_{11}|^2}.$$

Now, we want to compute the value of $d \in \mathbb{C}$ that minimizes $\overline{\sigma}(DT_uD^{-1})$. Let $f(x)$ denote the function of $x > 0$ defined by:

$$f(x) = |S_y|\sqrt{|G_1K_{11}|^2 + |\tilde{G}_1K_{21}|^2 + x|G_1K_{21}|^2 + \frac{1}{x}|\tilde{G}_1K_{11}|^2}.$$

We have:

$$\inf_{d\in\mathbb{C}} \overline{\sigma}(DT_eD^{-1}) = \inf_{x\in\mathbb{R}^+} f(x).$$

$f(x)$ is differentiable and its derivative is:

$$\frac{\mathrm{d}f}{\mathrm{d}x} = \frac{|S_y|^2}{2f(x)}\left(|G_1K_{21}|^2 - \frac{|\tilde{G}_1K_{11}|^2}{x^2}\right).$$

$f(x)$ therefore reaches its minimum for $x_0 = \frac{\tilde{G}_1K_{11}}{G_1K_{21}}$, and we have:

$$\mu_\Delta(T_u) = f(x_0) = |S_y|(|G_1K_{11}| + |\tilde{G}_1K_{21}|).$$

This is the result we wanted to prove.

For comparison purposes, in the unstructured case, we have:

$$\overline{\sigma}(T_u) = f(1) = |S_y|\sqrt{|G_1K_{11}|^2 + |\tilde{G}_1K_{21}|^2 + |G_1K_{21}|^2 + |\tilde{G}_1K_{11}|^2}.$$

Therefore, in the case where $|G_1| \approx |\tilde{G}_1|$:

$$\overline{\sigma}(T_u) \approx |S_y|\sqrt{2}\sqrt{|G_1K_{11}|^2 + |\tilde{G}_1K_{21}|^2}.$$

In that case, using unstructured perturbations leads to underestimate the robustness with respect to structured perturbations.

F.1.2 Structured Singular Value of S_u

In a similar way, we want to compute the singular values of $M = DS_uD^{-1}$, or the eigenvalues of MM^*.

Since $S_u = I - T_u$, the fact that T_u has an eigenvalue equal to zero implies that S_u has an eigenvalue equal to 1. Therefore 1 is a singular value of $M = DS_uD^{-1}$. The other singular value is equal to:

$$\overline{\sigma}(DS_uD^{-1}) = \sqrt{|S_y|^2(|1+G_1K_{11}|^2+|1+\tilde{G}_1K_{21}|^2+|dG_1K_{21}|^2+|d|^{-2}|\tilde{G}_1K_{11}|^2)-1}.$$

We now look for the value of $d \in \mathbb{C}$ that minimizes $\overline{\sigma}(DS_uD^{-1})$. Let $g(x)$ denote the function of $x > 0$ such that:

$$g(x) = |S_y|\sqrt{|1+G_1K_{11}|^2+|1+\tilde{G}_1K_{21}|^2+x|G_1K_{21}|^2+\frac{1}{x}|\tilde{G}_1K_{11}|^2-|1+G_1K_{11}+\tilde{G}_1K_{21}|^2}.$$

We have:

$$\inf_{d\in\mathbb{C}} \overline{\sigma}(DS_uD^{-1}) = \inf_{x\in\mathbb{R}^+} g(x).$$

The function $g(x)$ is differentiable and its derivative is given by:

$$\frac{\mathrm{d}g}{\mathrm{d}x} = \frac{|S_y|^2}{2g(x)}\left(|G_1K_{21}|^2 - \frac{|\tilde{G}_1K_{11}|^2}{x^2}\right).$$

$g(x)$ therefore reaches a minimum for $x_0 = \frac{\tilde{G}_1K_{11}}{G_1K_{21}}$, and this yields

$$\mu_\Delta(S_u) = g(x_0) = |S_y|\sqrt{|1+G_1K_{11}|^2+|1+\tilde{G}_1K_{21}|^2+2|G_1K_{11}\tilde{G}_1K_{21}|-|1+G_1K_{11}+\tilde{G}_1K_{21}|^2}.$$

Permissions

In the book we have used copyrighted material with permission from the copyright owner.

Figures 4.20 and 4.21 were reprinted from the *Journal of Hydraulic Engineering*, Vol. 130, No. 8, X. Litrico and V. Fromion, Frequency modeling of open channel flow, pages 806–815, Copyright (2004), with permission from ASCE.

Figures 5.3, 5.7 5.8, 5.10, 5.9 and 5.11 were reprinted from *Applied Mathematical Modelling*, Vol. 28, No. 7, X. Litrico and V. Fromion, Analytical approximation of open channel flow for controller design, pages 677–695, Copyright (2004), with permission from Elsevier.

Figure 6.3 was reprinted from *Automatica*, Vol. 45, No. 3, X. Litrico and V. Fromion, Boundary control of hyperbolic conservation laws using a frequency domain approach, pages 647–656 (2009), with permission from Elsevier.

Figures 6.4 and 6.5 were adapted from *Automatica*, Vol. 42, No. 6, X. Litrico and V. Fromion, Boundary control of linearized Saint-Venant equations oscillating modes, pages 967–972 (2006), with permission from Elsevier.

Figures 7.5, 7.6, 7.7, 7.8, 7.9, 7.10 and 7.11 were reprinted from the *Journal of Irrigation and Drainage Engineering*, Vol. 132, No. 4, X. Litrico and V. Fromion, Tuning of robust distant downstream PI controllers for an irrigation canal pool: (I) Theory, pages 359–368, Copyright (2006), with permission from ASCE.

Figures 11.4, 11.7, 11.8, 11.10, 11.9, 11.11, 11.16, 11.14, 11.12, 11.15 and 11.17 were reprinted from *Control Engineering Practice*, Vol. 13, No. 11, X. Litrico, V. Fromion, J.P. Baume, C. Arranja and M. Rijo, Experimental validation of a methodology to control irrigation canals based on Saint-Venant equations, pages 1425–1437 (2005), with permission from Elsevier.

Figures 11.19, 11.20, 11.21, 11.22, and 11.23 were reprinted from the *IEEE Transactions on Control Systems Technology*, Vol. 14, No. 1, X. Litrico and V. Fromion, H_∞ control of an irrigation canal pool with a mixed control politics, pages 99–111 (2006), with permission from IEEE.

Index

MIX
Papier aus verantwortungsvollen Quellen
Paper from responsible sources
FSC® C105338

If you have any concerns about our products,
you can contact us on
ProductSafety@springernature.com

In case Publisher is established outside the EU,
the EU authorized representative is:
Springer Nature Customer Service Center GmbH
Europaplatz 3, 69115 Heidelberg, Germany

Printed by Libri Plureos GmbH
in Hamburg, Germany